Electric Airplanes
and Drones

ALSO BY KEVIN DESMOND
AND FROM McFARLAND

Electric Motorcycles, Bicycles and Scooters: A History (2018)

Electric Boats and Ships: A History (2017)

Innovators in Battery Technology: Profiles of 95 Influential Electrochemists (2016)

Gustave Trouvé: French Electrical Genius (1839–1902) (2015)

Electric Airplanes and Drones

A History

KEVIN DESMOND

Foreword by IVO BOSCAROL

McFarland & Company, Inc., Publishers
Jefferson, North Carolina

LIBRARY OF CONGRESS CATALOGUING-IN-PUBLICATION DATA

Names: Desmond, Kevin, 1950– author.
Title: Electric airplanes and drones : a history / Kevin Desmond ; foreword by Ivo Boscarol.
Description: Jefferson, North Carolina : McFarland & Company, Inc., Publishers, 2018 | Includes bibliographical references and index.
Identifiers: LCCN 2018023833 | ISBN 9781476669618 (softcover : acid free paper) ∞
Subjects: LCSH: Hybrid electric airplanes—History. | Drone aircraft—History.
Classification: LCC TL685.35 .D47 2018 | DDC 629.133/3404—dc23
LC record available at https://lccn.loc.gov/2018023833

BRITISH LIBRARY CATALOGUING DATA ARE AVAILABLE

ISBN (print) 978-1-4766-6961-8
ISBN (ebook) 978-1-4766-3341-1

© 2018 Kevin Desmond. All rights reserved

No part of this book may be reproduced or transmitted in any form or by any means, electronic or mechanical, including photocopying or recording, or by any information storage and retrieval system, without permission in writing from the publisher.

Front cover: Antares DLR-H2 aircraft (courtesy of Lange Aviation GmbH); (insets, left to right) Tissandier airship (Musée EDF Electropolis); DJI Phantom 3 Quadcopter (photograph by Marek Uliasz)

Printed in the United States of America

McFarland & Company, Inc., Publishers
Box 611, Jefferson, North Carolina 28640
www.mcfarlandpub.com

To
Ἴκαρος (Icarus),
to Ἠλέκτρα (Elektra)
and to Ἥλιος (Hēlios)

Acknowledgments

The author would like to thank the following for their kind help in this book:

TajaBoscarol; André Borschberg; Robert J. Boucher; HeinoBrditschka; Musée Air + Espace; Chard Museum (Vince Lean); Julie Conti; Valerie Crosby (Haringey Archive and Museum Service); Leonard DeGraaf; Deutsches Museum (Christian Schlafner); Graham Dorrington; John Emms; Didier Esteyne; Karl Fauhaber; Randall B. Fishman; Benjamin Franklin House; Stefan Gehrmann; Stefan Graupner; Ute Holder (Archives, Andor Holtsmark (Lange) Kirchheimer Zeitung); Stephen Howard (Bristol Aerospace Centre); Dan Johnson; Ted Kemp (Institution of Engineering and Technology); Kirchheimunter Teck (Archives); Jonas Larsson (SAAB); René Meier; Walter Ray Morgan; The Institution of Engineering and Technology; Steve Ptacek; RAF Museum, Hendon (Andrew Renwick); Eric and Irena Raymond; Brian Riddle (National Aerospace Library); Cornelia Ruppert; Musée de l'Air et de l'Espace (Jean-Marc Lombarde); Musée EDF Electropolis (Damien Kunz); Carolin Scheumann (née Rochelt); Shuttleworth Aviation Trust; Siemens AG; Dean Sigler; Janice Sullivan (née Brown); Tine Tomažič; Willi Tacke; Jean-Marie Urlacher; Chip Yates; Eric Zunino, Novadem.

Thanks also to Alexandra Desmond (my long-supporting wife) and Kathryn Cooper (my indexer and proofreader).

Table of Contents

Acknowledgments vi
Foreword by Ivo Boscarol 1
Preface 3

ONE	Origins	5
TWO	If It Flies, It Must Be Electric!	35
THREE	Electricity Goes to War	48
FOUR	The Aeromodelers	64
FIVE	"Here Comes the Sun"	74
SIX	The Lithium Advantage	105
SEVEN	"H" Is for Hybrid, Hydrogen, Helium	129
EIGHT	Towards Commercialization	140
NINE	UAVs (Generation Gas)	171
TEN	UAVs (aka Drones) Go Electric	182
ELEVEN	Dronomania!	211
TWELVE	Flying Cars	231
THIRTEEN	Into the Future	243

Appendix A. The Birth of Drone Racing 267
Appendix B. The Adventure of Solar Impulse 270
Appendix C. Speed and Altitude Records 282
Appendix D. Timeline of Notable Electric Aircraft Flights 291
Chapter Notes 293
Bibliography 299
Index 301

Foreword
by Ivo Boscarol

At the beginning of the 20th century, three men, independently of one another, were tackling the question of how to convert mechanical energy into alternating current; the company cofounded by one of them, Siemens, is still developing that technology today. At the time none of them knew that they held in their hands the key to the largest revolution in the history of mankind after the inventions of the wheel and fire. Today, 100 years later, nobody can imagine life without alternating current and all its benefits.

In the last 140 years the production of electric energy has been one of the greatest pollutants of the atmosphere, but today, in the third millennium, we have finally started to practice cleaner means, without coal-fired power plants and dams which destroy nature. For example, did you know that in only two minutes the Sun sends to Earth the same amount of energy as the entire human race uses in one year?

One of the biggest advantages that electricity has over alternative means of energy is the ease of its transportation—equally simple on level ground, down or uphill, wherever there are wires. And hopefully mankind will one day obtain the secrets of Nicola Tesla for wireless high-power energy transfer. But until then the use of electricity for transport has great limitations—and the greatest are of course still in aviation.

Battery energy density is still too low for trans–Atlantic flights, but it already offers the perfect substitute for combustion engines in powered gliders and training aircraft, where the endurance needed is about one hour—the same as for transport in megacities. While waiting for more efficient batteries, different hybrid options such as hydrogen fuel cells are an ideal solution, producing no exhaust apart from pure water.

When, 20 years ago, some of us visionary aircraft producers believed in electric aircraft propulsion, we were labeled as weirdoes. Today even the sworn kerosene users such as Airbus or Boeing are slowly getting integrated into the mindset of global players such as Siemens, Tesla, Amazon and Uber, who are seeing the world greener every day. With their integration, support and potential, electric-powered general aviation and commercial passenger aircraft are becoming more and more a reality with every passing day.

For this book, appearing at a crucial point in history, Kevin Desmond must be commended.

Ivo Boscarol is the founder of Pipistrel, electric and hybrid-electric aircraft manufacturers. Since 2007, Pipistrel has made and flown more than five dozen electric aircraft.

Preface

I spent the first four months of my infancy in 1950 in a house adjoining the London Terminal Airport, then in Croydon, Surrey, and must have been awakened many a time by the noise of airliners taking off and landing overhead. My godfather worked for the British Overseas Aircraft Corporation. As a schoolboy, I built model aircraft powered by wound-up or stretched-out rubber bands, except for the toy my dad gave me one day: a tethered "Electronic Falcon Plane that flies itself," made by Remco Industries, Inc. Its power came from a battery enclosed in a pocket flashlight device. Like a flashlight, you switched it on, the prop turned, and the plane went round and round in circles, until you became bored! From then on, like everybody else, I have traveled the world in avgas-fueled aircraft, piston or jet. I use the reading lamp above, watch the screen on the back of the seat in front of me, watch the wing flaps move and hear the undercarriage retract—knowing they are electric or electro-hydraulic.

A definition is perhaps necessary. An electric aircraft is a vehicle which, with or without a pilot or passenger(s), is regularly capable of taking off from the ground, rising to a height of no less than 100 feet and no more than 80,000 feet, and then flying for between 4 minutes and 4 years, using electric or hybrid-electric propulsion. This definition covers everything from a fingertip drone to an airship; this also makes provision for a roadworthy vehicle or flying car.

Avgas airplanes dumped 700 million metric tons of carbon dioxide into the air in 2013, according to a 2015 report from *National Geographic*. The aircraft industry is expecting a sevenfold increase in air traffic by 2050, and a fourfold increase in greenhouse gas emissions unless fundamental changes are made. The crucial next step towards ensuring the aircraft industry becomes greener is the full electrification of commercial aircraft. This means zero CO2 and NOx emissions, with energy sourced from power stations that are themselves sustainably fueled; a step-change in fuel efficiency is crucial to maintain emission levels promised in the Paris Agreement—a 30 percent improvement in aircraft efficiencies is required by 2035. As this book is published, the world of aviation is bravely accelerating into a new, more silent and less polluting era of electric propulsion. Described as the "Third Revolution" in aviation (after heavier-than-air flight and jet engines), the introduction of hybrid-electric aircraft could be a massive breakthrough for sustainable aviation.

From the stratosphere to door-to-door, differing prototypes have now entered into their series-production phase, be they airships, or airplanes carrying up to eight passengers or training would-be pilots, be they vertical takeoff drones which can carry a single passenger across a city, or those for delivery, filmmaking or sport, or merely toys which can be hand-launched and piloted using virtual reality.

Remarkably, an approach to electrical power can be traced back to over 120 years ago, when our ancestors dreamed of this very era. But they were only held back by their then limited technology. This book tells how their dream is now becoming our reality.

Electrical airplanes are sure to change the look of aviation, but if humans plan to continue to fly in the future, we will have to embrace this new era of flight. According to some reports, the world contains only enough petroleum resources to last us through the year 2100. And as we get closer to that date, fuel prices are likely to rise higher and higher. Eventually, we will need to wean ourselves off internal combustion engines and the aircraft that use them. To do that, we'll need to see innovation in aircraft design, battery technology, solar cells and electrically powered engines themselves.

During the period I have been researching this book, international research and development of electric aircraft have been making giant strides: a solar airplane has circumnavigated the planet, electric aircraft have crossed the English Channel, electric airships and helicopters are being tested, electric drones are everywhere, and the major aircraft manufacturers are working on hybrid-electric passenger craft for the future. But the origins, going back more than 120 years, are just as fascinating.

In presenting this book, I am reminded of two classics published over a hundred years ago, both by French aviation pioneers. Octave Chanute's 308-page tome *Progress in Flying Machines* was published in 1894 by the Courier Corporation, while Joseph LeCornu's *La Navigation Aérienne Histoire documentaire et anecdotique* was published by Vuibert and Nony in 1903. Although these authors presented up-to-date detailed accounts of aviation, neither was able to record the exponential progress which would take place in the three decades which followed. Where electric airplanes are concerned, with the cut-off point of this book as of spring 2018, it will be the same for me, unless I live to one hundred years old—and why not!

ONE

Origins

The ongoing story of electric aviation as a love-hate relationship between mythology and hard scientific fact goes back a very long way. Let us enjoy some mythology first.

While the sun god Aton (14th century BC) proclaimed by the monotheistic Egyptian pharaoh Akhenaton is depicted as a solar disk emitting rays of light terminating in human hands, certain images sculpted by ancient Egyptians between 3,000 and 4,000 years ago might, according to some, suggest their knowledge of both airplanes and electricity. A unique group of hieroglyphs found in Sethi I's temple in Abydos, Egypt, are said to depict nothing less than a helicopter, a submarine, a glider, and another unknown type of aircraft among the usual insects, symbols and snakes. The initial carving translates to "He who repulses the nine [enemies of Egypt]." The rational explanation is these images were created by a startling coincidence—two overlaid Pharaoh names, Sethi I and Ramses II. As time went by, the plaster came off, and the pictures were created *by accident*. There is no ancient technology here.

A 6-inch wooden artifact with a bird's head, dating back 2,000 years and discovered in 1898 during excavation of the Pa-di-Imen tomb near the Saqqara Pyramid in Egypt, presents both a wing and a vertical tail fin more akin to those of a glider than a bird. Furthermore, the hieroglyphs on the "model" airplane read "The Gift of Amon," and three papyruses found near the artifact include the phrase "I want to fly." Egyptian physician Khalil Messiha has claimed that the Saqqara Bird has aerodynamic qualities and that the only thing missing is the tail wing stabilizer. To support his claims, Messiha built a balsa wood model six times the size of the original and added the tail, surprised to see that the model indeed could fly. Recent tests in modern wind tunnels have validated Messiha's test. In spite of these claims, however, no full-scale ancient Egyptian aircraft have ever been found, nor has any other evidence suggesting their existence come to light.

Elsewhere in the Hathor temple at the Dendera Temple complex, three stone reliefs depict what appear to be electric lights linked to a text proposing "high poles covered with copper plates." Critics suggest these images do not relate to electricity or lightning, pointing out that no evidence of anything used to manipulate electricity had been found in Egypt. They suggest that this was a magical and not a technical installation.

The Ark of the Covenant (in Hebrew: אָרוֹנְהַבְּרִית, ʾĀrônHabbərît), also known as the Ark of the Testimony, was described in the Book of Exodus as containing the two stone tablets of the Ten Commandments. Built in Sinai about 1400 BC by Bezalel, son of Uri, son of Hur, it served as a portable temple used during the Exodus in the desert and then the conquest of the land of Israel. It was made of acacia wood covered with gold, with two winged cherubim, one at each end. The Bible states that the Ark was carried on poles inserted in rings at the four lower corners carried by strong men from the tribe of Levi.

There is, however, a theory that it was able to levitate vertically and then move horizontally through the air; that its design was well known to match that of an electric capacitor, its wood being the perfect insulator between two charged plates of gold. In addition, the Ark was infamous for its deadly energy discharges. Those unqualified to touch, approach, or even look at the Ark would be struck dead. In this way, the following narrative becomes ambiguous: "And they departed from the mount of the LORD and the ark of the covenant of the LORD went before them in the three days' journey, to search out a resting place for them."[1] In addition, God's presence is frequently seen in the guise of a cloud in the Bible (Ex. 24:16). He appeared as a pillar of cloud (Exodus 33:9) and the Ark is constantly accompanied by clouds. When God spoke from between the Cherubs, there was a glowing cloud visible there (Ex. 40:35); when the Jews traveled, they were led by the Ark and a pillar of clouds (Num. 10:34). Was this pillar some form of aircraft?

It has again been suggested that the Biblical prophet Ezekiel (Yechezkel), living in Babylon around 800 BC, witnessed the arrival of an electric flying machine:

> And I looked, and, behold, a whirlwind came out of the north driving a great cloud, and whirling fire surrounding it, with the gleam of polished brass (chashmal) in the center of the fire. Now as I looked at the living beings, behold, there was one wheel on the earth beside the living beings, for each of the four of them. The appearance of the wheels and their workmanship was like sparkling beryl, and all four of them had the same form, their appearance and workmanship being as if one wheel were within another. Whenever they moved, they moved in any of their four directions without turning as they moved…."[2]

In modern Hebrew, "chashmal" חַשְׁמַל means electricity.

Staying with pseudo–Biblical legends, one elaborated across the ages has it that the Queen of Sheba gifted King Solomon a green and gold flying carpet sixty miles long and sixty miles wide studded with precious jewels, as a token of her love. It is said that a flying carpet was woven on an ordinary loom, but its dyes held spectacular powers. These were made from a special type of iron-rich clay procured from mountain springs and untouched by human hands. The clay was superheated at "temperatures that exceeded those of the seventh ring of hell" in a cauldron of boiling Grecian oil, and then acquired antimagnetic properties. So impregnated, the superconducting carpet, at an altitude of several hundred feet, could fly anywhere by following both the thermal air currents and the trillions of magnetic lines crossing the Earth from the North to the South Pole. When Solomon sat upon the carpet he was caught up by the wind, and sailed through the air so quickly that he breakfasted at Damascus and supped in Media.[3]

Known manuscripts of the Arabian *One Thousand and One Nights* date back as far as the ninth century. One of these relates how Prince Husain, the eldest son of the Sultan of the Indies, travels to Bisnagar (Vijayanagara) in India and buys a magic carpet. This carpet is described as follows: "Whoever sitteth on this carpet and willeth in thought to be taken up and set down upon other site will, in the twinkling of an eye, be borne thither, be that place near at hand or distant many a day's journey and difficult to reach."[4]

The Royal Library in Alexandria, Egypt, was one of the largest and most significant libraries of the ancient world. It was dedicated to the Muses, the nine goddesses of the arts. It flourished under the patronage of the Ptolemaic dynasty and functioned as a major center of scholarship from its construction in the 3rd century BC until the Roman conquest of Egypt in 30 B.C., with collections of works, lecture halls, meeting rooms, and gardens. According to a Jewish scholar named Isaac Ben Sherira, this library also kept a large stock of *flying carpets* for its readers. The carpets were handed out, traded for the visitor's slippers, and used to glide back and forth, up and down, among the shelves of papyrus manuscripts. The library was housed in a ziggurat that contained forty thousand scrolls of such antiquity that they had been transcribed by three hundred generations of scribes. The ceiling of this building was so high that readers often preferred to read while hovering in the air. The manuscripts were so numerous that it was said that not even a thousand men reading them day and night for fifty years could read them all! ... *except*, that this wonderful tale, including Isaac Ben Sherira, is the fictitious elaboration of Pakistani-Australian author Azhar Ali Abidi in his *The Secret History of the Flying Carpet*, published as recently as 2002 AD.

It was in the 4th century AD that the first concept of rotary-wing aviation came from the Chinese. A book called *Pao Phu Tau* (also *Pao Phu Tzu* or *Bao Pu Zi*, 抱朴子) tells of the "Master" describing flying cars (feichhe) made from wood from the inner part of the jujube tree with ox-leather straps fastened to returning blades that set the machine in motion (huancheiniyhichhi chi). This is the first recorded evidence of what we might understand as a helicopter.[5]

Such myths only show how from very long ago, man desired to ride around in the air. These myths, long after this book has been printed, may well become daily reality.

In 1410, *Le Livre de bonnesmoeurs* (*Book of Morality*) was presented to the Duke of Berry by an Augustinian monk, Jacques Legrand. Towards 1490, an artist from the Valley of the Loire decorated a copy of this work with 53 miniatures. One of these shows a pregnant woman turning a vertical wheel, and above whom hovers a golden disc. At virtually the same time, Domenico Ghirlandaio of Florence, Italy, painted his *The Madonna with Saint Giovannino* in which Mary the mother of Jesus looks down while in the background a man on a ledge blocks the sun with his hand and stares at the strange flying object in the sky. By his side is a dog also looking up with its mouth slightly open. The object appears as a dark oval structure with light rays projecting out from all angles. When Aert de Gelder of Amsterdam painted the baptism of Jesus by John the Baptist, instead of the Spirit of God descending as a traditional dove as he appears in hundreds of depictions of this event, de Gelder painted an oval object beaming down rays of light.

Who was indeed the first to fly using electricity? Reference might be made to one tale from Greek mythology, recounted by Pseudo-Apollodorus of Athens in his *Epitome of the Biblioteca*, a compendium of Greek myths and heroic legends, arranged in three books, generally dated to the first or second century AD From Epitome 1:

> On being apprised of the flight of Theseus and his company, Minos shut up the guilty Daedalus in the labyrinth, along with his son Icarus, who had been borne to Daedalus by Naucrate, a female slave of Minos. But Daedalus constructed wings for himself and his son, and enjoined his son, when he took to flight, neither to fly high, lest the glue should melt in the sun and the wings should drop off, nor to fly near the sea, lest the pinions should be detached by the damp. But the infatuated Icarus, disregarding his father's injunctions, soared ever higher, till, the glue melting, he fell into the sea called after him Icarian, and perished.

Modern physiological research suggests that electrical signals from Icarus's brain, during his attempt to fly, were telling his shoulders and arms to flap, even if in his case solar energy proved destructive. Similar electrical signals would be used by flying horses such as Pegasus and the whole world of birds, bats and insects, currently being observed and copied with the biomimicry-inspired drones of the 21st century. Daedalus was not a mythological figure; he was an aeronautical designer, one of the engineers of Knossos. They constructed water-chutes in parabolic curves to conform exactly to the natural flow of water—streamlined chutes. Streamlining could only be produced by long years of scientific development and is an essential part of aerodynamics, which Daedalus must have mastered.

In 1505, Leonardo da Vinci, the Italian polymath genius, took a new notebook and began writing down in his coded reverse script observations of bird flight, the nature of air and flying machines. This *Codex on the Flight of Birds* runs to more than 35,000 words with 500 margin sketches. Leonardo noted: "A bird is a machine working according to mathematical laws. It lies within the power of man to reproduce this machine with all its motions, but not with as much power.... Such a machine constructed by man lacks only the spirit of a bird, and this spirit must be counterfeited by the spirit of man."[6] Leonardo makes such observations as how the tips of a bird's feathers are always the highest part of the bird when its wings are lowered, and how the bones in the wing are the highest part of a bird when its wings are raised. Later on he begins to examine how bird flight could be applied to a man-carrying machine.

Eight years later, while staying in the Vatican, Leonardo predicted how humanity would use the sun's energy. In a series of notes scribbled on blue paper he draws and describes "a pyramidical structure which brings so much power to a single point that it makes water boil in a heating tank like they use in a dyer's factory."[7]

Mythical stores of flight continued into the modern era when in the 1620s, Francis Godwin, Bishop of Hereford, wrote a Utopian-style book, *The Man in the Moone, or a discourse of a voyage thither*. He was inspired by the swans that regularly flew low above the River Wye beside his cathedral. In his book, the hero, Domingo Gonsales, flies to the moon in a chair pulled by the *gansa*, a species of wild swan able to carry substantial loads. Gonsales discovers the *gansa* on the island of St. Helena, and contrives a device that allows him to harness many of them together and fly around the island. Gonsales resumes his journey home, but his ship is attacked by an English fleet off the coast of Tenerife. He uses his flying machine to escape to the shore, but once safely landed he is approached by hostile natives and is forced to take off again. This time his birds fly higher and higher, towards the moon, which they reach after a journey of twelve days. *The Man in the Moone* was first published posthumously in 1638 (John Norton, London) under the pseudonym of Domingo Gonsales and enjoyed a popularity across Europe. By the second edition (1657), the pseudonym had been replaced by "F.G. B. of H." ("Francis Godwin, Bishop of Hereford").

There were those reputed to have used mechanical means to make an object fly. Johannes Muller von Konigsberg was a German mathematician, astronomer, astrologer and inventor who also frequently went by the Latin pseudonym Regiomontanus when he published his works. In the early 1500s Regiomontanus is said to have built an eagle out of wood and metal, and having fitted it with the cogs and wheels of clockwork, to have it fly out of the city of Nuremburg to greet the Holy Roman Emperor, then fly back again![8]

But there were some seriously interested in magnetic effects and their use in flight. William Gilbert, the royal physician to Queen Elizabeth I, devoted much of his time, energy and resources distinguishing the magnetic effect of the lodestone from static electricity produced by rubbing amber, which he named "electricus." In his book, *De Magnete*,

Magneticisque Corporibus, et de MagnoMagneteTellure, published in 1600, Gilbert invented a terrella (Latin for "little earth"), a small magnetized model ball representing the Earth. This gave rise to all sorts of fanciful theories which were published in *The Philosophical Transactions* of the Royal Society of London, including speculations about flying chariots.

In 1709, Bartolomeu Lourenço de Gusmão, a Portuguese priest and naturalist, presented a petition to King João V of Portugal, seeking royal favor for his invention of an airship, in which he expressed the greatest confidence. Gusmão wanted to spread a huge sail over a boat-like body like the cover of a transport wagon; the boat itself was to contain vacuum tubes through which, when there was no wind, air would be blown into the sail by means of bellows. The vessel was to be propelled by *the agency of magnets*, which were to be encased in two hollow metal balls. However, the public test of the machine, which was set for June 24, 1709, did not take place.

One book dared to poke fun at such projects: *Travels into Several Remote Nations of the World. In Four Parts. By Lemuel Gulliver, First a Surgeon, and then a Captain of Several Ships*, commonly known as *Gulliver's Travels*, is a prose satire by Irish writer and clergyman Jonathan Swift. It was published in 1726 and amended in 1735. In Part Three, "A Voyage to Laputa," Gulliver sees:

> ...a vast Opaque Body between me and the sun, moving forwards towards the island; it appeared to be about two Miles high, and hid the Sun for six or seven minute ... the Reader can hardly conceive my Astonishment, to behold an Island in the Air, inhabited by Men, who are able (as it should seem) to raise, or sink, or put into a Progressive Motion, as they pleased. At the Center of the Island there is a Chasm about fifty Yards in Diameter containing at bottom a dome extending 100 yards into the adamantine surface. This dome serves as an astronomical observatory.... The greatest Curiosity, upon which the fate of the Island depends, is a Loadstone of a prodigious size, in shape resembling a Weaver's Shuttle. It is in length six Yards, and in the thickest part at least three Yards over. This Magnet is sustained by a very strong Axle of Adamant passing through its middle, upon which it plays, and is poised so exactly that the weakest Hand can turn it.... *By means of this Loadstone, the Island is made to rise and fall, and move from one place to another.* For, with respect to that part of the Earth over which the Monarch presides, the Stone is endured at one of its sides with attractive Power, and at the other with a repulsive. Upon placing the Magnet erect with its attracting end toward the Earth, the Island descends; but when the repelling Extremity points downwards, the Island mounts directly upwards. When the Position of the Stone is oblique, the Motion of the Island is too. By this oblique Motion the Island is conveyed to different Parts of the Monarch's Dominions.

Swift therefore entered into science fiction when he described this dome-shaped airplane which could fly around the kingdom using a lodestone. Magnetic levitation? He adds that is the custom of the inhabitants to throw rocks down at rebellious cities on the ground. Aerial bombing? While at the Grand Academy of Lagado, great resources and manpower are employed on researching completely preposterous schemes such as extracting sunbeams from cucumbers. Solar power?

But some scientists produced real effects. On January 20, 1746, Pieter van Musschenbroek of Leyden University's Theatrum Physicum announced in a letter to the Paris scientist René Réaumur that he had come up with a technique using linked jars of *storing* the transient electrical energy which could be generated by friction machines. The letter, written in Latin, was translated by the scientist-clergyman Abbé Jean-Antoine Nollet, who named it the "Leyden jar." The name stuck. On page 46 of his book *L'Essai sur l'Electricité des Corps*, published in Paris by Frères Guerin, Abbé Nollet recounts how he first sent a discharge from a Leyden jar through a company of 180 soldiers holding hands.

In Part Three of Jonathan Swift's satire *Gulliver's Travels*, published in 1727, Laputa is in fact a floating island, powered and directed by a magnetic lodestone—or in other words, an electrically propelled airship (Wikimedia).

This demonstration was before King Louis XV at Versailles. The King was both impressed and amused as the soldiers *all jumped into the air* simultaneously when the circuit was completed. The King requested that the experiment be repeated at the Royal Academy of Sciences in Paris. So Nollet next gathered about two hundred Carthusian monks into a circle about a mile (1.6 km) in circumference, with pieces of iron wire connecting them. He then discharged a battery of Leyden jars, which had been charged from a glass globe design of a generator, through the human chain, and once again observed how the monks, swearing and contorting, physically *jumped into the air like one man*, in sharp response to the discharge from the Leyden jars.[9]

Maybe the first practical experiment with a flying object and electricity is reported to have been made in 1750 by Benjamin Franklin of Philadelphia with his kite in a thunderstorm. According to the legend, during the thunderstorm, Franklin kept the string of the kite dry at his end to insulate him while the rest of the string was allowed to get wet in the rain to provide conductivity. A key was attached to the string and connected to that latest technology, a Leyden jar, which Franklin assumed would accumulate electricity from the lightning. The kite wasn't struck by visible lightning (had it done so, Franklin would almost certainly have been killed), but Franklin did notice that the strings of the kite were repelling each other and deduced that the Leyden jar was being charged. He reportedly received a mild shock by moving his hand near the key afterwards, because as he had estimated, lightning had negatively charged the key and the Leyden jar, proving the electric nature of lightning. Fearing that the test would fail, or that he would be ridiculed, Franklin only took his son William to witness the experiment, and then published the accounts of the test in the third person. It was Franklin who coined the word "battery" and assigned a positive sign (+) for a gain in electricity and a negative sign (−) for a loss of electricity.[10]

In 1774, Abbé Pierre-Nicolas Berthelon, Professor of Experimental Physics for the States of Languedoc in Montpellier, a friend and admirer of Franklin, sent up a number of paper kites, to which he had attached pieces of metal, long and narrow, and terminating in a cylinder of glass, or other substance suitable for the purpose of isolation. In this way, he obtained sufficient electricity to demonstrate the phenomena of attraction and repulsion, as well as electric sparks.

In northern France, in 1775, Louis-Guillaume de Lafolie, a physicist and chemist of the Academy of Rouen, annoyed his fellow members when he published a novel he called *La Philosophe sans pretention ou l'Homme rare. Ouvrage physique, chymique, politique et moral, dédié aux savants* (*The Unassuming Philosopher or the rare Man. A physical, chemical, political and moral work, dedicated to scientists*) (Paris: Clousier). During this period, de Lafolie's fellow academicians were trying to distinguish scientific fact from fiction. So to write a tale, told by an Arab about Scintilla, a former resident of the planet Mercury, who narrates her experience of space flight in an elaborate, newly invented ship which she has crash-landed on Earth, was bound to upset them as much as Jonathan Swift had upset the Royal Society of London.

De Lafolie wrote: "I imagined a machine with wings.... But what was my surprise when arriving on the platform, I observed two glass globes three feet in diameter, mounted above a small convenient enough seat. Four wood studs covered with glass plates supported these two globes. The lower room that served as base support and the seat was a camphor-coated tray and covered with gold leaf. The whole was surrounded by metal." De Lafolie was suggesting that his fictional flying contraption would be *powered by static electricity*, produced when, using a system of hand-cranked wheels, its two glass globes were rubbed together, their powerful light changing the air pressure and enabling the alien operator to navigate. In 1705, Francis Hauksbee of London had demonstrated a glass globe (that could be evacuated) rotated rapidly by a pulley and rubbed with the hand became electrified, allowing various observations on electrostatic attraction and repulsion and on electrical discharges in a vacuum. By the 1770s such frictional machines had become very popular for electrostatic therapy. Strange that the Mercurian flying machine did not use Leyden jars.

In 1779, de Lafolie made amends for his planetary fantasy when he invented a varnish composition suitable for protecting copper-plated ships from the corrosive effects of seawater. He had been appointed Inspector of the Royal Factories by King Louis XVI, in recognition of his laborious and useful works, when he died at forty-one, following a slight injury that he suffered when a chemically charged flask he was holding in his hand shattered.

"La Fumée Electrique" ("Electric Smoke") was the phrase erroneously used by the Montgolfier brothers, Joseph-Michel and Jacques-Etienne, to describe the hot smoky air which, in the summer of 1782, raised their pioneering balloons to a height of 300 m (980 ft.) into the skies above Annonay, in France's Ardeche region. The brothers had begun experimenting with balloons made of paper and early experiments using steam as the

In 1775, Louis-Guillaume de Lafolie, a physicist and chemist of the Academy of Rouen, published a tale about the visit to Earth by Scintilla from the Planet Mercury in a space ship powered by static electricity generated by rubbing two globes together (Musée EDF Electropolis, Mulhouse).

lifting gas were short-lived due to an effect on the paper as it condensed. Mistaking smoke for a kind of steam, they began filling their balloons with hot smoky air which they called "electric smoke." The French Academy of Sciences soon invited them to Paris to give a demonstration.

On December 12, 1783, Horace-Bénédict de Saussure published a letter in the *Journal de Paris* to prove that, "contrary to what scientists are thinking, it is not the 'lightness' of this air pushing up aerostatic machines; but the air rarified by the heat of the flames. The evidence is very easy to check: by introducing a piece of metal white hot in a paper bag, it is easy to propel it to the ceiling. The strong smell of electrical smoke remains!" In a letter he sent from Geneva on March 26, 1784, de Saussure mentions his having made some experiments on atmospheric electricity, with a tethered aerostatic machine, which, raised by means of the combustion of spirits of wine, was fastened to a long string. Although it was a cloudy day, he obtained a positive electric charge strong enough to create sparks.

Some fourteen years before, a young Italian called Alessandro Volta had written a treatise *On the forces of attraction of electric fire* (*De vi attractive ignis electrici ac phænomenis independentibus*) in which he put forward a theory of electric phenomena. His fame soon spread, and in 1781–1782, Volta, inventor of the electric pile or battery, traveled to Switzerland, Alsace, Germany, the Netherlands, Belgium, Paris and London.[11] The Montgolfiers most surely had heard of Volta when they used the phrase "electric smoke."

In 1783, Benjamin Franklin, then living in Passy as Commissioner for France, had witnessed a Montgolfier ascent above the gardens of the King's hunting lodge in the Bois de Boulogne, on the outskirts of Paris. Of this, he made the following general observation in a letter recounting the Montgolfier Brothers' demonstration to the president of the Royal Society, Sir Joseph Banks, dated July 27, 1783: "I am pleas'd with the late astronomical Discoveries made by our Society. Furnish'd as all Europe now is with Academies of Science, with nice Instruments and the Spirit of Experiment, the Progress of human Knowledge will be rapid, and Discoveries made of which we have at present no Conception. I begin to be almost sorry I was born so soon, since I cannot have the Happiness of knowing what will be known 100 Years hence!" Franklin believed hot-air balloons "to be a discovery of great importance, and one which may possibly give a new turn to human affairs. Convincing sovereigns of the folly of wars may perhaps be one effect of it; since it will be impracticable for the most potent of them to guard his dominions."[12] But Franklin made no specific statement about the potential application of electricity to human flight.

On July 18, 1803, a Voltaic pile and an electrometer were taken aloft for five hours in a Montgolfier balloon high above Hamburg to measure the existence or nonexistence of electrical matter. The experiment was organized by a Belgian physicist and keen balloonist called Etienne-Gaspard Robert, otherwise known as "Robertson" or "Arago," and his assistant L'Hoëst. In his nacelle, Robertson had taken "a Voltaic pile composed of sixty couples, silver and zinc; it worked very well when we started and gave, without condenser, one degree to the Volta electrometer. At our greatest altitude, the pile was giving no more than 5/6th of a degree to the same electrometer. The galvanic light seemed to me far more sensitive than on the ground; this effect appears contradictory." The following year, Robertson proposed his plans for the *Minerva*, "an aerial vessel destined for discoveries, and proposed to all the Academies of Europe." He dedicated his project to none other than Alessandro Volta. It would be propelled by sails, but was never built.[13]

Flight inspired great poets. In 1835 Queen Victoria's Poet Laureate, Alfred, Lord Tennyson, in his poem "Locksley Hall," wrote a passage called "Prophecy":

> For I dipt into the future, far as human eye could see,
> Saw the Vision of the world, and all the wonder that would be;
> Saw the heaven fill with commerce, argosies of magic sails,
> Pilots of the purple twilight, dropping down with costly bales....

One electrochemist who was passionate about aviation was John Stringfellow of Chard, Somerset, England. In 1856 Stringfellow, who earned his money making bobbins for the local lace-making industry, patented what he called his Electro-Voltaic Pocket Battery, a device measuring only 3 in × 4 in, to be used to treat a whole range of medical ailments. If Stringfellow had subscribed to a publication called *The Annals of Electricity, Magnetism and Chemistry*, in the April 1837 edition, he may have read how his fellow countryman William Sturgeon had succeeded in propelling a boat and also a locomotive carried by the power of electromagnetism. But Stringfellow's prior passion was flight, and the same year, he and his associate William Henson had planned a monoplane powered by the latter's patented lightweight steam engine, not with a series of his own pocket batteries. The first hop of the 10-foot wingspan *Aeriel*, achieved in a lace factory, gave them the confidence to form the Aerial Transit Company to raise money to construct a 50-foot wingspan Aerial Steam Carriage, which would take people to exotic locations like Egypt, India and China; in short, an airline.

However, attempts involving a larger model with a 20-foot wing span were unsuccessful. While Henson emigrated to the USA, Stringfellow, still financing himself by

In 1843, to raise funds for their project, Henson and Stringfellow envisaged their Aerial Steam Carriage taking people as far as India, using an ingenious system for taking off (Chard Museum).

bobbin-making, continued his experiments. An unmanned demonstration flight of some 40 feet was achieved in August 1848 under a canopy specially erected for the purpose in Cremorne Gardens, London. In 1871 Stringfellow, now in his seventies, had built a steam powered triplane, which was exhibited at the Crystal Palace in London. The old man had planned to eventually build a flying machine which would carry him aloft, and equipped a building for just that purpose. Age and illness intervened, however, and that machine was never built. He died in 1883. All that was missing was that improved but elusive power source which he knew so well: electricity.

On New Year's Day, 1871, Thomas Alva Edison, 24 years old, of Newark, New Jersey, who had already invented the electric vote recorder, the automatic repeater, and other improved telegraphic devices, speculated: "A Paines engine[14] can be so constructed of steel & with hollow magnets. .. and combined with suitable air propelling apparatus wings … as to produce a flying machine of extreme lightness and tremendous power."[15]

Nine years later, in 1880, a journalist working for the *New York Daily Graphic* obtained a scoop from the world-famous inventor of the phonograph and the electric light. In an interview given at Menlo Park, Edison is supposed to have confided to the journalist his latest project to build a featherweight "flying canoe," borne aloft by two bat-like wings and operated by *electric motors*. Silk, stretched over a bamboo framework, would form the lifting surfaces, and the aerial boat would be steered from side to side by means of a bird-like tail at the rear of the cigar-shaped body. The craft would take two passengers and would be able to come down on land or water. Later on, the article recounts, Edison wanted to build a giant six-winged craft which could land on rough seas if necessary for flights across the ocean. Such a machine, driven by silent electric motors, would rush through the sky at one hundred miles an hour and would carry passengers from New York to Paris or Moscow nonstop in less than thirty hours. Weird skyscraper landing towers were included as a possible feature of such an airline. These huge pylons of steel, rising up above the lower clouds, would carry at their tops waiting depots and strange spoon-like landing cradles into which the big machines would settle at the end of a flight and from which they would take off again, loaded with fresh cargo and passengers. Lit with electric lamps and surrounded by a railed-in promenade, the octagonal waiting room at the top of the tower would provide shelter and refreshments for incoming and outgoing passengers. Electric elevators would make rapid trips up and down the tower.

Perhaps significantly, this interview as published on April 1, 1880, known to some as "April Fool's Day," or in French "Le Poisson d'avril" (April's Fish), celebrated every year for its published hoaxes. Indeed, some months later, Edison stated that the *Daily Graphic* reporter had rather exaggerated the plans he had in mind. But the news was out and Europe was caught up with the idea that the Wizard of Menlo Park was constructing a flying machine to tour the world. In June 1880, Abel Hureau de Villeneuve of Paris, editor of *L'Aéronaute: moniteur de la Société generale d'aérostation et d'automotion aériennes* (*The Aéronaute: Monitor of the General Society of Aeronautics and Air Automation*) translated and reprinted the interview. De Villeneuve then pointed out that electricity simply would not have the power to make the Edison Air Ship work like bird flight.[16]

In reality, on 7 July 1880, Edison did make a rough sketch of a helicopter. He built a test stand and tested several different propellers using an electric motor. He deduced that in order to create a feasible helicopter, he needed a lightweight engine that could produce a large amount of power.[17]

Ironically, it was the French who started more modest and realistic experiments

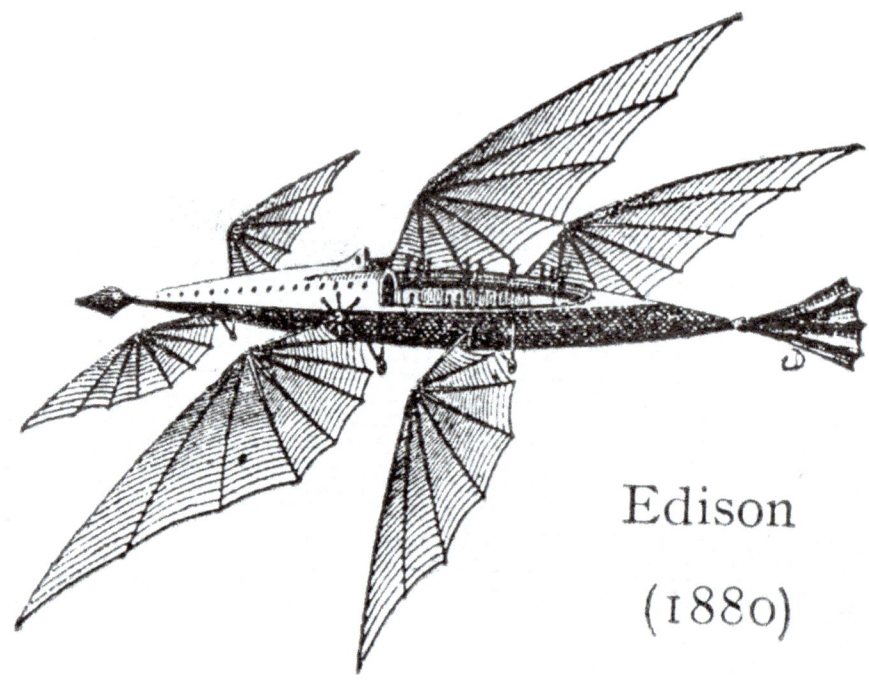

In 1880, inventor Thomas A. Edison is said to have disclosed in an interview with a journalist from the *Daily Graphic* that he had built an electric flying canoe capable of crossing the Atlantic, except that the hoax article was published in April 1, known as "April Fool's Day" (Musée EDF Electropolis, Mulhouse).

with electric airplanes. In the fall of 1881, an International Electrical Exhibition was held in Paris, showing the applications of electricity to medicine, entertainment and transport. Floating under the roof of the exhibition gallery was the oblong aerostat of the Tissandier brothers, Gaston and Albert, experienced balloonists. It measured 3.5 meters (11 ft.) long, 1.30 meters diameter at the middle. It had a volume of 2 cubic meters (71 cubic feet) and 200 grams (7 oz.). Filled with pure hydrogen, it had a lifting force of 2 kg (4 lbs.). The lower part of the nacelle of the little balloon was equipped with a minuscule *electric* motor, built by Parisian electrical engineer Gustave Trouvé and weighing just 220 grams (7.7 oz.). What made the motor unique was that to make it lightweight, parts of it had been machined in the then-revolutionary aluminum.[18]

The shaft of this little machine was connected to a two-bladed propeller, made of wood and textile and turning at six and a half revolutions per second. Energy came from two small secondary lead-acid batteries supplied by the Tissandiers' and Trouvé's friend, Gaston Planté. The motor and the batteries had a weight inferior to the lifting force of the balloon and would be raised by this when it was filled with hydrogen.

Before arriving at the exhibition, the team had tested out their aerostat in the rooms of the Conservatoire des Arts et Métiers, with the enthusiastic approval of its director, Hervé Mangon. Further experiments were then made in the workshops of M. Lachambre, in the rue de Vaugirard.

For the exhibition, in full view of the public, the little aerostat was tethered to a guide rope which towed it to and fro like a circus horse. It had a stern rudder to enable

During the International Electrical Exhibition, held in Paris in 1881, this electrically propelled tethered aéronef gave regular demonstrations to visitors, some of whom went away determined to build the full-scale version, while others were inspired to write fictional adventure stories (Musée EDF Electropolis, Mulhouse).

it to move right or left. Throughout the exhibition, demonstrations of the aerostat were given twice a week—on Thursdays at 4:30 p.m. and on Saturdays at 9 p.m. Visitors were entranced. Tissandier now calculated that with Trouvé's motor, a scaled-up aerostat, in calm weather, could reach between 20 and 25 km/h (12 and 15 mph)!

Among those hundreds who had admired the tethered aerostat was William Delisle Hay, an English writer and member of the Royal Geographical Society. In 1881, Hay's science fiction novel *Three Hundred Years Hence, or, a Voice from Posterity* was published by Newman of London. In this book, set in the year 3001, Hay describes numerous types of flying machines; there are balloons whose canopies contain the extremely light "lucegen" gas; that canopy is positioned below rather than above the car. "The car was thus immersed, as it were, in a bladder covering it externally but leaving it open above; it sat in its balloon just as it might in water." Even the author concedes that such a design might tend to lead to instability and turning turtle. To prevent this, he uses a powerful magnet which is attracted to the Earth's magnetic field. Thus stabilized, the "Lucengenostat" is able to carry considerable weight of freight and passengers.

Yet these are superseded by even more powerful aerial machines, working on new principles. The lift is provided by "basilica-magnetism" for greater power and greater safety. The locomotive power may come from "generated heat and electricity," which causes a pair of fans modeled on birds' wings extending along the sides of the craft from stem to stern to flap. This type is known as the "alamotor" and is used for small utilitarian craft. Then there is the "spiralmotor," driven by one or more propellers of the "pusher"

The 220-g (7.7-oz.) motor of the aeronef, using aluminum parts to make it lighter, was designed and built by Gustave Trouvé, electrical engineer of Paris, who would also innovate the electric tricycle and electric boat using the same engine design powered by rechargeable lead-acid batteries (Musée EDF Electropolis, Mulhouse).

variety normally placed at the stern. "Heat and electricity give the motive power, and this form is the most generally employed on aircraft." For really heavy loads, the "zodiamotor" is available, being powered by the naturally occurring "zodiacal electricity." When Hay says that "it is that which holds together the elements of air," he can be construed (just) as referring to atomic energy. At any rate, his various flying machines are able to lift any weight and to travel at a thousand miles an hour, "though seldom employing more than half that rate."[19]

While the English public marveled at Hay's almost visionary imagination, back in Paris, Gaston Tissandier had taken out his seminal patent "New Application of Electricity to Aerial Navigation": "Electric motors offer the following advantages: 1st their weight is constant, so that the balloon can stay balanced in the air ... 2nd absence of fire, which offers a considerable danger under an aerostat inflated with hydrogen gas ... 3rd the electric motor offers the advantage of the ease of starting and stopping, and that of the mechanical simplicity."

Tissandier also formed a company with the intention of scaling up his aerostat to man-carrying dimensions. During the two years that followed, the Tissandier team assembled the full-scale 28-meter (92-foot) version of the tethered Trouvé-engined aerostat they had demonstrated at the Electrical Exposition. Their nacelle was a cage made of bamboo lashed together with ropes and copper wire covered in guttapercha (rubber), while the floor was made of walnut planks surrounded by basketwork. It was equipped with a D4 long-coil motor specially built in their workshops by Siemens Frères and weighing just 54 kg (119 lb.). It was mounted on a wooden chassis with special transmission. Energy came from heavy-duty dichromate of potash batteries. The two-bladed canvas and bamboo pusher propeller turned at 180 rpm and could be warped by pulling on steel wires. The rudder was made of unvarnished silk also stretched over a bamboo framework. The envelope weighed 170 kg (374 lbs.), the motor and batteries 208 kg (460 lbs.), and with crew, the whole weighed 1,240 kg (2,734 lbs.).

On October 8, 1883, at Auteuil, southwest Paris, the airship rose up into the sky with the intrepid brothers on board.[20] [W]hen we got our motor to function at great speed, with the help of the 24 elements, the effect produced was completely different. The movement of the aérostat was becoming suddenly appreciable, and we felt the cold wind produced by our horizontal movement. Our aérostat regularly planed at a height of 4–500 meters. When she turned into the wind, with her forward point headed for the Auteuil bell tower near to our departure point, she held her head to the oncoming wind and stayed motionless...."[21] This was the first time a real electric airship had taken to the skies! Thousands of Parisians looked upwards and gasped in awe and wonder.

The airship flew at a gentle speed of 10 kph (6 mph.), passing over the Bois de Boulogne. It held its head against the wind but the rudder had little effect, and after 20 minutes airborne, at 4:55 p.m., the Tissandiers had to land in a large field near Croissy-sur-Seine. Painters and photographers arrived to portray the aerial ship, among a big and friendly crowd that the novelty of the spectacle had attracted from everywhere. They left her inflated the whole night. They would have made a second ascent, but the cold of the night had crystallized the dichromate of potassium in the ebonite reservoirs, and the battery, far from being exhausted, was unable to function. So the Tissandiers took their aerostat back to the banks of the Seine close to Croissy Bridge and deflated it.

They continued to improve their electric airship or aerostat. On the afternoon of Friday, September 26, 1884, to the clapping and cheering of a large surrounding crowd,

October 8, 1883, Auteuil, southwest Paris: the Tissandier brothers ready for takeoff in their Siemens-engined 28-meter (92-foot) airship (©D.R./Coll. musée de l'Air et de l'Espace, Le Bourget).

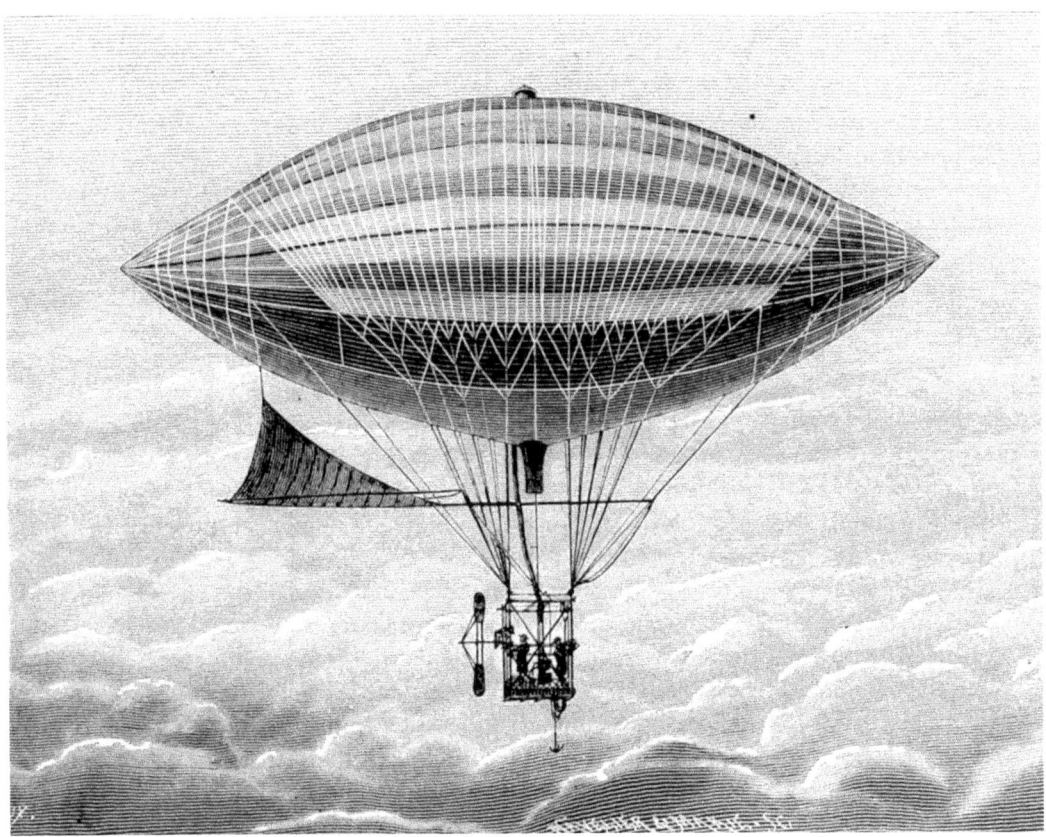

The Tissandier airship in the skies over Paris. Lack of rudder control forced it to land after a flight of only twenty minutes (Musée EDF Electropolis, Mulhouse).

it rose vertically into the skies again, with the brothers and a retired sailing captain called Lecomte on board. With its improved rudder, it flew for two hours, passing over Grenelle, the Luxembourg Observatory, Bercy Bridge, the Vincennes woods, Varenne-Saint-Maur, and Sucy-en-Brie. But without being able to go against a strong headwind, it had to land at 6:20 p.m. at Marolles-en-Brie, near Servon woods in the Seine-et-Oise region. In a letter sent three days later to Madame Hervé Mangon, Tissandier writes:

> Albert and I were really tired on the day of our ascent. Nobody can imagine how much it costs to make such experiments, trouble and frustration. It required all our efforts on the previous night in our workshop to be ready to prepare everything for 5 o'clock in the morning. The workers who were with us turned pale from our demands; they fought and we had to separate them! I had to work almost alone with big gas appliance; I was covered with acid stains and splashes. On the descent, I got a few liters of dichromate of potassium acid we were carrying on my legs, and when I returned in the evening to Paris, I had the appearance of a convict. Albert remained the following day in Marolles where we had touched down, to bring back all the equipment.

"La Scala," 13 boulevard de Strasbourg, Paris 10ème, named after Milan's opera house, was famous for its operettas and popular songs. With the Tissandiers' flight the talk of the town, it was rather fun that one of the songs was "Le Ballondirigeable" (steerable balloon), with words by Messieurs Lafaurié and Bourges and music by Monsieur Giraud-Malteau. The illustration on the front of the score was inspired by the Tissandiers' aerostat.

Some weeks before, on August 9, 1884, a second cigar-shaped balloon had graced the skies, to the delight of the Parisians. Its battery was what is now called a flow battery—electrical energy was stored by pumping chlorine from one chamber to another containing zinc, and so generating power to a motor. Its creator was one Charles Renard. Louis-Marie-Joseph-Charles Clément Charles Renard was born in Damblain in the Vosges regions of France in November 1847. A clever child, he won a scholarship that eventually led him to graduate with honors from one of the leading polytechnics (France's so-called "grandes écoles" or top universities).[22]

Renard was only 23 when the Franco-Prussian War broke out in 1870. Given command of a section of the 15th Army Corps on the Loire, he took part in the battles of Artenay, Cercottes and Orléans, and was awarded the Cross of the Légion d'honneur for his bravery in leading a defense against the Prussians. Extraordinarily, at the same time, this brilliant young engineer presented a way to adjust numerical values used in the metrical system for mechanical construction. The interval from one to 10 was divided into five, 10, 20 and 40.[23] Incidentally, the Renard series, in one of the most obscure facts about ballooning, helped the French army to reduce the number of different balloon ropes kept in its inventory from 425 to 17!

Three years later, Renard, promoted to lieutenant in the 3rd Regiment of Engineers, started working on flying machines, with what he called a "directional parachute" or aéride. The idea was for a 10-winged glider, without pilot and weighing just 7.5 kg (16 lb.), to be launched from a balloon to transport messages. It was tested with success in 1873 close to Arras, from one of the towers of the Mont Saint-Eloi Abbey.

In 1877, Renard, assisted by Captain la Haye, founded the Central Establishment of Military Aérostation in the park of the old Château de Chalais at Meudon outside Paris. First, he modernized the existing equipment. This included the building of a powerful, continuous-circulation hydrogen generator, designed by Charles Renard and built under the supervision of his brother Paul.

They were next joined by Arthur Constantin Krebs. Three years younger than Renard, his senior officer, Krebs had also fought against Prussia. Transferred to the workshops of the Chalais-Meudon Aérostation, the innovative Krebs worked for Renard and la Haye, resulting in a direct circulation steam generator, called the Renhaye.

One of Krebs's contributions was the development of the engine that would power the airship and the building of an electric boat, the *Ampère*, to measure the resistance of airship models. At the end of 1881 Renard had begun to research the electric

Charles Renard, who led the project to build *La France*, the world's second airship (Musée EDF Electropolis, Mulhouse).

generator indispensable for his project. His previous experiments showed he needed to create a generator capable of giving 10 hp for two hours, and weighing less than 480 kg (1060 lb.).

Working on the new theory of chemical affinity, Renard retained as his anodic couples chlorine and bromine. For the cathode: magnesium, aluminum, calcium and zinc. After he abandoned bromine for safety reasons, and magnesium, aluminum and calcium as being too expensive, the chlorine-zinc couple remained. The positive electrode retained was a silver-plated leaf of a thickness of one-tenth of a millimeter, the negative being a rod of pure zinc. After a great deal of experimentation he chose as his electrolyte a mix of hydrochloric acid at 11° Baumé and chromic anhydride, likely to release the chlorine. To avoid excessive overheating during the discharge, a tubular form was created, the container serving as a thermal radiator.

At the start of 1883, the construction of the definitive battery began. It had a specific capacity of 44 kg (97 lb.) per horsepower. Krebs built the electric motor: a rotor of two crowns of eight electromagnets, supplied on average by eight brushes. The whole weighed 88 kg (194 lb.) for 8.5 hp. On the side of the airship gondola, a steering wheel controlled the power of the batteries by pushing the zinc cathodes in and out of the batteries.

On August 9, 1884, the balloon airship *La France* took to the air at Meudon. It then flew above a farm at Villacoublay, then made a controlled return to Meudon—the world's

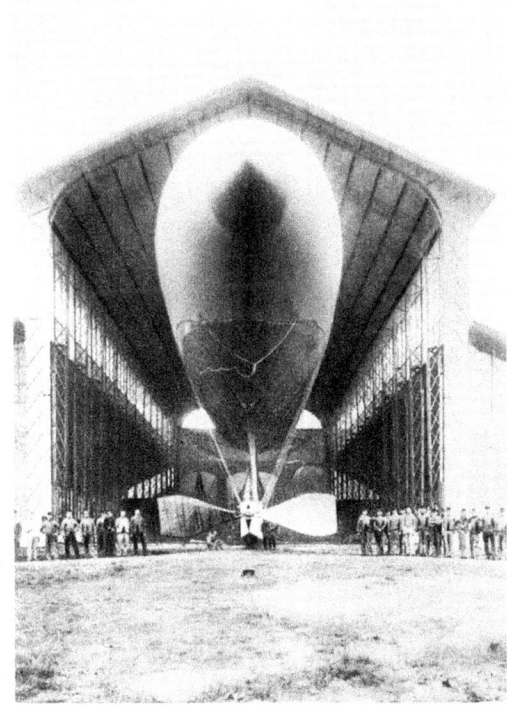

In 1884 and 1885, in making several controlled flights, *La France*, seen here at the Chalais-Meudon Aérostation, earned the title of "dirigible" (steerable). Its 88 kg 8.5 hp motor was a rotor made up of two crowns of eight electromagnets, supplied on average by eight brushes. Energy came from a revolutionary chlorine-zinc flow battery (Musée EDF Electropolis, Mulhouse).

first closed-circuit flight. The flight lasted 23 minutes for a circuit of 8 km (5 mi), giving "le dirigéable" (steerable) *La France* an impressive speed of 19.8 kph (12 mph). On its seven flights in 1884 and 1885, *La France* returned five times to its starting point.

The following year, Renard, now a colonel, persuaded France's Minister of Finance to invest the then-staggering sum of 200,000 francs in the project, including the erection of a hangar necessary for the construction and sheltering of balloons and airships. This would become the cradle and home for *La France*, which had demonstrated that controlled flight was possible if the airship had a sufficiently powerful lightweight motor. In 1889 it was proudly put on show at the military pavilion in the Place des Invalides, Paris, during the Universal Exhibition held that year. Among those impressed by the technology was the Marquis Henri de Graffigny, who by coupling the Renard flow battery to an 8 kg Trouvé motor, was able to increase the range of his pioneer electric-pedal tricycle to five consecutive hours and a distance of 95 km.

But Renard also realized that lighter-than-air, battery-powered, electric-engined dirigibles had their limitations. And for the last 20 years of his life he would move away from electrical research and spend his time creating more efficient gasoline engines. In 1902, working with the engineer Léon Levavasseur, he developed a revolutionary V8 aero-engine capable of 80 hp.

But Renard's last years were not happy ones. He had financial difficulties and grew tired of fighting the inertia within an army whose enthusiasm for aeronautics was lackluster. He was also a bachelor and his only relaxation was to visit his brother Paul's family and play and compose on the piano. Then his health deteriorated. Early in 1905 he suffered a bout of flu, which lowered his resistance. On April 13 he was found dead in the office of his chalet in the park. He was 58. Although the official diagnosis was a heart attack, rumors circulated—and still persist to this day—that while depressed, Renard had committed suicide. The rumors were never substantiated.

Among those impressed by the 1884 flight of *La France* was Augustin Henri Hamon of Boulogne-sur-Seine. In 1885, Hamon wrote a book, *Aerial Navigation*, published in Paris by Marpon and Flammarion. The following year, he was awarded patents in France (350,303), Germany and England for a dirigible aerostat that incorporated an electric motor with battery, and in particular a propeller-wheel with feathering blades whose blade angle could be altered for ascent and descent. Although never built, Hamon's ingenious design would later be seen again in ailerons and variable pitch propellers.

In March 1886, another aeronautical engineer called de Latour announced details of his battery-electric dirigible balloon. It would be propelled by a propeller turning at 300 rpm, giving it a speed of 4 meters (13 ft.) per second. Steering would be carried out using sails that could be furled and unfurled using a magnet, a prototype of the electric winch.[24] Three months later, Abel Clarin de la Rive of Chalon-sur-Saône, a senior journalist for various French newspapers, published a pamphlet in which he reviewed: "All the attempts which have been made to control balloons … up to Captains Krebs and Renard. Having noted the inherent faults in each system, the author announces a new dirigible aerostat of his invention also equipped with an electric motor."[25]

Another engineer was Constantin Senlecq, a notary from Ardres, northern France, and founding member of the International Electricity Company of Paris. Senlecq wrote a 12-page brochure, also titled "Aerial Navigation," in which he proposed a system combining the advantages of lighter-than-air (balloon using hydrogen gas for the ascent) and heavier-than-air (char of the nacelle with a horizontally rotating propeller driven by a motor powered by electricity). Dated January 1886, this pamphlet was registered at the Academy of Sciences meeting of August 27, 1886. Senlecq's machine was never built, nor was his "telectroscope," the forerunner of television.

In the late summer of 1886, France's popular science fiction writer Jules Verne serialized his novel *Robur the Conqueror* (French: *Robur-le-Conquérant*), also known as *The Clipper of the Clouds*.[26] Robur's *Albatross* is a slender, clipper-shaped hull made of hydraulically compressed paper or cellulose, 100 feet (30 m) long and 12 feet (3m 60) wide. Above its flat deck stands a veritable forest of slender masts, 37 of them, each with twin contra-rotating propellers at the top. At the bow and stern are two more propellers. A large rudder steers the "clipper of the clouds," and spring-loaded shock absorbers cushion its landings.

As for its propulsion:

> He employed electricity, that agent which one day will be the soul of the industrial world. But he required no electro-motor to produce it. All he trusted to was piles and accumulators. What were the elements of these piles, and what were the acids he used, Robur only knew. And the construction of the accumulators was kept equally secret. Of what were their positive and negative plates? None can say. The engineer took good care—and not unreasonably—to keep his secret unpatented. One thing was unmistakable, and that was that the piles were of extraordinary strength; and the accumulators left those of Faure-Sellon-Volckmar very far behind in yielding currents whose amperes ran into figures up to then unknown. Thus there was obtained a power to drive the screws and communicate a suspending and propelling force in excess of all his requirements under any circumstances.[27]

Often inspired by Verne, the Cuban-born Brooklyn writer Luis Senerans, alias "Noname," devoted one adventure in the weekly U.S. dime-novel series to *Frank Reade Jr. and his Queen Clipper of the Clouds: A Thrilling Story of a Wonderful Voyage in the Air*. Again propulsion was by electricity. This was serialized between February and July 1889. Another story, *The Electric Island; Or, Frank Reade, Jr.'s Search for the Greatest Wonder on Earth with His Air-Ship, the "Flight,"* is a complete novel. The Electric Island itself, southeast of Kerguelen, is a gigantic storage battery, and to land on it, one must wear rubber boots and insulated clothing. The flora and fauna are electrified. The explorers must land with a land-dwelling eel and an electric tortoise. Natural spark gaps provide illumination. When a storm arises, the atmospheric electricity reacts with that of the island; things are too perilous for the men and they leave. The island sinks beneath the sea, for it was ultimately volcanic. The rotascope-lifted airship cracks up in Australia.

In his novel *A Fortnight in Heaven: An Unconventional Romance*, published by Henry Holt and Company, New York (1886), Harold Brydges (aka James Howard Bridge) describes the voyage to Jupiter of an English sea captain's "spiritual double." Here he finds gigantic humans populating an alternative futuristic America. One of the first spectacles to confront him is a new Chicago, transformed into a city of crystal. The main force behind this transformation is electricity, powering "electric pedestrianism" (through a kind of accelerated bicycle) and "aerial ships."[28]

From 1886, perhaps inspired by the pioneering attempts of his compatriots, Dr. Arthur DeBausset, of French origins, resident of Chicago, ambitiously attempted to raise funds to construct his "vacuum-tube" airship design, which he called the "aero-plane." Instead of being filled with lighter-than-air gas such as hydrogen or helium like a dirigible, DeBausset's "aero-plane" gained buoyancy by having air *removed* from the cylinder by a vacuum system. The absence of air would make it float, with the huge size of the airless chamber negating the weight of a passenger cabin and the weight of the tank itself. Its

Among those inspired by such flights was the author Jules Verne. In 1886 Verne wrote *Robur the Conqueror* (French: *Robur-le-Conquérant*), also known as *The Clipper of the Clouds*. The 37 masts making up the distributed electric propulsion of the *Albatross* used accumulators to voyage through the skies (Bibliothèque Louis Aragon, Amiens: Agence Roger Viollet).

riveted steel cylinder would measure 236 meters (774 ft.) long, with a cylinder diameter 44 meters (144 ft.). Its conical ends would be constructed of ¼₄" steel plates. Power would come from six 60 hp engines; six 45 kW dynamos; *24 electric motors*; 40 propellers; and 4 quadruple pneumatic pumps. It would have a top speed of 120 mph (200 kph) and a cruising speed of 100 mph (160 kph). This would promise, for example, two-hour trips between Chicago and New York, and overnight trips to Europe.

DeBausset's pamphlet *Aerial Navigation* (a popular title for such books), printed by Fergus, was published under the auspices of the Transcontinental Aerial Navigation Company of Chicago, duly incorporated under the laws of the State of Illinois, February 18, 1886. In this Dr. DeBausset stated that the first notice of his invention was given at Saint Louis, Missouri, April 8, 1884, at a public conference he gave at Bomberger's Hall, and reported by the *Globe Democrat*. On November 6, 1886, an article titled "A New Air-Ship to be Propelled by Electricity" appeared in *The Electrical World*, New York.

An article, "Mammoth Airships," published in the *Chicago Herald* on March 20, 1887, begins:

> Above the entrance to the hallway at 236 State Street is a gilt sign reading "Transcontinental Aerial Navigation Company." In a pleasant office upstairs, its door bearing the same words, is found a short, round-headed Frenchman who answers to the name of Dr. A. DeBausset, president of the Aerial. He is a keen-eyed, energetic little man, who puffs at innumerable cigarettes placed one after another in a long amber holder, the while talking most rapidly in a language which is imperfect English with a full French pronunciation. Dr. DeBausset is the inventor of the aero-plane....

Writing in the "Chronique" column of *XIXème Siècle, Journal Républicain*, published in Paris, Raoul Lucet reported: "DeBausset's ship can carry 200 passengers and is destined to explore the Polar regions. It will leave New York on June 10, 1888 (the departure time is not given) and head for the Pole, stopping successively at Philadelphia, Washington, Toledo, Chicago, Omaha, San Francisco, Jeddo, Canton, Constantinople, Rome, Paris, Berlin, Copenhagen, Stockholm and St Petersburg."

In 1888, DeBausset published another pamphlet, titled *The Arctic Explorer*, in which he claimed that tickets for the maiden voyage of the *AS Artemis* had already sold out and the ship was still being fitted for passengers, while the demand had seen tickets changing hands for handsome sums. He even suggested that a four-strong fleet of "aero-planes" could take passengers across the Atlantic to the forthcoming Paris Universal Exhibition in thirty hours. At the time, a Cunard Line steamship held the record for a transatlantic crossing of 7 days, or 168 hours.

Promotional materials included testimonials by physicists backing the feasibility of the concept. DeBausset attempted to sell $100,000 in shares at $100 each to finance his venture, the U.S. government expressed interest for defense purposes, and a huge tract of land was secured in outlying Worth to build it. But the general public was skeptical. His patent application was eventually denied on the basis that it was "wholly theoretical, everything being based upon calculation and nothing upon trial or demonstration."[29]

Not only was *AS Artemis* never even built, but DeBausset disappeared from Chicago, briefly resurfaced in New York City, then disappeared for good from the aviation and business world.[30]

At the same time, in February 1887, Dr. Martin Braun, of German origins, residing in Cape Vincent, New York, obtained Patent U.S. 356743A for an "Electro-Dynamic Air Ship." Articles appeared in both *Scientific American* and *Manufacturer and Builder*. However, in his patent, Braun states, "Motion may be imparted to the shaft 24 by any suitable motor," not specifically electric. There is no reference that the Braun craft was ever built or flown.[31]

Rather like the Tissandier brothers, Professor Peter Carmont Campbell of Rhinebeck, Duchess County, New York, was no stranger to balloons. Following his first ascension in 1857 in the *Brooklyn Eagle* inside the Crystal Palace, New York City, Campbell went on to patent five different air ships. In 1887, with advice and encouragement from Samuel Morse and Horace Greeley, Campbell submitted his latest, more controllable design to aeronautic engineer Carl Edgar Myers for examination. The craft comprised an oblong cylindrical envelope 60 feet (18.3 m) in length and 42 feet (12.8 m) across, filled with coal gas. A long metal rod beneath the gasbag served as a keel. The keel was directly tied to the bag at the center. A web of cords extended from the bar to the ends of the bag. A boat-like car was slung from the keel. Large enough to house a pilot and three passengers, the car sported two large birdlike wings on each side. The wings were not designed to be flapped, but could be raised or lowered to control the direction of motion. A forward rudder was also employed. A large, multi-bladed propeller was located on the underside and at the front of the car. Campbell had originally planned *to use an electric motor and batteries*, but considering them too heavy, he resorted to pedal power. He received the patent for this invention in May 1887 (U.S. No. 362.602) and the Dirigicycle was built by the Novelty Air Ship Company in the months which followed for $ 2,500.

On its first flight in December 1888, with James K. Allan at the pedals, the Dirigicycle rose 30 m; then, called down by Campbell for a photo, it easily descended. On the next flight it climbed to a height of about 150 m (500 ft.) and took half an hour to cross the sky between Coney Island and Brooklyn. On July 16, 1889, Professor W.M. Hogan's brother Edward set out on a third flight from Brooklyn, New York. One of the propellers fell off, and the Dirigicycle was finally blown out to sea and lost. It was last seen 150 miles (250 km) out to sea heading for Africa. Edward Hogan was never seen again. It was later theorized that Hogan had been asphyxiated by coal gas escaping from the balloon.

Sometimes it paid to be less ambitious. Perhaps the first use of a flying machine to electrically transmit messages appeared at this time. A 30-year-old British meteorologist, Eric Henry Stuart Bruce, MA Oxon, conceived of a remarkably simple way to send message on a clear night in Morse code using a tethered balloon. Inside the balloon, made of transparent cambric, a little ladder was headed with six incandescent light bulbs, which were connected to a battery on the ground by a wire that ran side by side with the cable tethering the balloon. By switching the lights on and off through the translucent balloon, using the dot and dash system, the operator could telegraph messages to distant points, the only requisite being that the night must be dark and clear enough. By 1885, Bruce's electrical war balloon had been adopted by the British, Belgian and Italian governments.

In 1888, Bruce was giving demonstrations to members of the Royal Society, the Royal Institution, the Institute of Chemistry, the Birmingham and Midland Institute, and in both the Town Hall in Kensington, southwest London, where he lived, and the Crystal Palace. He even produced a sales catalogue, giving as an advantage the absence of danger in introducing the electric light inside a gas balloon; detailing how to put a red-hot poker inside a balloon without setting fire to the gas; instructing how to signal over hills and woods, as well as coast signaling, and describing successful experiments at Chatham, Aldershot and Antwerp. During the 68th Meeting of the British Association for the Advancement of Science held in Bristol in 1898, Bruce gave a talk on "The Use of Electric Balloon Signaling in Arctic and Antarctic Expeditions." From 1899, Bruce was Honorary Secretary of the Royal Aeronautical Society. Bruce soon found out, however, that his electric balloon signaling had been made redundant by the Morse wireless telegraphy developed by the Italian physicist Guglielmo Marconi, a system that could be sent in any weather conditions and at any distance. This did not stop Bruce's inventiveness, which also resulted in such ideas as meteorological kites, about which he published a paper in *Quarterly Journal of the Royal Meteorological Society* in 1909.

Over in Paris, Gustave Trouvé, at his workshop in 14 rue Vivienne, had continued to invent compact electrical machines for medicine, theatrical effects and transport. Toulouse is a town some 680 km (423 miles) south of Paris, and it was here that the French Association for the Advancement of Sciences held its congress on September 26, 1887. Among those things demonstrated by Trouvé was his 90 gram (3 oz.) Lilliputian electric motor which, fitted with an aerial propeller and attached to one end of a scale, once electrified, lifted the scale arm up. Occupying less than a 3 cm. (1 3/16 inch) cube, the motor could rise to a height of 22 meters (72 feet) in one second and would enable future experimentation with electric helicopters and aeroplanes.[32]

In the summer of 1891, twenty years after his first attempt, Trouvé resumed his ambition to build and fly a mechanical bird. After a summer of trials, he wrote a document, *Study of Heavier-than-air Aerial Navigation. Tethered Electric Military Helicopter. Aviator Generator-Motor-Propelling Unit*. Examining which motor was best qualified for aerial navigation, combining great power with lightness, Trouvé eliminated pure electricity: "I am rejecting the electric engine for the moment, for this reason: that with its generator and propeller, it goes beyond the weight of 8Kg per horsepower that I have demanded of myself...."[33]

Science fiction continued. In 1890, Robert Cromie, a Belfast journalist and novelist, wrote *A Plunge into Space*. In the book, Henry Barnett discovers, after 20 years of experimenting, how to control the ethereal force, "which permeates all material things, all immaterial space" and that combines electricity and gravity. Barnett succeeds in his experiments, and a large black globular spaceship, 50 feet (15.24 m) in diameter, called the "Steel Globe," is secretly built in an inaccessible region in Alaska for a flight to destination: Mars. The book was first published in 1890 by Frederick Warne & Co. of London and New York. It was prefaced by Jules Verne, to whom Cromie penned this dedication: "To Jules Verne, to whom I am indebted for many delightful and marvelous excursions—notably, a voyage from the earth to the moon, a trip twenty thousand leagues under the sea and a journey round the world in eighty days—and who, in return, has now courteously consented to accompany me to the planet Mars, at the rate of fifty thousand miles a minute. Robert Cromie, Belfast, February, 1891." It became a best-seller.

Hiram Maxim's first patent concerned with electricity was in 1878, and he joined the first electric lighting company ever formed in the United States. He became their chief engineer, and as their representative he went to Europe for the Paris Electrical Exposition in 1881, displaying his machine for regulating the pressure of an electrical system. For this he was awarded the *Chevalier de la Legion d'Honneur*. One imagines that, like the other visitors to that exposition, Maxim must have looked upwards and admired the Tissandier tethered electric aerostat.

When, in the early 1890s, as a successful machine-gun inventor and wealthy industrialist, Maxim built his unsuccessful prototype helicopter, he powered it with two lightweight naphtha-fired 360 horsepower (270 kW) steam engines driving two 17 ft. 10 in.

(5.5 m) diameter laminated pine propellers. But before starting design work, Maxim carried out a series of experiments on airfoil sections and propeller design.

In October 1891, Maxim wrote an article "Aerial Navigation, the Power Required," that was published in *The Century* magazine.[34] In this, he began with a critical look at recent French attempts to build powered balloons (Tissandier, Renard), judging them a clumsy form of flight. "Look at birds," he says. "A bird weighs 600 times more than the air it displaces." He shows that a goose in flight never exerts more than a tenth of a horsepower. "Heavier-than-air flight is surely the way to go. Yet birds combine lift and propulsion in the wing, and that's too subtle for us to mimic."

Knowing he would have to separate the wing from the propeller, Maxim built a central tower with a 32-foot (10 m) rotating arm to measure the effectiveness of propellers and wing surfaces. A steam engine drove the arm. At the end of the arm was a propeller with a streamlined engine pod and a short section of a wing. That test configuration circled the tower at speeds up to 60 mph (100 kph), while an electric motor inside the pod drove the propeller. The apparatus offered a means for measuring power input to both the propeller and the rotating arm.

Maxim's instruments let him separate out lift, thrust, and drag. He found that, at 60 mph, the propeller might use 16 horsepower to lift the wing and another 35 horsepower to overcome drag and its own inefficiency. With such detailed preliminary researches, Maxim had done superb work on the power inventory of flight, but he had not solved the crucial problem of controlling a moving airplane.

Over in Austro-Hungary, since 1884, Georg Wellner, professor of mechanical engineering at the TechnischeHochschule in Brünn (today's Brno), had developed what he called a navigable sail balloon. He took the example of a paddle wheel ship, then placed the paddlewheel horizontally to the wedge-shaped balloon, otherwise a two-seater sail wheel flying machine, with power from a Siemens-Halske electric motor. The French called this a hélicoptère. Obtaining patents in 1895, Wellner experimented with a small model at the Hochschule; these efforts proved successful and promising, so work began on a 50 ft. (15 m) wheel. But when, by 1895, Wellner's full-scale flying machine was given its first test, it was only able to lift 350 kg (110 lbs.) with an expenditure of 12 hp; the weight of the engine was 90 kg (200 lbs.), and it lacked a force of 540 kg (1,190 lbs.) to raise itself. Wellner continued to work on his sail-wheel concept, but without conclusive results.

On the ground and afloat, practical electric transport had become popular. On May 1, 1893, President Grover Cleveland pushed a button to inaugurate the World's Columbian Exposition. During the six and a half months of the exposition, a fleet of 55 electric launches, built by the Electric Launch & Navigation Company (ELCO), made 66,975 trips on the lagoon, carrying 1,026,346 passengers 200,925 miles (323,357 km) and earning $314,000 for the World's Fair organizers. Their greatest test came on Chicago Day, when 622 trips of three miles were successfully made by fifty boats. In London, Walter Bersey of the London Electric Cab Co. Ltd. ran a fleet of fifteen taxicabs, while over in Paris, a similar service of hackney electric cabs, fitted with EPS batteries, had been inaugurated by the Compagnie Générale des Voitures, its fleet rising to no fewer than 100 cabs.

In this context, in 1893, French author Albert Robida wrote a science fiction novel, *La Vie électrique (Vingtième siècle). (The Electric Life [The 20th Century])*.[35] In this, he looked at everyday life in the years 1952–1953, depicting machines made by an illustrious French scientist Philox Lorris, such as the "téléphonoscope" (an interactive television)

ONE. Origins

and "les helicopters électriques," vehicles that serve both for individual transportation and for military reconnaissance and attacks. These are put to the very widest use, day and night, moored to many rooftops. Public buildings like Notre Dame Cathedral served as balloon interchange stations.

> The month of September 1952 was drawing to a close. Summer had been magnificent; the sun, cooler now, bathed the golden days of autumn with a soft and caressing glow. Airship omnibus B, whose route went from the central Tube station on boulevard Montmartre to the aristocratic suburbs of Saint-Germain-en-Laye, was following the winding lines of the outer avenues and cruising at the statutory altitude of 250 meters. The arrival of the train at the Brittany Tube had quickly filled a dozen airbuses parked above the station. A swarm of aircabs, veloces, skiffs, flashes, and baggage tartans (whose heavy-winged tugs can barely do 30 kilometers an hour) bustled to and fro....

Robida's fellow countryman, Camille Flammarion, a respected author and astronomer, and friend of Gustave Trouvé, wrote and published *La Fin du Monde* (*The End of the World*). In a story set in the 25th century, a comet made mostly of carbon monoxide (CO) could possibly collide with the Earth. The plot is concerned with the philosophy and political consequences of the end of the world. In his description of the Earth in five centuries' time, Flammarion projects: "We traveled, especially during the day, preferably in airships, in electric aircraft, airplanes, helicopters, aerial devices, some heavier than air, like birds, others lighter, like aerostats." In the Hungarian illustrated edition of this book, with a bevy of drawings done by famous French illustrators, one sees two lovers profiting from the intimate seclusion of their wing-flapping electric carriage!

The same year, George Chetwynd Griffith wrote and published his novel *The Angel of the Revolution: A Tale of the Coming Terror* (1893). The story begins on September 3, 1903, with twenty-six-year-old scientist Richard Arnold, devoted heart and soul to the

In his science fiction novel *The End of the World* (1893), Camille Flammarion envisions the skies filled with "electric aircraft, airplanes, helicopters, aerial devices, some heavier than air, like birds, others lighter, like aerostats," even with room for romantic trysts! (Musée EDF Electropolis, Mulhouse).

invention of a flying machine, finally realizing his dream in the form of an airship that can fly on its own.

Colston made his first inspection of the interior of the airship, under the guidance of her creator. What struck him most at first sight was the apparent inadequacy of the machinery to the attainment of the tremendous speed at which Arnold had promised they should travel. There were four somewhat insignificant-looking engines in all. Of these, one drove the stern propeller, one the side propellers, and two the fan-wheels on the masts. He learnt as soon as the voyage began that, by a very simple switch arrangement, the power of the whole four engines could be concentrated on the propellers; for, once in the air, the lifting wheels were dispensed with and lowered on deck, and the ship was entirely sustained by the pressure of the air under her planes.

Using this airship, a crippled, brilliant Russian Jew and his daughter, the "angel" Natasha, set up "The Brotherhood of Freedom" to establish a "pax aeronautica" over the Earth.[36]

Charles Dixon joined the genre with his *Fifteen Hundred Miles an Hour*, published in 1895 in London by Bliss, Sands and Foster. Professor Heinrich Hermann, FRS, FRAS, FRGS, has discovered that the space between planets is not airless but filled with a rarefied atmosphere that can be traversed by a ship with *electric propellers*, driven by a petroleum fuel cell of his own devising: "First, as to my means of conveyance. I have here a design for an air carriage, propelled by electricity, capable of being steered in any direction, and of attaining the stupendous speed of fifteen hundred miles per hour. It can be made large enough to afford all necessary accommodation for at least six persons, and its attendant apparatus is capable of administering to their every requirement." Taking off in the *Sirius*, Hermann and some boys head off at fifteen hundred miles per hour to an adventure on the planet Mars!

The idea of supplying electrical power to a transport by cables was also in vogue. Only ten years after the first experimental trolleybus, in 1893, Frank W. Hawley, a wealthy entrepreneur and a director of the Cataract General Electric Company of Niagara Falls, converted a steamboat called *Ceres* into a trolley boat. Electricity was taken from the Rochester Street Railway station, and two 25 hp Westinghouse street railway–type motors using 500 volts were installed on board, each with its own shaft and prop. An initial line of trolleys was set up above New York's Erie Canal, to which the boat was linked by two flexible wires attached to an overhead traveler. The first public demonstration of the renamed *Frank W. Hawley* trolley-boat was made on November 19, 1893, in the presence of New York Governor Roswell P. Flower and many distinguished guests, and was pronounced completely successful. The Financial Panic of 1893, political squabbling over the choice of electrical contractor, and the problems of positioning trolley wires for boats traveling through locks, over wide water stretches, or under raising and lowering drawbridges put a damper on an enterprising venture.

If you could cable a boat, then why not a flying machine? Such a system was described in "The Electric Motor in Aeronautics," published in *The Electrical Review* on 23 March 1894:

> There would probably be little difficulty in arranging rails, trolleys and wires so that the current for operating the electric motor placed in the aerial machine could be conveniently generated and used. Thus we might, at first, realize a sort of electric railway with an aerial machine attached to the cars, and moving (not pulled) in the direction determined by the line of rails. Such, at least, would be a step from which more independent flights might ultimately be made.... It is the present state of our

In 1893, French author Albert Robida wrote the science fiction novel *La Vie électrique (Vingtième siècle)* (*The Electric Life [The 20th Century]*). In this, Robida looked at everyday life in the year 1952–1953, including this electric aircab. Is this the precursor to tomorrow's flying taxis? (Musée EDF Electropolis, Mulhouse).

knowledge that the electric motor would be the easiest of all to develop the required degree of lightness and power. We honestly think that attempts to provide steam or explosive motors will prove costly, if not futile, and that the most promising direction in which to look for the realization of Tennyson's much-quoted prophecy will be towards the development of the electric motor.[37]

Indeed, Joseph Marie Augustin Deydier of Nyons in the Drome region of France was even awarded French Patent 653,221 for an "electric traction balloon."

On July 31, 1894, Hiram Maxim had tested his own heavier-than-air flying machine, weighing two tons, measuring 105 ft (32 m) wide with 17-ft (5 m)-long propellers, along railway tracks at his home in Baldwyn's Park, Bexley, England. The aircraft successfully lifted itself off the track before it crashed safely to the ground. *The Electrical Review* had only this to say on Maxim's hop:

> The simple mind of an electrical engineer concludes that since failure and grave personal risk has hitherto attended the exploitation of Mr. Maxim's ideas, an attempt with electric motors should be made. In this case the element of personal danger could be practically eliminated, for the trials of the aerial machine would be made without the necessity of an operator being actually on it. Cables, capable of being rapidly paid out, would supply the current to the motors, and the whole machine could be actuated and directed from a safe distance by electrical means. The use of electric motors is of comparatively recent date, yet such results have already been attained as to suggest that there is at least as much likelihood of placing upon a flying machine horse-power per unit weight, supplied electrically to anything that can be obtained by other means. Did not Monsieur Trouvé construct last year an electric motor which developed energy at the rate of 1 hp for each 7 lbs. weight? The electric motor is a mere baby in experience compared with the steam engine, yet the latter is already being compelled to "look to its laurels." Not only is it possible that the electric motor may be the means of giving the large horsepower per unit of weight, which is claimed to be a *sine quâ non* in aerial navigation, but, from what it has already achieved, it is reasonable to anticipate that it will—and that very shortly.[38]

TWO

If It Flies, It Must Be Electric!

Captain Charles A. Smith of San Francisco was a merchant, physician, miner, mechanic and inventor. After three years of study, Smith took out his fifteenth patent—for his version of an Air Ship (U.S. 565805A)—in August 1896. The captain's previous patent had been for a tethered aluminum balloon equipped with arc lights for street lighting; four such balloons could light up the whole of San Francisco Bay. His 100-ft (30 m)-long × 30-ft (9 m)-wide lighter-than-air machine would also be built of aluminum, pushed by an immense screw propeller, and would use a system of chains and pulleys to flap its 3,700 sq. ft. (344 m^2) bird-like wings and its 200 sq. ft. (18.5m^2) tail to navigate through the air. "I propose to use a single engine or motor, one having the least possible weight, to drive all the operating machinery. It is only necessary that suitable machinery and connections be employed for operating the wings. In practice I propose to employ either a naphtha-engine *or a small electric motor*, according to the purposes for which the ship is to be used and the distances to be traversed." Smith's *Airship* would also possess an anchor for mooring.

The company set up to finance Smith was called the Atlantic and Pacific Aerial Navigation Company and its superintendent was none other than Hiram Maxim. Smith, it is said, even had a scale model of his invention, which was on display in the Mechanics' Pavilion window in San Francisco, where it was visible to all who passed by.[1]

Maxim became superintendent of the Atlantic and Pacific Aerial Navigation Company, with Smith as president, for the realization of the captain's dream. The machine was built in New Kensington aluminum sheeting supplied by the Pittsburgh Reduction Company. Motive power was a 105 hp Maxim naphtha engine. A battery would serve for the on-board electric lighting. From late 1896, the Smith-Maxim Air Ship went on trials, and in the spring of 1897, they flew it across the U.S. from San Francisco to Cleveland, Ohio. Smith announced that his company would build a number of these airships of diverse designs.[2] The press provided eyewitness reports of the ship. In November 1896 an object was reported in the night sky over Sacramento, California. It was described as a light with a dark body of some kind above it. It was seen a second time about a week later. Similar reports, usually describing only the light, came from other cities, from northern California to Michigan to Louisiana. Reported movement of the light indicated a slow motion. The dark body seen above it was variously held to be "cigar-shaped," "egg-shaped" or "barrel-shaped." This was undoubtedly the Smith-Maxim Air Ship. In an article published on December 3, 1896, in the *Evening Times* (Washington, D.C.) and titled "Professor Langley Scouts it. Puts No Faith in the Alleged California Airship": "The airship, although it appears from the working drawings to be an entirely unworkable device, corresponds closely to the description of the big ship now said to be nightly wandering around over the Pacific slope."

In Timothy C. Parrott's biography, *The Secret Life of Dr. Charles Abbott Smith: San Francisco's 19th Century Airship Inventor*, he reports: "When Smith's plans for his airship were first revealed in the pages of *The San Francisco Call* in August of 1896, he was said to be entering into ruinous competition with the trans-continental railway companies. However, it is equally plausible that the political tensions leading to the Spanish-American War diverted the attention of both the public and investors, creating a sudden loss of financial backing and, ultimately, the collapse of this fledgling airship industry."

Responding to sightings previously reported in the *Dallas Morning News*, on April 17, 1897, one respected Erath County farmer, C.L. McIlhany, discovered such a craft had landed on his property in Stephenville, and reported two human operators, a pilot and an engineer, who gave their names as "S.E. Tilman" and "A.E. Dolbear." The two operators performed minor repairs on their *electrically-powered* lighter-than-air craft, then again flew away. Yet Samuel Escue Tillman, aged fifty, was an astronomer, geologist and resident educator at the United States Military Academy at West Point, New York. Amos Emerson Dolbear of Somerville, Massachusetts, aged sixty, was a physicist better known for his device for communicating over a distance of a quarter of a mile without wires in the Earth. Neither of them had the slightest interest in aerial navigation, nor had they ever met!

At this period, the U.S. government is said to have offered a bonus of $100,000 for the invention of a practical airship if completed within a certain time, and this new corporation would claim the prize if its venture proved a success.

In 1898, Joseph M. Gaites, a rising 26-year-old impresario and playwright from Pittsburgh, wrote a musical farce comedy called *The Air Ship*, in which one of the most realistic stage scenes ever presented would be the flight of a real airship with fifteen passengers on a Klondike expedition, with a view of Dawson City in winter. The show originally opened at the Masonic Temple in New York City in 1899, where it received rave reviews and hilarious approval by a very large audience. From there, the actors hit the road, with *The Air Ship* being performed in places as diverse as Trenton, New Jersey, and Fort Wayne, Indiana. It was still being performed as late as 1911. Gaites went on to work with the Marx Brothers in vaudeville.

Marketing took to the air. From 1899, the French chocolate manufacturer Gaston Menier produced a series of colored postcards, predicting what life would be like in Paris "En l'An 2000," a century later. For each packet of chocolates, the buyer received a card to collect. One of these cards shows a helicopter airbus called *Sylphe* flying between Paris and Bordeaux. One imagines that an optional propulsion system was electric.

Between April and December 1897, English science fiction writer H.G. Wells's novel *The War of the Worlds* was serialized in the UK by *Pearson's Magazine*. It tells of a Martian invasion of southeast England by towering three-legged fighting machines equipped with lethal heat rays. On January 9, 1898, several months after the story was officially serialized in the U.S. in *Cosmopolitan Magazine*, the *Boston Post* began publication of an entirely revised version of *The War of the Worlds*, in which the action was relocated to the local area under the somewhat cumbersome title, *Fighters from Mars—or The Terrible War of the Worlds as it was waged near Boston in the year 1900*.

Then there was a sequel, *Edison's Conquest of Mars*, written by Garrett P. Serviss, an astronomer and an established popular science writer.³ In this, Earth's leaders, including U.S. President William McKinley, Queen Victoria, Kaiser Wilhelm II, and Emperor Mutsuhito, unite the world against the common threat and plan an attack on Mars. American inventor Thomas Edison leads a group of scientists studying derelict Martian equipment; they are able to develop an anti-gravity device powered by electric repulsion as well as a disintegration ray. A stupendous fleet of electric spacecraft heads off toward the red planet and makes a series of brutal engagements with the enemy. Edison himself, however, had nothing to do with the story other than to authorize the use of his name.

In 1901, when the Englishman Wells decided to write *The First Men in the Moon*, he came up with an ingenious antigravity material, Cavorite, named after its inventor Mr. Cavor. When a sheet of Cavorite is prematurely processed, it makes the air above it weightless and shoots off into space. Cavor hits upon the idea of a spherical spaceship made of "steel, lined with glass," and with sliding "windows or blinds" made of Cavorite, by which it can be steered. He persuades a businessman called Bedford to undertake a voyage to the moon. Published in 1901 by George Newnes, the book became an immediate success.

In 1903, Octave Chanute, a French-born aeronautics researcher based in Chicago, was looking for a safer way to launch his man-carrying biplane glider. He met William Avery, a carpenter and electrician, and together they designed and built a winch to be powered by an electric motor. In September 1904 Chanute then patented their invention: "Means for aerial flight."[4] Their chance to demonstrate this came at the St. Louis World's Fair held in early October 1904, which included an International Aeronautical Congress. There, leading scientists and aeronauts from four countries came together, in the words of Professor C.M. Woodward of Washington University, "To demonstrate that progress in aerial navigation was possible [and] to learn, through the failure and experiments of others and a comparison of notes and results, that success cannot be hoped for except through careful study and scientific investigation."[5]

Instead of returning home, most of the international aeronautical experts remained on the ground to witness, and in some cases to participate in, the aerial events at the stadium, including the Chanute glider. To launch himself into the air, Avery mounted the glider on a small flatcar, which ran on a miniature railroad track about 100 feet (30 m) long. One end of a long rope was attached to the glider and the other to a large drum, 400 feet (120 m) away, which was powered by a 10 hp electric motor. At Avery's signal, the motor rotated the drum at a rapid rate, and rapidly taking up the slack in the rope, it pulled the glider forward and launched it like a kite. When the pilot was safely airborne, he released the towrope. Launched by the electric winch, William Avery made three flights at the stadium on October 7. At 2:30 p.m. he took off, reached a height of 35 feet (10 m), and glided through the air for 175 feet (53 m). On his two other efforts, altitudes of 18 to 35 feet (5 to 7 m) were reached, and glides of 90 to 100 feet (27 to 30 m) were attained.

For the next two weeks, Avery made glides almost daily in the stadium. On October 25, however, he moved his equipment to the Plaza St. Louis because he felt he needed more room to maneuver. At 3:00 p.m. October 26, he made his first trial of the biplane glider at the plaza. He reached a height of 30 feet (9 m) when the rope, apparently defective, parted before his signal. The machine plunged to the ground, throwing Avery to the asphalt pavement and causing him to break his ankle. His career at the fair was ended, but he had earned a creditable record. In all, William Avery made 46 flights in Octave

Chanute's glider, and he was the only man to fly a heavier-than-air machine at the exposition.

Samuel D. Mott of Passaic, New Jersey, had begun his professional career as an electrical engineer in Thomas Alva Edison's Menlo Park laboratories. In 1902, Mott proposed a machine to raise self-registering meteorological instruments into the higher atmosphere by ornithological flight. Constructed of aluminum, steel, wood and silk, Mott's electrically-operated "aerodrome" or "air runner" as he called it, would combine the continuous thrust of a screw propeller in close mechanical imitation of wing flight. However, Mott's ornithopter designs were soon eclipsed by the wing warping of the Wright Brothers.

In 1903, Frederick Walker of Oxford and Frank Barton, a well-known aeronaut, developed what they called an Electrical Equipose for Air Ships:

> The great desideratum in successful aerial navigation by means of air ships combining an aerostat with a system of aero-planes, is to provide an automatic equipoise at any desired angle or plane. Birds and insects possess an instinctive sense of equilibrium in flight, which is anticipatory in action. The only method by which a mechanical equivalent can be applied to an air ship is by using an electrical current and this has been successfully done by Fred Walker of Oxford and Dr. F.A. Barton—using a Hart accumulator set (motor-car new type). The twin screw-propellers rotate under triple aerocurves, and the action of the regulator is to diminish the speed of the motor on the lowest side by putting resistance in. The longitudinal equilibrium is controlled by a similar regulator set fore and aft. This

In 1905, in order to enthuse the public about heavier-than-air flying machines, Sir Hiram Maxim patented and built a fairground attraction called Captive Flying Machines, propelled by twin 50 hp electric motors (Blackpool Council).

acts by reversing a small electro motor, which alters the angle of a long tube centrally pivoted, containing about 60 lbs. of mercury. This regulator is specially designed for the new airship being built for the War Office.[6]

Sir Hiram Maxim remained convinced about heavier-than-air machines. Despite his newfound status and connections in high places, he was having great difficulty raising interest for his flying machine project. In an attempt to relieve his bronchitis, he spent the winter in the south of France, where he came up with an idea: a fairground roundabout ride in what he called "Captive Flying Machines" to simulate the experience of flying and enthuse the public. To do this, he adapted the mechanism he had been using—but had not invented—to explore the efficiency of wing designs for planes.

He gathered a consortium of backers and formed the Sir Hiram Maxim Electrical and Engineering Company to handle the construction and operation of the new ride. A series of cars ("flying machines") attached by arms to a central rotating upright would "fly" through the air. At the bottom of the central column, in the machine room, was a large-toothed horizontal gear wheel driven by beveled pinions. Two identical 50 hp electric motors made by Lister Ltd. of Dursley with 990mm (39 in.)-diameter drive wheels, delivered drive via six ropes to the main 3.6 m (12 ft)-diameter fly wheels. Modifications had to be made to his prototype, which was constructed at Thurlow Park in Norwood. Maxim had wanted the cars to have working wings and aerofoils, allowing riders the ability to control their "flight path," making the car sweep and dive at will, but this proved too dangerous. One of the first riders during the testing phase was Maxim's chief assistant, Albert Thurston, who later recorded: "Speed was increased until the centrifugal force was 6.47 times gravity. After a mighty mental struggle, I fell consciousless to the bottom of the car." Obviously, knocking your passengers unconscious would not be good for ticket sales, and so the speed was decreased on subsequent runs.

Sadly, the problems continued. Like all the rides at Earl's Court for the summer 1904 season, the Captive Flying Machines would have to be inspected and declared safe by London County Council before it could be opened to the public. While council inspectors were watching a demonstration of the ride in action, a strong wind caught one of the cars, and sent it soaring far higher than anyone expected. While Maxim's team insisted that such an event could never happen when passengers were aboard, the council refused to allow the ride to open until the wings were removed. As a result, the final version of the ride was a bitter disappointment to Maxim, who said that it had become "Simply a glorified merry-go-round."

Although the ride was popular, it broke down partway through the season and didn't make enough money. Its projected construction cost had been £3,000, but the actual cost was more like £7,000. Maxim decided that more would have to be constructed to make the venture profitable. Maxim had originally deemed that no more than two Captive Flying Machines rides would need to be built in order to fund his research, and the loss of money from the Earl's Court breakdown may well have influenced the decision to build more. These included ones for Crystal Palace, Brighton, Southport, and Blackpool.[7] In 1905, Willow Grove Park in Pennsylvania opened another version of the ride, where no doubt Maxim would have been less than flattered to discover the park promoting the ride by describing him as "The Edison of England"! The ride had a 100 ft. (30 m)-high central structure, around which miniature airships attached to 10 extending arms. The arms lifted the ships about 600 feet (180 m) from the ground, as they circled around the central structure.[8]

Maybe inspired by Maxim, in 1905 Horace A. Lockwood of Kansas City filed a patent (U.S. 864317 A) for what he called an Air-Wheel "designed for pleasure purposes." A series of "aero planes" would revolve around an electrically powered central shaft. "The aero planes will by their resistance to the air and by reason of their inclination, cause the car supporting members to rise and soar through the air like a bird on the wing."

As for those attempting to build and fly heavier-than-air machines, it had become evident that current electric motors and batteries were too heavy for constructions made from bicycle wheels, lightweight wood and canvas. This all changed with two brothers, Orville and Wilbur Wright of Dayton, Ohio.

One of the key men behind the Wright Brothers' success was Vincent G. Apple of Dayton. At the age of 18, in 1892, Apple had founded the Franklin Electric Company, later the Dayton Electric and Manufacturing Company. In 1902, he introduced what is thought to be the first electric self-starter for an automobile. The Wright brothers, unable to find an off-the-shelf engine to power their *Flyer*, borrowed one of Apple's machinists, Charles E. Taylor. The aluminum water-cooled unit took Taylor only six weeks to build. The cast aluminum block and crankcase weighed just 152 pounds (69 kg). The Wrights needed an engine with at least 8 horsepower (6.0 kW). The engine that Taylor built produced 12. But for the ignition system, Taylor had recourse to his employer, Vincent Apple. The latter's solution involved the opening and closing of two contact breaker points in the combustion chamber of each cylinder via a camshaft. The initial spark for starting the engine was generated with a coil and four dry-cell batteries, not carried on the airplane. A low-tension magneto driven by a 20-pound flywheel supplied electric current while the engine was running. Using Taylor and Apple's engine, the Wrights were able to make the world's first controlled, sustained flight from the sand at Kill Devil Hills on December 17, 1903, four miles south of Kitty Hawk, North Carolina. In the historic photo of that flight, one observes the starting battery, with its kinky, stiff wire sticking out of the wooden box.[9]

Simultaneously, in France, Léon Levavasseur, sponsored by Jules Gastambide, who owned an electricity-generating station in Algeria, had come up with a 24 hp (18 kW) V-8 unit called the Antoinette and weighing just 55 kg (121 lb.). It was shown at the Paris Aero Salon held in December 1908 and was first flown in 1909. This was the world's first rotary engine produced in quantity. Its introduction revolutionized the aviation industry.[10] Another advantage to the pioneer airmen was that it took a matter of minutes to refuel these units. These and subsequent motors began to relegate electric propulsion to the realms of research and of science fiction.

Where electric propulsion was concerned, science fiction continued. Take the remarkable short story, "Sultana's Dream," published in 1905 in *The Indian Ladies Magazine* of Madras, India, then part of the British Empire. It was written by Begum Rokeya Sakhawat Hussain, a Muslim feminist, writer and social reformer. Hussain's story depicts "Lady Land," a peaceful, crime-free utopia ruled by women, where men are secluded in a purdah-like system. The women use advanced technology that makes possible laborless farming, having solved the problem of solar energy, and able to control the weather. The women go around in flying cars:

> Then she screwed a couple of seats onto a square piece of plank. To this plank she attached two smooth and well-polished balls. When I asked her what the balls were for, she said they were hydrogen balls and they were used to overcome the force of gravity. The balls were of different capacities to be used according to the different weights desired to be overcome. She then fastened to the air-car two wing-

like blades, which, she said, were *worked by electricity*. After we were comfortably seated she touched a knob and the blades began to whirl, moving faster and faster every moment. At first we were raised to the height of about six or seven feet and then off we flew. And before I could realize that we had commenced moving, we reached the garden of the Queen. My friend lowered the air-car by reversing the action of the machine, and when the car touched the ground the machine was stopped and we got out [emphasis added].

The same year, the British Empire's popular fiction writer, Rudyard Kipling, in his science-fiction stories set in the year 2065, "With the Night Mail" (*The Windsor Magazine*, December 1905; *McClure's Magazine*, November 1905) and "As Easy as ABC" (*The London Magazine*, 1912), presents the Aerial Board of Control, a supranational organization created to manage air traffic for the whole world. The organization is able to limit the influence of national states and create a *de facto* world government. Technology has moved on, and there is an array of sophisticated machines and weapons based on electricity and radio.

Towards the end of his life, Emilio Salgari of Verona, Italy, already a best-selling writer of action-adventure swashbucklers, turned his mind to the sky and the future. He began with *King of the Sky* and *The Children of the Sky*, published in 1904 and involving *lo Sparviero*, a fantastic electric flying machine with four giant wings and a propeller, piloted by the enigmatic Capitano. Soon after, Salgari wrote *Le Meraviglie del Duemila* (*The Wonders of Two Thousand* [AD]). The novel was published in 1907 by Bemporad in Florence, but for contractual reasons, it was signed with the pseudonym "Guido Altieri," Salgari's fictional grandson, although later expanded and republished under Salgari's name. It is the story of two men, Brandok and Toby, who, thanks to the discovery of an active substance of a strange exotic plant suspending vital functions, can travel through time for a hundred years, moving from 1903 to 2003. They find themselves living in a profoundly changed society: a world of *flying cars*, fast trains, underground and underwater cities, and many other technological marvels. In the end, they perish from the excessive bustle of life in the future and the electrification of the sky. Salgari envisaged a cigar-shaped metal passenger airship which at high speeds passes from and through electromagnetic rings, each mounted on a tall tower. On each ship, there are only six men: the commander, two mechanics, a helmsman, a steward and a doctor. The interior of the gallery is divided into four compartments. One is reserved for the machine and the mechanics; one is a bedroom compartmented into small cabins; the third is the dining room; the fourth a library with a conversation room and an electric organ to divert the voyagers.

On September 6, 1906, a 24-year-old physicist called Robert H. Goddard, studying at Worcester Polytechnic Institute in Massachusetts, made an entry in his personal notebook concerning the problem of producing a reaction with electrons moving with the velocity of light: "At enormous potentials can electrons be liberated at the speed of light, and if the potential is still further increased will the reaction increase (to what extent) or will radioactivity be produced?"[11] In fact, Goddard, who became known for his rocketry experiments, would later record how his earliest recollection of a "scientific experiment" at the age of five had involved the use of electricity for propulsion. Robert believed he could jump higher if the zinc from a battery could be charged by scuffing his feet on the gravel walk. But, holding the zinc, he could jump no higher than usual. Goddard halted the experiments after a warning from his mother that if he succeeded, he would "go sailing away and might not be able to come back."[12]

The idea of an ion engine first appeared in Donald W. Horner's *By Aeroplane to the Sun: Being the Adventures of a Daring Aviator and His Friends*, published by the Century Press (London: Bennett & Co., 1910). Horner was a UK astronomer, meteorologist and writer who specialized in popular-science texts. His science fiction novel offers a numerate and complex vision of a high-tech near future, featuring picture phones, television and electric cars, and describing the protagonists' usual ion-engined tour of the Solar System with prescient realism.[13]

During the period between 1906 and 1914, would-be aviators, "those magnificent men in their flying machines,"[14] while all dependent on the power-weight advantage of the gas engine, took a wide and diverse approach the challenge of wing design: bird's swept or dart wings, biplane, triplane, multiplane, staggered, corrugated, curved, circular wing, tail plane at the front (canard) or at the back; and also that of propeller design—with two, three or four blades. Eventually these were whittled down to the Blériot tandem wing design—which became the conventional norm. It was only one hundred years later that those planning to fit lithium batteries and electric motors to their airplanes would seriously rethink the wings and propellers configuration again.[15]

One of the main dangers with those early flying machines was their stability in flight. In 1907, J.A. Colquhoun obtained a patent (17366) that stated in part, "[W]hen the flying machine becomes tilted, a pendulum, owing to its inertia, makes contact with one or two of four insulated plates which are electrically connected respectively to the driving mechanisms of four propellers, so that the tilting of the machine causes less power to be transmitted to one or two of the propellers. The propellers rotate about verbal axes and are so disposed that the reduction in speed of rotation of any one propeller, or any two adjacent propellers, will tend to right the machine."

As late as 1909, there were would-be airplane designers who still made provision, even if mistakenly, for electric propulsion. In filing his patent for "an aeroplane flying machine," Peter Levins of San Francisco described the wing design of his biplane, suggesting that for propulsion it could be fitted with a gas engine or "a storage battery, where the motive power is electricity, the motor is connected with the battery on the platform." Levins was never to know which system would be used because he died in an accident, although his widow Grace L. Levins eventually obtained the patent in 1912 (U.S. 1043781 A).

Another approach was taken by Joseph Danziger of Chicago, who in March 1910 filed his "Automatic Equilibrating Device for Aeroplanes," using electricity to correct the flight angle. The same year, Dighton B. Ellsworth of Portland, Oregon, filed for his "Equilibrator," or automatic balancing system, not only for airplanes but for airships: U.S. 1024398 was granted in April 1912.

Perhaps the best way was to test a design before building it. Gustave Eiffel, the engineer whose iconic tower in the heart of Paris bears his name, was fascinated by aeronautical research. He would often tell journalists, "Wind had always been my concern. It was my enemy when building the tower." In 1909, to determine aerodynamic laws, he built his first open-return wind tunnel (soufflérie in French). It had an airstream 1.5m (5 ft.) in diameter and 3 meters (10 ft.) long. Airspeed could vary from 5 to 20m/sec, using a 50 kW electric motor. As the tunnel was built on the Champ de Mars at the foot of Eiffel's Tower, the motor obtained its energy from the tower's electric generating station. Since 1900, the tower's electric lighting (5,000 bulbs) had been entirely electric. Between 1909 and 1912, working in a wooden shed, Eiffel ran about 4000 tests in his wind tunnel, and his systematic experimentation set new standards for aeronautical research. Those airplane wings and propellers benefiting from the Eiffel Wind Tunnel included the Wright, Voisin, Farman, and Blériot, followed by completed airplane models conceived by Esnault-Pelterie, Nieuport, and Levasseur. By that time, Eiffel had moved to Auteuil, a suburb of Paris, where he built an improved wind tunnel with a 2-meter test section and a speed of 30 m/sec (100 ft/sec). He donated this to the French government (Services Techniques de l'Aéronautique) in 1920 when he was 88 years old. It has been operated without interruption since its inauguration and continues to be used for testing airplanes, buildings, and Formula 1 racing cars.[16]

Among those reporting on Eiffel's experiments were Messrs. Fernand Barrés, Eugene Brémaud, and Adolphe Schœller in their book published in 1910: *Les transformateurs d'énergie, générateurs, accumulateurs, moteurs avec les plus récentes applications à la navigation aérienne* (Energy transformers, generators, batteries, motors with the most recent applications to aerial navigation), published in Paris by A. Quillet.

An interesting electrical propeller-testing device was developed in 1911 by Hermann Scheit, Professor of Mechanical Engineering and Strength of Materials, with his assistant Bobeth, at the Royal Sächs Institute of Technology in Dresden-Strehlen, Imperial Germany. The propeller to be tested was mounted on the shaft of an electric motor. The shaft was stretched in one direction by a wire fixed at one end, and in the opposite direction by a wire passing over a pulley, and supporting a set of weights. The work of the air-propeller relieved the pull of the weights, and the reduction in the length of the wire thus obtained was registered automatically. To find the thrust of the propeller when the aircraft is in motion, an artificial wind was created by a subsidiary propeller. Scheit and Bobeth determined the thrust for different powers and wind velocities, adding, in some experiments, screens to represent the air resistance of the aircraft. For a consumption of 25 hp there was a falling-off in the thrust amounting to 8 percent, on running the motor in a wind of 10 m (33 ft.) per sec.[17]

Across the English Channel, also in March 1909, George Crosland Taylor FRGS of the Crosville Motor Company in Chester, England, innovated a lightweight electric welding system for airplane construction (*Flight*, March 27, 1909). For the next century, electric welding for airframes became the norm.

Arguably the world's first electrically driven tethered *model* airplane was tested in North London in 1907. Two years before, Baron Alfred Lord Northcliffe had offered £150 for the best design of a flying model. One of the rules stipulated that a model must weigh not less than two pounds and not more than fifty pounds (1–22 kg). From a starting altitude of not more than five feet, the model airplane had to fly at least 50 feet (15.2 m). The contest was held on April 15, 1907, at Alexandra Palace, Wood Green, North London, organized by the new Aero Club. A crowd of one thousand people turned up to watch the 29 entries, 15 competing for the prizes. No plane at the contest could win the first prize, so the prize money was held to be awarded at future meets. However, a second prize of £75 was awarded to a young marine engineer, Edwin Alliott Verdon Roe, for the flight of his 9 ft. 6 in. (2.85 m) wing-span rubber-powered model "Avroplane," which outflew everything else, reaching 100 ft. or more. Roe went on to use his winnings to build his first man-carrying triplane and later founded the Avro company. Two years later, on July 25, 1909, Frenchman Louis Blériot successfully flew across the English Channel in his Anzani-engined monoplane and won the £1,000 *Daily Mail* prize offered by Lord Northcliffe.

Unknown to Blériot, to Northcliffe or to Roe, in the months that followed, an aero-modeling enthusiast called Joseph L. Cannon, a builder and plumber resident in Bowes Park, North London, decided to build and fly an electrically-driven model airplane. He supplied the details of his experiments to *The Model Engineer and Electrician*, which published them on October 21, 1909:

> The design is of the monoplane type, made of American white wood and American oak. The main plane is made of cardboard, 18 ins [46 cm.[across, while the rear plane is 11 ins [28 cm.] across. It is powered by a compound wound electric motor at 4 volts pressure, current being supplied by the refill dry batteries for a pocket electric lamp and attached direct to the motor spindle. The weight of machine and motor is 1.5 lbs. or with batteries just under 2 lbs. [1 kg]. Our readers will notice that the machine is similar to the Blériot monoplane, but it has no vertical rudder. Steering to right or left is accomplished by either of the two small end planes situated at the rear.
>
> When the machine is to be used, a short length of strong rubber is attached to the underside of the frame, a light line is attached to the rubber, the object being to absorb any sudden jerk when pulling down. An assistant holds the line loosely, and the operator launches the machine off a tall pair of trestles. It then flies in a large circle, plenty of line being played out at discretion. When the batteries are exhausted the machine descends slowly, and is caught by the operator before it reaches the ground. The highest flight yet accomplished is 35ft [10 m], duration 8 minutes approximate. Mr. Cannon calls his machine an electroplane, and considers that an electrically driven model is better than an India rubber–driven model to experiment with. He states however that the batteries become exhausted very quickly so that convenience is obtained at some expense if many flights are made. A drawback to electric propulsion is also heaviness but this does not matter if a high number of revolutions per minute is used. In a recent experiment a higher voltage was tried with the result that the motor winding fused somewhere.
>
> It is gratifying to find an experimenter having the courage to use an electric motor and battery to drive his model instead of the almost universal twisted rubber cord. That electricity is not hopelessly outclassed seems evident by Mr. Cannon's sad story of the loss of an electrically propelled model of the box type which, imitating the unfortunate dirigible La Patrie, broke away one evening in the early part of this year and flew into the unknown, the course being from Belsize Avenue over the Alexandra Palace. The lifting surface of this machine was about 12 sq. ft. [1 m²] and it carried an electric accumulator weighing about 6 lbs. [2.7 kg]. Mr. Cannon although highly gratified at the flying capacity of his machine, still deplores his loss. If any of our readers have found a derelict aeroplane which appears to be the long lost one, perhaps they will be kind enough to communicate with Mr. Cannon.[18]

No more was heard of Joseph Cannon, and model airplanes continued to be powered by twisted rubber bands for decades to come.

Marconi's wireless communication via electricity was now being tested for air-to-ground communication from heavier-than-air airplanes. On August 27, 1910, Canadian James A.D. McCurdy transmitted a Morse code message to Henry M. Horton while flying over the Sheepshead Bay race track in Brooklyn. The same year, Englishman Robert Loraine, also a highly successful actor, sent an air-to-ground message while flying over Salisbury Plain. In January 1911, Major Lefroy, flying in the airship *Nulli Secundus 1*, made and received wireless communication with a ground station.[19]

In 1911 Nikola Tesla, Croatian-born U.S.–based inventor of alternating currents, granted an interview to the *New York Sun* in which he announced that he was working on what could broadly be defined as an electric airplane design. It would use an electromagnetic field lift-and-drive system, drawing its energy from external power stations and Tesla's "wireless transmission of energy-system," or with a power generator inside the craft. Tesla humorously likened the saucer-shaped design of his high-speed machine to a kitchen gas stove, which would have the same weight. He indicated that it would have neither a lighter-than-air gas bag, nor wings, nor propellers, nor landing wheels.

Nor would it be affected by winds or holes in the air, and it could be held absolutely stationary in the air, even in a wind, for great length of time. Tesla's short-range machine would use a low-voltage motor to turn the generators, occasionally recharging by hovering next to high voltage power lines and using antennas mounted on the outer hull to take in the electricity. The short-range machine could also have electricity beamed to it from a generating plant on a long-range aircraft or spacecraft, or on the ground.

Tesla's circular electric airplane would hold 480 magnets with 480 field coils wired in series surrounding it in close tolerance. At 50 revolutions per minute, it would produce 19,400 cycles per second. The electricity would be fed into a number of large capacitors, one for each metal sheet. An automatic switch was adjustable in timing by the pilot so that as the electricity jumped back and forth, it raised its own frequency; a switch was used for each capacitor. The electricity would then go into a Tesla transformer; again, one transformer for each capacitor.[20] Due to his pacifist sympathies, Tesla originally contemplated giving his electric flying machine to the Geneva Convention or League of Nations, for use in "policing the world" to prevent war. Later disillusioned after World War I, he changed his mind. Was Nikola Tesla's flying machine concept an "antigravity" craft? Some have even suggested that it is a prototype flying saucer, and that it was built; they have linked it to UFO conspiracy theories.

From the sublime to the everyday, ironically, the world's first air freight was a package of electric light bulbs, flown by Horatio Barber in his gasoline-engined ASL *Valkyrie* from Shoreham to Hove in 1911! Over in France, Monsieur Albert Moreau, a salesman of printing equipment and the father of 6 children, spent most of his modest income on the design and construction of his *Aérostable* flying machine. With his brother André, Albert had started in 1902 with models of gliders, observing the reflexes of flying pigeons. One of the features of their *Aérostable* was its automatic stabilization. Indeed, in 1912 Moreau innovated an *electro-motor* clutch servo elevator.

The use of electricity on board flying machines was not far off. During the Tennessee State Fair at Cumberland Park, Nashville, on June 22, 1910, an estimated 23,000 people witnessed the world's first nighttime airplane flights. That evening, the intrepid Charles K. Hamilton made two flights in his Curtiss biplane. The first at 7:30 p.m. was in pitch darkness, at a speed of 45 mph (72 kph), at an altitude of 240 feet, and for 11 minutes. Hamilton then fitted an acetylene tank and a searchlight to his machine, and between 10:57 and 11:12 p.m., he astounded and dazzled the crowds with a second flight, despite a forced landing. For this feat, he received a diamond-studded gold medal.[21]

Eight months later, at 2:20 a.m. on February 11, 1911, Robert Grandseigne of Le Havre, Normandy, took off from Issy les Moulineaux aerodrome in his Anzani-engined Caudron biplane, its wings equipped with electric lights, and made an historic circuit in the night skies above Paris. He passed above Grenelle, the Champ-de-Mars, wing-tipped to the Eiffel Tower, then moved on to Passy and Auteuil, landing at 3:32 a.m. Tens of thousands of Parisians were woken up by the engine noise, running to their windows and looking down at their street to discover the origin. But then they looked up and caught sight of the tiny lights on Grandseigne's airplane.[22]

Two years later, in June 1913, at the Grahame-White Flying School grounds in Hendon, North London, thousands of spectators marveled as a flying machine, fitted out with multi-colored electric light bulbs, flew around the night sky. The Maurice-Farman biplane was piloted by Claude Grahame-White's instructor Louis Noel, who wowed the crowd every time he switched his lights off and on.[23]

In June 1913, at the Grahame-White Flying School grounds in Hendon, North London, thousands of spectators marveled as a flying machine, fitted out with multicolored electric light bulbs, flew around the night sky. The Maurice-Farman biplane was piloted by Claude Grahame-White's instructor Louis Noel, who wowed the crowd every time he switched his lights off and on! (©RAF Museum).

Less successfully, the following month the Royal Flying Corps No. 4 Squadron based at Farnborough was tasked with testing electric landing lights (Experiment 34). A Maurice Farman S7 Longhorn (serial 306) was fitted with a 60-candlepower Blériot acetylene lamp. It was considered unimpressive as it did not pick out the ground until only 25 feet (7.6 m) off the ground. Holt flares were also tried during this experiment. These were small flares fitted with a parachute and were equally unsuccessful.[24]

Inspired by the news from Hendon, on July 23, 1913, H.W. Blakeley flew through the Canadian night in his biplane, again equipped with electric lights on the wings, over the Dominion Livestock Show and Fair, in Brandon, Manitoba. Bonfires guided his way from and to the landing strip.

Claude Graham-White continued to develop Hendon Aerodrome as "London's Modern Rendezvous," and in September 1913 the millionth spectator passed through the turnstiles.[25] On November 5, traditionally "Guy Fawkes Night" for fireworks and bonfires, Marcus D. Manton made a 15-minute flight of several circuits around Hendon, again switching on and off his colored lights positioned along the wings like a Christmas tree.[26]

In Imperial Russia, the four-engined *Ilya Muromets* (Sikorsky S-22) was designed and constructed by Igor Sikorsky at the Russo-Baltic Carriage Factory (RBVZ) in Riga in 1913. It was first conceived and built as a luxurious aircraft. For the first time in aviation history, it had an insulated passenger saloon, comfortable wicker chairs, a bedroom, a lounge, and even the first airborne toilet. The aircraft also had heating and electrical lighting. The first flight was from St. Petersburg to Kiev in June 1914, a flight lasting more than 14 hours and carrying 16 people. With the outbreak of war, the Sikorsky S-22 became one of the world's first heavy military bombers, using electricity for other purposes.[27]

THREE

Electricity Goes to War

> What more ingenious devices than those of M. Trouvé. How not to marvel at them! The knowledgeable inventor showed me an aerial propeller, which could indeed, in its application to aerial navigation, one day lead to the abolition of frontiers and militarism....

So wrote the great French astronomer Camille Flammarion in *Le Voltaire* in February 1881. Trouvé's accidental death in 1902 spared the brilliant electrical engineer the knowledge that twelve years later, aerial navigation would be used to kill and to bomb, assisted by electricity. In 1892, with a war between the USA and Chile threatening, Thomas Edison had expressed different views to the *Straits Times* of Singapore:

> Mr. Edison is very sanguine about his new flying machine; he assures us that it can be projected into space at any given angle and with *the aid of an electric motor* and revolving fans for about fifty miles, dropping dynamite by the way on the heads of an enemy. The experiments I have made lead me to think that I can carry on this machine 500 lb. of explosive material, and drop it from aloft at any point I choose. Of course, I must allow for the state of the atmosphere, as one does with artillery; but my experiments make me feel that I can come within 20 per cent of my object [emphasis added].[1]

At the start of World War I, airplanes were only sports machines, powered by V or rotary gasoline engines. Of course, every one of these was fitted with magnetos for starting and the pilot's legendary command "Contact!" For redundancy, virtually all piston engine aircraft are fitted with two magneto systems, each supplying power to one of two spark plugs in each cylinder. A magneto is an electrical generator that uses permanent magnets to produce periodic pulses of alternating current in airplane engines, in which keeping the ignition independent of the rest of the electrical system ensures that the engine continues running in the event of alternator or battery failure. All the engines for the British Naval and Military Aeroplane Engine Competition in 1914 were using the Bosch ignition system.

But other than that, pilots went off on reconnaissance, hunting with a cavalry rifle or an automatic pistol. This only lasted for a short time. From December 1914, the first wireless transmitters appeared, which aided aircraft in carrying out some artillery settings. After two years of R&D, Royal Flying Corps Major Herbert Musgrave and his team had devised a system whereby pilots could use wireless telegraphy to help the artillery hit specific targets. The aircraft observer carried a wireless set and a map, and after identifying the position of an enemy target was able to send messages such as A5, B3, etc., to the artillery commander. Unfortunately the early transmitters weighed 75 lb. (34 kg) and filled a seat in the cockpit. This meant that the pilot had to fly the aircraft, navigate, observe the fall of the shells, and transmit the results by Morse code by himself. Also, the wireless in the aircraft could not receive. Originally only a special Wireless Flight attached to No. 4 Squadron RFC had the wireless equipment.

One wireless ace who was prepared to test each new development was Lieutenant B.T. James, who made flight after flight in B.E.2a. James brought the science of wireless in aircraft to a high state of efficiency, testing the Sterling lightweight wireless. By 1915, each corps in the British Expeditionary Force had been assigned an RFC squadron solely for artillery observation and reconnaissance duties. The transmitter filled the cockpit normally used by the observer, and a trailing wire antenna was used which had to be reeled in prior to landing. The Royal Flying Corps began research into how wireless telegraphy could be used to help home-defense aircraft during German bombing raids. In 1916 the RFC developed a lightweight aircraft receiver and a Marconi half-kilowatt ground transmitter. These transmitters were located on aerodromes in raid-threatened areas. The aircraft receiver was tuned in advance, and the pilot had to unreel a 150 ft. aerial from its drum and switch on. Trials started in May, and pilots reported that signals were clearly heard up to ten miles, but at longer distances they weakened.

In April 1915, Captain J.M. Furnival was the first person to hear a voice from the ground when Major Prince said, "If you can hear me now, it will be the first time speech has ever been communicated to an aeroplane in flight." In June 1915, the world's first air-to-ground voice transmission took place at Brooklands Motor Course (England) over about 20 mi (32 km). Ground-to-air was initially by Morse, but it is believed 2-way voice communications were being achieved by July 1915. In early 1916, the Marconi Company (England) started production of air-to-ground radio transmitters/receivers which were used in the war over France. By November 1918 the newly established RAF had some 600 aircraft fitted with the new Mark III choke-controlled telephone set, operating in conjunction with 1,000 ground stations and manned by over 18,000 wireless operators.

Over in the USA, in 1917 AT&T invented the first American air-to-ground radio transmitter. They tested this device at Langley Field in Virginia and found it was a viable technology. In May 1917, General George Squire of the U.S. Army Signal Corps contacted AT&T to develop an air-to-ground radio with a range of 2,000 yards (1800 m). By July 4 of that same year, AT&T technicians achieved two-way communication between pilots and ground personnel. This allowed ground personnel to communicate directly with pilots using their voice instead of Morse code. Though few of these devices saw service in the war, they proved this was a viable and valuable technology worthy of refinement and advancement; therefore, further models had this technology installed into biplanes on airstrips in France by 1919.

Airships also benefited. In 1917, *Rigid No. 9* was fitted with a valve transmitter at Howden, the first such radio installation in a rigid airship; earlier spark sets were thought to be too dangerous because of the stored inflammable hydrogen. Subsequently all British dirigibles carried Marconi valve sets, and in late 1917, RFC long-range night bombers were also fitted with Marconi equipment. For the electrical power required to transmit, the British Coastal class blimps, one of several types of airship operated by the Royal Navy, carried a 1.75 horsepower (1.30 kW) ABC (Anglo-Belgian Corporation) auxiliary gas engine. These powered a generator for the craft's radio transmitter and, in an emergency, could power an auxiliary air blower.

Parallel development had been taking place for nocturnal flight. Initial nighttime test flights with particular lighting, by causing the loss of several crews, brutally demonstrated that an aircraft cannot count on the light alone to land. The question was abandoned for almost one year. In 1914 No. 1 Squadron, Royal Naval Air Service fitted an 80hp Avro 504 with a 50-watt searchlight mounted in the undercarriage structure. This was powered by a 12-volt battery mounted in the fuselage. By 1915 the RNAS fitted a Royal Aircraft Factory BE2c with instrument panel lighting which had variable intensity. The aircraft was also fitted with two 5⅜ inch (13 cm) headlamps, one on each outer interplane strut and set to converge on the ground 30 yards (27 m) ahead of the aircraft. The installation only weighed 16 pounds (7 kg). The Royal Naval Air Service at Burgh Castle fitted a Lucas headlamp to an aircraft which proved "very useful."[2]

The British RFC Supermarine Nighthawk, an anti–Zeppelin night fighter, used a trainable nose-mounted searchlight, a 1½-pounder (37 mm) Davis gun mounted above the top wing with 20 shells, and two .303 in (7.7 mm) Lewis guns. Power for the searchlight was provided by an independent gasoline engine–driven generator set made by ABC Motors, possibly the first instance of a recognizable airborne auxiliary power unit hybrid. Although touted as being able to reach 75 mph (121 km/h), the P.B.31E prototype only managed 60 mph (97 km/h) at 6,500 ft (1,981 m) and took an hour to climb to 10,000 ft (3,048 m), which was totally inadequate for intercepting Zeppelins.

When at the end of 1915, daytime flying becoming dangerous for the heavy and clumsy bomber planes, these were oriented towards night flights. They were equipped with lightweight lighting installations, with small 300W generators constant tension–activated by a wooden reel and supplying energy to a ramp of three small adjustable parabolic lamps placed at the front of the airplane. These summary installations, just like those of the wireless transmission, underwent successive improvements (back-up batteries, more powerful lamps, supple warming clothing, etc.)

So electricity became part of the way of life of the military pilots. In France, special services were created at the S.T. Aé to the S.F. Aé; squadrons and aerodromes were equipped with electricians specially recruited and trained in a school perfectly organized at the Retrenches Park of Paris. Entirely new equipment was put into service. Military telegraphy brought out a series of transmitters and

1915: The only disadvantage when you lit up your bombing target from above was that you were a "sitting duck" for enemy ground fire. The wind-powered turbine gave electrical power to the searchlight (©RAF Museum).

receivers specially developed for the applications to which they were aimed. The Technical Section of the Aéronautique researched and developed landing lights, searchlights of optical telegraphy, various body warmers, and a whole range of equipment for generating and distributing electricity. Aircraft that benefited from these developments were the Farman F-50 night bomber, Spad fighter plane, Nieuport, Breguet and Salmson.

Combat pilots flying at night, in order to see their dashboard, had small, shielded incandescent lamps placed next to the dials. Importantly, these lamps could be turned on and off as needed. However, engine vibrations often caused light bulbs or electrical connections to fail. Even the small amount of light emitted by such lamps could cause great difficulty for a novice since it could be too dazzling; later, luminous paint composed of zinc sulfide and radium was applied to the dials.[3]

There was, however, one flying machine that went further than using electricity for heating and lighting. Its innovator was a forty-year-old Austro-Hungarian pilot officer called Stephan von Petróczy. A graduate of the Theresian Military Academy in Wiener Neustadt, Petróczy trained to fly in 1910, obtaining Pilot Diploma No. 13, despite crashing on a flight from Wiener Neustadt to Fischamend, suffering a broken arm. He became a flying instructor at Wiener Neustadt, the first flying school in the Empire. On the outbreak of World War I, he flew combat planes first on the Serbian front and then the Russian front. But in 1915 he was withdrawn from service to establish a flight training battalion to compensate for the high pilot fatalities, and then promoted to Commander of the Aviation Arsenal. Well aware, from his own experiences, that the hydrogen-filled observation balloons were too clumsy and dangerously inflammable, Petróczy conceived of a tethered device, deriving its energy from the land. For his project, in April 1916, the Imperial War Ministry allotted Petróczy 100,000 crowns to conduct experiments on a prototype. In this he was assisted by two fellow Hungarians: an aerodynamics expert, Lieutenant Dr. Theodore von Kármán; and engineer Wilhelm Zurovec, who built a 6 hp Austro-Daimler electric-generating motor at 2400 rpm, weighing 4 kg (9 lbs.) which could power a 35 kg (77 lbs.) flyable model.[4]

This very rare 1918 photograph shows the *PKZ 1* (Petróczy-Kármán-Zurovec), the world's first tethered quadcopter, deriving its electric power from an Austro-Daimler gasoline engine. Designed for use as an aerial observation machine, the prototype was abandoned after the 1918 Armistice (courtesy Reinhard Keimel).

The full-scale *PKZ1* rotorcraft was built by Mátyásföld. A 190hp Austro-Daimler-generated electric motor was used to drive two propellers in front of the observer and two behind, as in today's quadcopter. The electrical power was transmitted through a cable but still the motor weighed 195 kg. In flight, the power would be transmitted through an 800-m-long aluminum cable. On its first test flight in Fischamend Hangar, the unit took off at a rotor speed of 700 rpm and rose up to the height limit of 5 meters (16 ft.). Approximately fifty flights took place between July 1917 and March 1918, with heights of more than 10 m (33 ft.) obtained. A second PKZ was built and tested to an altitude of 50 meters (164 ft.), but then, in a demonstration flight on June 10, 1918, in Fischamend, in front of a high-level military delegation, the machine crashed. Five months later, Armistice was declared and the project abandoned. But at least the potential of an electrically-powered quadcopter had been demonstrated. In the years which followed, Stephan von Petróczy was involved in the construction of the Hungarian Air Force—gasoline-engined, of course.

While the PKZ aircraft were being test flown, in May 1917, an article titled "Electrifying the Aeroplane" appeared in *The Electrical Experimenter* describing some recent advances. It begins: "Electricity is being rapidly introduced in the new art of Aeronautics." The article refers to "the aerial limousine the Autoplane, recently exhibited at the Aeroplane Show in New York." The Autoplane had just been invented by Glenn Curtiss of Seattle as a roadable aircraft. It was a triplane, using the wings from a Curtiss Model L trainer, with a small foreplane mounted on the aircraft's nose. Its aluminum body resembled a Model T and had three seats in an enclosed cabin, with the pilot/chauffeur sitting in the front seat and the two passengers side-by-side to the rear. It used a four-blade pusher propeller and a twin-boom tail. An electric start 100 horsepower (75 kW) Curtiss OXX engine drove the propeller via shaft and belts. It was shown at the Pan-American Aeronautic Exposition at New York in February 1917. Although the vehicle was capable of lifting off the ground, it never achieved its projected 65 mph (104 kph) in full flight. The entry of the United States into World War I in April 1917 ended development of the Autoplane. Electrically, at least it had electric heating and cabin lighting.

The Electrical Experimenter article continues with devices innovated by Elmer Sperry, then working with Peter Hewitt on an Automatic Airplane (see Chapter Nine). An adapted Curtiss N-9 Seaplane, which included the Sperry automatic gyroscope, serves to control and maintain an airplane in any desired position and also free a pilot for aiming and dropping bombs. The article also mentioned an Incidence Indicator,

> which warns the aviator before he stalls and enables him to get the best climbing and gliding angles out of his machine mounted on a forward strut. The lamp bank and indicator is on the instrument cowl, always visible to the pilot observing other essential instruments. The red light warns the aviator before he stalls as well as when he begins climbing at a dangerous angle. The white lamp signals wherever the pilot dives at too steep an angle. The green light indicates the best climbing angle. Being of low voltage as well as low current consumption, the lights can be operated by a dry battery, encased in metal and installed wherever most convenient. The signals are regulated by a vane operated by the air stream.

To provide the electrical power necessary to gyroscopically motor-control his airplane, Sperry also invented an aerodynamically-shaped wind-driven generator, mounted on the upper wing in the slipstream of the airplane's propeller. In his patent, filed in December 1916, Sperry stated:

> As has become the case with automobiles, electricity has been found to be an indispensable asset for aeroplanes also. It is being used for lighting signals and other lights, operating wireless transmission sets, operating the aeroplane controls through servo-motors, actuating various kinds of instruments for signaling to other machines or between the pilot and observer, operating gyroscopes or other stabilizing apparatus, and for charging engine starting batteries. The object of this invention is to provide a practicable means whereby a sure supply of electricity at a constant voltage may be obtained.

Unknown to Sperry, in January 1915, Bruno Rosenbaum of Berlin had filed a similar patent for propeller-driving wireless sending equipment on aircraft, and particularly referred to the case where the propeller was driven by the air current generated by the main propeller and not by the air current created by the flight of the aeroplane. When this arrangement was adopted, it would be possible to drive the generator of the sending equipment when the craft is not in flight, as the main propeller may be kept in motion in order to drive the smaller propeller. With Armistice, Rosenbaum would sell his invention to Westinghouse.

By this time, airship bombing raids were terrifying city dwellers. Japan was on the side of the Entente Powers, and although their air force was nonexistent, in October 1916 Toshio Yoshida of Tokyo, "subject of the Emperor of Japan," applied for a patent for "a novel system for dropping a bomb or bombs or the like from an airplane or airship." This was "a telescopic aiming device using electrical circuit including electromagnets for operating the said electrically controlled means to release the bomb from the case when the airplane or airship reaches a proper point above an object to be attacked." (U.S. 1290858 was granted in 1919.) Count Oscar Wilcke, who had served as a Zeppelin airship pilot based in Friedrichshafen, Germany, carrying out those notorious raids on the British Isles, also obtained an almost identical patent for an electromagnet-powered bomb-throwing apparatus, including a device for testing the electric circuits. Wilcke had applied for this patent in 1916 at the height of the Zeppelin raids, and ironically obtained it in March 1921 (Patent No. 316584) after Germany had been defeated and all combat Zeppelins had been destroyed.

In 1917, Godfrey L. Cabot, a middle-aged American industrialist and airplane pilot and organizer of the Naval Militia Aviation Unit based in Marblehead, invented and patented:

> An apparatus for launching aircraft such as airplanes and hydroaeroplanes, the principal object being to provide electro-magnetic means for holding the aircraft upon the launching apparatus until the required speed has been attained, and for then releasing it. A further object is to provide electric means for operating the launching apparatus and means for simultaneously cutting off the current from said operating means and the electro-magnetic holding means when the ship is traveling at a safe speed for flight. Yet another object is to provide a track for the catapult carriage upon which said carriage and aircraft move forward under the power of the latter or of both for a predetermined distance, the electric propelling means of the carriage and the electric magnet being then automatically thrown into action so that the craft is gradually driven faster until a safe flying speed is attained.

Cabot filed his patent in 1917 and obtained it the following year. In June 1918, Bernhard E. Fernow, Jr., engineer for the Cutler-Hammer Manufacturing Co. of Milwaukee, obtained U.S. Patent 1410395 A for "a system of electro-magnetic brakes to stop airplanes."

Throughout the war, pilots had been able to fire their machine guns through their spinning propellers without splintering them, thanks to a mechanically or hydraulically synchronized timing system. By the end of the war, German engineers were well on the way to perfecting a gear using an electrical link between the engine and the gun, with the latter being triggered by a solenoid. When the guns entered a forbidden zone, electrical power was cut off to disable the trigger.

Following the Armistice, a number of bomber planes had the space previously taken up by bombs replaced with luxurious cabins which had to be heated and lit for passenger comfort. The other challenge was navigation and various approaches were used, all indirectly dependent on electricity. The April 1922 issue of *Aerial Age Weekly* reported the first nighttime flight from the UK to the Continental Air Route. A British Air Ministry airplane took off from Croydon Airport, London, at 9:20 p.m., landed at an aerodrome at St. Inglevert, France, and returned to Croydon at 11:30 p.m. On board were a couple of Air Ministry personnel who handled RDF wireless communication and lighting, as well as the crew of the plane itself. No passengers were on board.

Unable to use the same lighthouses for ships as for navigation, the French also decided to build lighthouses that airliner navigators could see at great distances. To stake out the Paris-Algiers air route, an electrically powered lighthouse was commissioned by the Technical Section of the Aéronautique at Issy-les-Moulineaux and built by the firm of Barbier, Bénard & Turenne. It had an intense luminosity of one thousand candles and was composed of 8 two-meter diameter lenses extending 180°; in the foyer of each one was an arc lamp absorbing 120 amps at 65 volts. Each lens in its optical system was made up of seventeen dioptric elements and catadioptric elements.

In 1925, the 11-meter (36 ft.) lighthouse was dismantled and transported for erecting on Mont Afrique, twelve kilometers from Dijon. It was inaugurated that summer by Gaston Gerard, mayor of Dijon, and General Jacques Theodore Saconney. Run and maintained by the Aerial Navigation Service, this was the first of over thirty terrestrial aeronautical lighthouses for aircraft, at a time when air navigation was still carried out by eyesight. Located at the crossroads of Paris-Switzerland-Italy and Paris-Mediterranean, the powerful beam of this aeronautical beacon was visible in certain conditions up to 400 km (250 mi). It is even said that it was "visible" from Rouen, Brussels, Antwerp and Turin.[5]

The same happened across the USA. After initial tests with simple bonfires, in 1922 two young U.S. Army Air Corps lieutenants, Donald L. Bruner of Iowa and Harold R. Harris of Illinois, based at McCook Field, Dayton, Ohio, proposed ideas for electrically lit airport boundaries, spot-lit windsocks, and rotating beacons on towers. Each beacon consisted of a revolving motor-driven light that sat at the top of a 60-foot (18 m) tower. At night, the beacon flashed in a certain sequence so that pilots could match their location to the printed guide that they carried in their aircraft. The first nighttime flight test was between Dayton and Norton Field in Columbus, a distance of 72 miles (116 km). Its financing had been thanks to Paul Henderson, Second Assistant Postmaster General, who had managed to secure funds for lighting a portion of the transcontinental, despite initial criticism. By 1926, all segments were in place; the Transcontinental Airway System's light beacons were brought under the authority of the Bureau of Lighthouses, and crossing the country by air (day or night) was considered much safer. In 1926 Bruner, promoted to captain, received the Distinguished Flying Cross for developing and perfecting night-flying equipment, thus making it possible for military and commercial airplanes to traverse the length and breadth of the United States during the hours of darkness. He went on to invent and patent airplane landing lights (U.S. 1554198), electrically warmed anti-frosting/fogging goggles (U.S. 2099464), and electrically retractable landing lights (U.S.

2124050). By this time there were about 1,550 electric light beacons stretching across 18,000 miles (30,000 km) of the USA.

In 1921, Nikola Tesla submitted his first patent: "An apparatus for Aerial Transportation," modifying it in an application made in October 1927, which was granted in 1928 (U.S. 1,655,114). In this, he described a "helicopter-plane" adapted for vertical and horizontal propulsion and change from one to the other attitude, the combination of means for tilting the machine in the air, a fluid pressure generator of a capacity several times greater than normally required in horizontal flight, a motor capable of carrying overloads adequate for support in all attitudes, and means for controlling the supply of the fluid to the motor in accordance with the inclination of the machine.

Far from these practical applications, in India, a mystic called Subbaraya Shastry was writing a book called *Vaimānika Shastra*, or *The Science of Aeronautics*. He is said to have obtained the information by psychic channeling with the ancient Saint Bharadvaja. The story goes that Subbaraya Shastry was believed to have contracted leprosy. He left his home and spent nine years living in the forest. During this time he is supposed to have spoken with the ancient saint (sage Bharadvaja) and was enlightened with this newfound knowledge of flying machines. He later returned home (as he also had been cured of leprosy), but Shastry could not read or write, so he dictated his new knowledge over the period of 5 years (25 years after the psychic experience itself). The dictated text was apparently discovered in 1952 by G.R. Josyer, who later translated it into English in 1973. This publication contains eight chapters claiming that ancient vimānas from the King Ravana legend were actually feasible flying machines, perhaps even similar in ability to rockets. The text indicates that propulsion was provided using rotating gyroscopes of electricity and mercury.[6]

Some visionaries were more practical. In 1929, Hermann Julius Oberth, an Austro-Hungarian-born German physicist and engineer, wrote a book, *Wege zür Raumschiffahrt* (*Ways to Spaceflight*). In this, he devotes a whole chapter, "Das elektrische Raumschiff" ("The Electric Spaceship"), to electric propulsion, advocating electrostatic acceleration of electrically charged gases that can be created from refuse on the orbiting space station that is a major theme of his book.[7] But not of this one.

On a more practical level, to deal with the problems of flying through thick fog or poor visibility, on April 3, 1928, French Captains Gérardot, Cornillon, Vigroux and Rey—respectively, pilot, navigator, mechanic and assistant pilot—took off in their Lorraine-engined Amiot from Le Bourget near Paris, to carry out a 68-hour, 10,000 km (6,000 mi) study flight across Africa, where they navigated by "radiogoniométrie électromagnetique" or radio direction finding. The flight path took in Colomb-Bechar, Timbuktu, Bamako, Dakar, and Casablanca, returning then to Villacoublay. During the flight, Cornillon was only able to use his wireless positioning until Oran, where, due to its malfunctioning, he had to resort to a traditional compass.

Such Marconi wireless direction-finding was also used by those in airships or solo airplanes risking their lives to make transatlantic or polar flights to transmit their whereabouts either to potential rescuers or to the press.

On September 24, 1929, Lieutenant James H. Doolittle, U.S. Army Air Corps, Ph.D. MIT, made the first completely blind airplane takeoff flight and landing, solely by reference to instruments on board his aircraft. Flying from the rear cockpit of a civil-registered two-place Consolidated NY-2 Husky training airplane, NX7918, Doolittle had his visual reference to earth and sky completely cut off by a hood enclosure over his cockpit. The experimental gyroscopic compass, artificial horizon and a precision altimeter were developed by Elmer Sperry, Jr., and Paul Kollsman, both of Long Island, New York. Funding for the Full Flight Laboratory at Mitchel Field was provided by the Daniel Guggenheim Fund for the Promotion of Aeronautics.

To train new pilots, in 1929 J.P. Buckley patented an aeronautical instructing device (or flight simulator) "comprising a body, means of mounting said body for universal movement, *electric energizers* for imparting movement to said body, circuits for said energizers, means for closing circuits for selected energizers operable by an occupant of the body."[8]

Since the beginnings of aviation, no device had been installed on airplanes that would indicate when one or both of the wheels had either been broken off in ascending or fallen off after a plane was in the air. Numerous accidents had occurred due to the pilot's ignorance of such a condition, because of his inability to see the landing gear. In 1930, Harry H. Semmes of the Bendix Aviation Corporation, South Bend, Indiana, applied for a patent for an electrical circuit and a mechanical device in or on the airplane for indicating when a wheel of an airplane has broken loose and fallen from its axle. (Patent U.S. 2025909 was published in 1935.)

In the same way, the hazards of flying caused by ice accumulating on airplane wings and stabilizers at high altitudes or in cold weather were well known, the additional load of the ice on the plane often seriously interfering with safety. Again in 1930, Archie F. Thompson of the Iceless Air Wing Corporation, Tulsa, Oklahoma, applied for a patent concerning a heater for airplane wings. In combination with aircraft including a source of electrical energy, heating wires were arranged on a selected wing surface, a layer of insulation between said surface and wires, feed wires leading from the source of energy to the heating wires, means interposed in the feed wires for controlling the amount of current flowing through those wires, and a covering for the insulation and heating wires substantially conforming in shape to the wing surface and spaced therefrom to form a dead air chamber. Thompson was granted Patent U.S. 1868468A in 1932, but the Wall Street crash had bankrupted him.

Electricity continued to power testing devices. In 1935 a 24-ft (7 m) wind tunnel, powered by a 2,000 hp DC electric motor, opened at the Royal Aircraft Establishment, Farnborough, England. Parallel to this a seaplane tank was commissioned and consisted of a 400-yard (366 m) tank with rails either side on the edges of the tank. The rails supported a trolley that ran above the tank carrying a model under test. The model was in contact with the water surface and the forces generated could be measured. The trolley was propelled by electric motors and the control gear was similar to that found in tram cars. A brake was applied automatically towards the end of each run.[9]

One piece of electrical apparatus regularly used on board aircraft during the interwar years was the Eagle Automatic Electric Aircraft Camera, developed by Colin M. Williamson of Willesden Green, London, and used throughout the world for air survey photography in all types of aircraft. In 1928, the RAF developed an electric heating system for the aerial camera. This allowed reconnaissance aircraft to take pictures from very high altitudes without the camera parts freezing.[10]

In 1932, similar to Hiram Maxim's fairground ride of thirty years before, brothers Victor and Joseph Stanzel of Schulenburg, Texas, developed a fairground ride called "Fly-A-Plane." They designed and built a full-sized two-passenger conventional high-wing cabin-type electric airplane on a supporting beam structure, able to carry up to four passengers at a time. When it was completed in 1932, they stationed it along a well-traveled road near town and offered rides for twenty-five cents. Joe served as pilot of the aircraft. Later, it was located in a Houston park and eventually sold to an amusement park in Kilgore, Texas. Victor Stanzel received his first patent for the "Fly-A-Plane Amusement Ride" in December of 1933 (U.S. 1941024 A).

During the 1930s a prevailing movement in the design of buildings and furniture was Art Deco, which extended from a cigarette case to the Chrysler Building in New York City. Art Deco pieces included electric desk lamps and ceiling fans in the form of airplanes made of Bakelite or chromed metal. Obviously switching on the light or the fan would not make the airplane fly!

In 1941 Arthur M. Young, an American inventor, designed Bell Aircraft Corporation's first helicopter, the Model 30, and innovated the stabilizer bar used on many of Bell's early helicopter designs. Although these were gasoline-powered, Young later recalled, "As a boy, in about 1915, I made a crane from 'Meccano Set' parts which would lift one of my little brothers into the air with the help of a motor and endless pulley. I also made the electric motor. That was my initiation into the electrical field." In 1927, as a Princeton University graduate, Young started work on model helicopters and used electric hover motors to drive the rotor head. By 1939, he decided to abandon larger scale models and the tip powered concept and concentrate on smaller models which were again powered by electric vacuum cleaner motors.[11]

If electricity could not be the main propulsion unit, might it at least be used to control propeller pitch? So reasoned a Canadian engineer, Wallace Rupert Turnbull. Coming from a wealthy family, Turnbull studied at Cornell University and in Germany until age 25, and then

In 1929, Rupert Turnbull stands beside his invention, the electrically powered variable pitch propeller, which improved takeoff and extended the range of many gas airplanes (Canada Aviation and Space Museum).

worked at the Edison Lamp Works, Harrison, New Jersey, for six years, learning about electricity. He then developed an electrically activated variable-pitch propeller that gave aircraft maximum power in takeoffs and landings, and economical cruising at speed. He built the first wind tunnel in Canada to test propeller designs under a constant wind speed. This was tested with an Avro 504K airplane on June 29, 1927, at Camp Borden, Ontario, Canada, then patented in 1929 (U.S. Patent 1,828,348).

That same year, Turnbull won a silver medal at the Inventions Show in New York City, then sold his patent to the Reed Propeller Company, a subsidiary of the Curtiss Airplane and Motor Company, a division of Curtiss-Wright Ltd., Montreal. While Turnbull went on to other inventions, his electrical propeller was further refined by Charles W. Chillson.

Chillson had received his B.S. in mechanical engineering from Stanford University in 1931 and gone on to work in chemical engineering at the California Institute of Technology (1931–36). At the same time, Chillson worked with C.K. Greene on a mechanical controllable-pitch propeller, which progressed through whirl-testing at the USAAF Engineering Division at Wright Field, Ohio. He then moved to the Curtiss-Wright Propeller Division as their engineer and project designer to concentrate on the electrical propeller. His patent (Serial No. 261,879), filed March 15, 1939, explains how the fluid pressure controlled by the governor actuates an electric switch, which in turn controls the operation of an electric motor to adjust the pitch of the propeller blades.

By 1940, the electric propeller had become more sophisticated. During World War II, Curtiss-Wright manufactured 146,468 electric propellers with variations, while the German Luftwaffe's Messerschmitt BF 109F was also equipped with the system (San Diego Air & Space Museum).

During World War II, Curtiss-Wright manufactured 29,269 airplanes, 142,840 aircraft engines, and 146,468 electric propellers with variations ingeniously developed by Chillson and his team. Between 1942 and 1945, the Curtiss-Wright propeller factory in Beaver, Pennsylvania, fabricated more than 100,000 new propeller blades for a variety of aircraft, such as the Lockheed YP-38 1940 twin engine bomber. Rotol Airscrews in the UK purchased a manufacturing license to produce the Curtiss Electric propeller at their Kilmarnock plant.

The German Luftwaffe was swift to adapt the electrical pitch propeller design to their warbirds. With a license originally obtained from Curtiss, the VDM (Vereingite Deutsche Metallwerke) electric propeller pitch control mechanism was fitted to the Messerschmitt BF 109F. Propeller pitch was changed electrically, and was regulated by a constant-speed unit, though a manual override was still provided. Designed by Kurt Tank, the Focke-Wulf Fw 190 Würger (English: Shrike) single-seat fighter aircraft, dubbed the "Butcher Bird," used an electric VDM control-pitch propeller. In addition, to minimize changes in the aircraft's trim at varying speeds, thus reducing the pilot's workload, the entire horizontal tailplane was tilted with an electric motor, with an angle of incidence ranging from -3° to +5°. The retractable undercarriage was operated by pushbuttons controlling electric motors in the wings, and was kept in position by electric up and downlocks. The armament was also loaded and fired electrically. The Fw 190 F-8/U4, created as a night fighter, was equipped with various electrical systems such as the FuG 101 radio altimeter, the PKS 12 automatic pilot, and the TSA 2 A sighting system.[12]

The increasing sophistication of warplanes meant that ground facilities also became more sophisticated. The high-powered engines in use could no longer be started by hand-swinging the propeller, but powered starting systems had to be provided. One of these, the Trolley Accumulator, was used for aircraft such as the Spitfire, which had an inbuilt electric starter motor. "Trolley accs" were produced in several forms, but they all had a bank of lead/acid batteries contained in a covered box, which had been wired "in series" so that each lead/acid cell (nominally producing about 2 volts) produced 12 volts and sufficient power to turn over the standard aircraft engines of the day. When not in use, these accumulators were connected to the main electricity supply at the RAF station they were on, to build up charge (usually overnight); they would need periodic servicing and topping up with distilled water, as they tended to lose that component of the electrolyte during the charge/discharge process.[13]

To move around the giant plants where these warplanes were being built was always a challenge. In September 1935, Consolidated Aircraft Corporation opened its new "Building 1," a 247,000-square-foot (22,900 m^2) continuous flow factory in San Diego, California, while Douglas Aircraft Company's largest facility was its Long Beach plant, totaling 1,422,350 sq. ft. How executives and technicians might get around these hangars silently and safely was solved by Consolidated's chief test pilot, William B. Wheatley. Bill Wheatley was as much an innovator as a pilot, having patented "Apparatus for arresting launching devices for airplanes"; "Launching airplanes from water" (flying boats); "Device for handling aircraft"; etc. His solution to mobility was the "Electricycle," a standard bicycle with the battery and single-speed electric motor mounted on the front fork. Wheatley's Electricycle was as streamlined as the airplanes he was test-flying, with production advantages secured through utilization of standard parts, foot-controlled throttle and brakes, and easy-to-exchange batteries. For manufacture, in April 1941, Wheatley joined forces with Newton C. Blood and O.L. Weaver of Blood Sales Co., Inc., with their plant in Long

In the later stages of the war, high-powered engines in use could no longer be started by hand-swinging the propeller, so powered starting systems had to be provided. One of these, the Trolley Accumulator, was used for aircraft such as the Spitfire and the Hurricane (©RAF Museum).

Beach, California. At the time, Consolidated was developing the XB-24 Pratt and Whitney–engined bomber. In June of 1941, Wheatley and his crew of four were killed undertaking the final test of the B-24 before the aircraft were to be delivered to the Royal Air Force in England. The crash into San Diego Bay was initially thought to be sabotage, but was later discovered to have been caused by a mechanical anomaly in which the elevator locked in the "up" position, rendering the crash unavoidable. While more than 18,000 B-24 Liberators were built in just over five years, making it the largest military production in U.S. history, nothing else was heard of the Electricycle.

The Avro Lancaster was a British four-engined Second World War heavy bomber designed and built by Avro for the Royal Air Force (RAF). The "Lanc," as it was affectionately known, thus became one of the more famous and most successful of the Second World War night bombers, delivering 608,612 long tons of bombs in 156,000 sorties. But with its rear gun turret, neither the mid-upper nor the rear gunner's position was heated, and the gunners had to wear electrically heated suits to prevent hypothermia and frostbite. Extremely low temperatures would occur above 10,000 ft. (3,000 m).

In September 1944, the first of thousands of B-29 Superfortress bombers rolled out of Boeing's assembly line in Wichita, Kansas. Based on the highly successful platform of the B-17 bomber, the B-29 became the largest aircraft operational during World War II, a combination of cutting-edge tech and devastating firepower. This was due to a battery of .50 Browning M2 machines linked to five interconnected analog computers, built at General Electric's Erie plant. These were located in the nose and tail positions and three Plexiglas blisters in the central fuselage, and required only one single gunner and a fire-control officer. This electrical system increased the weapons' accuracy by compensating for factors such as airspeed, lead, gravity, temperature and humidity.

THREE. Electricity Goes to War

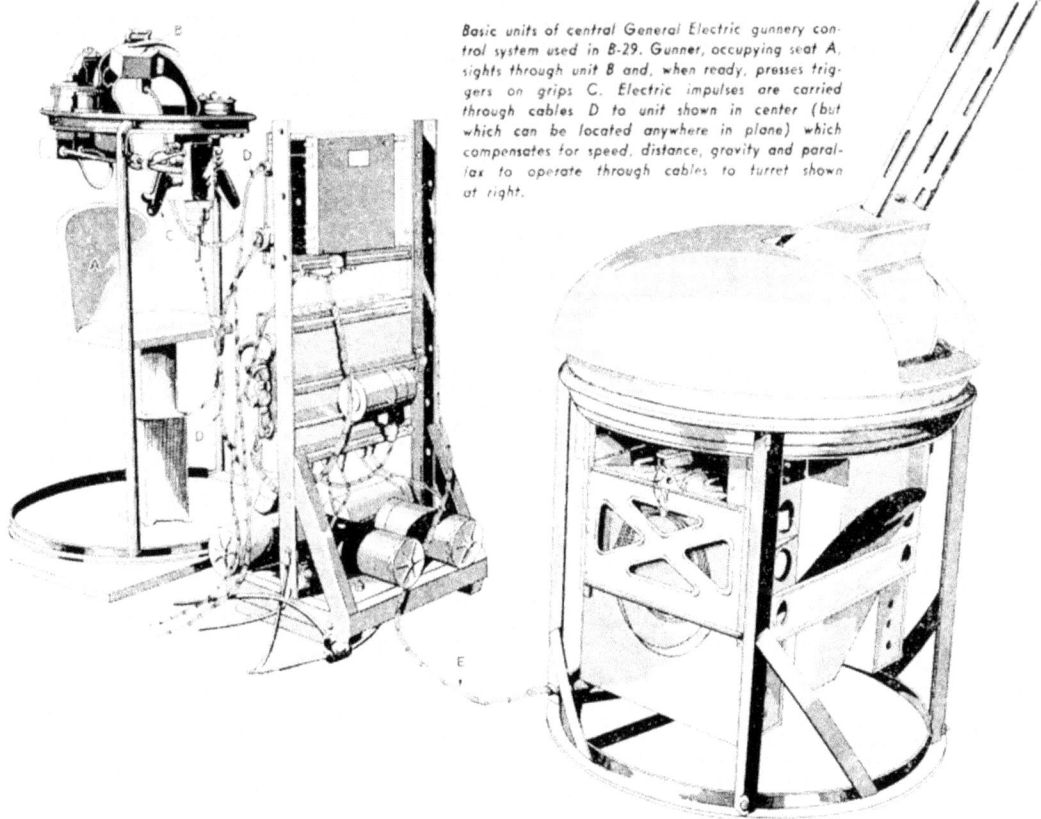

Basic units of central General Electric gunnery control system used in B-29. Gunner, occupying seat A, sights through unit B and, when ready, presses triggers on grips C. Electric impulses are carried through cables D to unit shown in center (but which can be located anywhere in plane) which compensates for speed, distance, gravity and parallax to operate through cables to turret shown at right.

1944: The USAF's B29 Superfortress bombers were equipped with an electrically activated, spherical-shaped, altazimuth mount gun turret, giving gunners a lethally precise firepower (author's collection).

Simpler, the ball turret, an electrically activated, spherical-shaped, altazimuth mount gun turret, was fitted to some American-built aircraft such as the Boeing-17E Flying Fortress, the B-24H Liberator, and the United States Navy's Liberator, the PB4Y-1. It was manufactured by Sperry, and to a lesser extent Emerson Electric.

Second World War aircraft needed large amounts of power, as they had many electrical systems and literally miles of wiring—a Wellington bomber had approximately 7 miles (11 km) of wiring, and the Halifax had no less than 12 (20)!

While the air war was going on, in 1943, some engineers still believed in electricity. Stanley Bizjak, a 31-year-old inventor living in the tiny village of Crivitz in Marinette County, Wisconsin, obtained a patent on

> an electrically powered glider, partially supported by buoyant gas, which could take off silently and operate under its own power during flight, and one which during aerial maneuvering can have its motive power manually disconnected so as to utilize the glider's propeller which is always directly connected to a generator to automatically regenerate the power batteries independently of the electrical power to compensate for loss of electrical energy consumed in rising into the air or else for power expended in flight in performing various aerial maneuvers which the electric motor per se is incapable of producing.

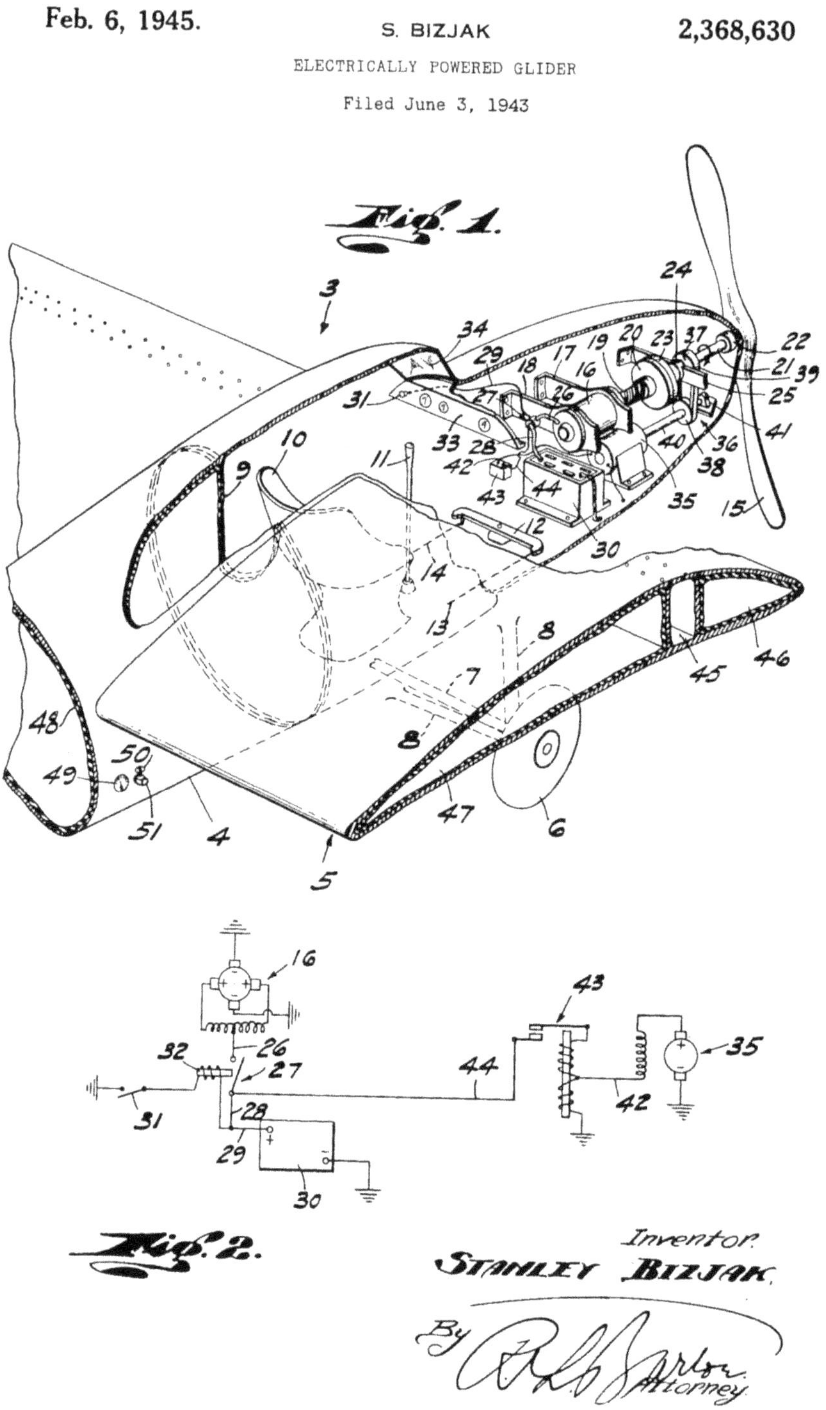

Feb. 6, 1945. S. BIZJAK 2,368,630
ELECTRICALLY POWERED GLIDER
Filed June 3, 1943

Fig. 1.

Fig. 2.

Inventor:
STANLEY BIZJAK,
By [signature] Attorney

In 1945, U.S. Army Corporal Stanley Bizjak, based at his family dairy farm in Crivitz, Wisconsin, received U.S. Pat. No. 2,368,639 for this electrically powered glider. He too had filed in June 1943, four months after Westinghouse.

Bizjak had also patented a tractor embodying a retractable mechanism whereby it could be propelled alone on four wheels, or could be propelled on two wheels when hitched to a plow, harrow or other soil-working implements. We do not know whether the Bizjak electric glider, for which he obtained U.S. Pat. No. 2,368,639 in 1945, was built.

A unique aircraft produced by the Luftwaffe, designed by Alexander Lippisch, was the rocket-powered Messerschmitt Me163 "Komet." In early July 1944, piloted by Heini Dittmar, the Komet reached 1,130 km/h (700 mph), an unofficial flight airspeed record unmatched by turbojet-powered aircraft for almost a decade. Less known is the fact that a small windmill generator on the extreme nose of the fuselage, and the backup lead-acid battery inside the fuselage that it charged, provided the electrical power for the radio, the Revil 6B, -C, or -D reflector gunsight, the direction finder, the compass, the firing circuits of the cannons, and some of the lighting in the cockpit instrumentation. This was later developed as the Ram Air Turbine (RAT), a propeller-driven generator which is moved into the airflow when all other means of generating electricity and hydraulic power have failed. With the exception of crop dusters, modern aircraft use RATs only in an emergency.

It is also reported that towards the end of the war, the Nazis were testing an electric airplane developed by an Austrian, Viktor Schauberger. In 1940, Schauberger had begun construction of his Repulsin(e) discoid motor in Vienna with help of the Kertl Company. He patented his idea on March 4, 1940, in Austria under Patent 146,141. But very soon afterwards he was reported by the Viennese Association of Engineers to the SS, who placed Schauberger in a mental hospital in Mauer-Ohling. Schauberger was then forced to work with Messerschmitt on liquid vortex cooling systems, and with Heinkel concerning applications of water towards aircraft engines. At this point Heinkel received reports on the early Repulsin A. It was at the Mauthausen Concentration Camp in Upper Austria, under orders from Heinrich Himmler himself, that Schauberger was to carry out research and development for the Third Reich war effort. He was given approximately 20 or 30 prisoner engineers to proceed with his research into what was termed "higher atomic energies." For this Schauberger was given special dispensations from the SS for both himself and fellow engineers.

The construction and perfection of the Repulsin A model discoid motor continued until one of the early test models was ready for a laboratory test that ended in disaster. The model was 2.4 meters in diameter with a small high-speed electric motor. Upon initial start-up the Repulsin A was set in motion violently and rose vertically, quickly hitting the ceiling of the laboratory, shattering to pieces. The SS were not pleased and even threatened Schauberger's life, suspecting deliberate sabotage. Despite this, according to Schauberger, a full-sized "flying saucer" was built in collaboration with the first-class stress-analyst and propulsion engineers assigned to the Austrian. It was apparently flight-tested on February 19, 1945, near Prague, when it attained a height of 15,000 meters (45,000 ft.) in 3 minutes and a horizontal speed of 2,200 km/hour (1,366 mph)! It was destroyed by the Nazis before it could be captured by the Allies. We do not know whether it too used an electric motor.[14]

FOUR

The Aeromodelers

As a child, I had a book called *The Wonder Book of the Air*. I loved to look at the photos of the RAF's latest jet fighter planes such as the Lightning and the Vampire, as built by English Electric Aviation Ltd. Ironic choice of company name? The company was formed in 1918 and made up of five businesses, including the United Electric Car Company of Preston and the Phoenix Dynamo Works. From World War II until the late 1950s, English Electric Ltd. built a sequence of bombers (the Hampden, the Halifax, the Lightning, the Canberra and the Vampire), but none of these were electrically propelled.

The real origins of electric aircraft as we know it today came from the aero-modelers of the late 1950s.

Control-line or tethered flying of electric model airplanes was not new. In 1958 Victor Stanzel of Schulenburg, Texas, filed a patent for a remotely-controlled propulsion and control mechanism for model aircraft. His patent U.S. 3,018,585: "The provision of electrically powered propulsion mechanism for model aircraft wherein the electric motor and power supply therefore is located at a distance from the craft so that the weight of such mechanism does not affect the craft in flight. A further object of the invention is the provision of an electrically powered model aircraft which is operated by an electric motor and power supply source located at a distance from the craft and drivingly connected thereto by means of a flexible shaft."

But free flight with an electric model airplane was a challenge. The first to take this on was Englishman Harold John Taplin. "Taps" was born in Stoke Newington, London, in January 1891. By the outbreak of World War I, Taplin had become a professional aero engineer and designer by working for the Empress Motor Car and Aviation Co. in Manchester. In August 1916 he was appointed a probationary 2nd lieutenant in the Royal Flying Corps. Embarked for France as an engineer officer in December 1916, Taplin returned to the Home Establishment to take his aviator's certificate (No. 4483) in a Maurice Farman in April 1917. He ended the war with a staff appointment in the temporary rank of captain at the Air Ministry, though he is believed to have served as a test pilot and instructor in the interim, and was demobilized in January 1919. He returned to his pre-hostilities profession as a designer, and as he worked for Gerrand Industries Limited in London, a wide variety of his work was registered with the UK and U.S. Patent Offices in the 1920s and 1930s.

By 1957, Colonel Taplin was running Electronics Developments Ltd. when on June 30, his *Radio Queen* made the first official recorded electric-powered model aircraft flight above Chalgrove Aerodrome in Oxfordshire, England. He was assisted by his son Michael.

Electrical power for the colonel's model was supplied by 20 Venner H-105 silver/zinc cells weighing a total of a little over 28 oz. (800 g). Motor weight was 30 oz. (850 g) and the total model weight was 8 lbs. The government-surplus Emerson D20 motor pulled 8 amps. Total weight, 8 lbs. (3.6 kg) (and just like now, a 10-minute flight was optimum).[1] *Radio Queen* managed an altitude of 10 m. The colonel died in 1969, but by then both Fred Militky of Germany and the Boucher brothers of the USA and Sanwa Denki Keiki Seisakusho of Japan were also experimenting.

Alfred Militky was born in 1922 in Jablonec nad Nisou, northern Bohemia. From his childhood, he devoted himself to making model aircraft. In this he was helped by a jeweler's son, Heinrich Brditschka, who made their propellers. In the 1930s Militky read about Professor Alexander Lippisch's Delta Wing aircraft gliders at the German Institute for Sailplane Flight, DFS. At the beginning of World War II, Militky of Gablonz, in his early twenties, was part of the Nazi Hitler Youth Movement, making and flying swept-wing duration-flight gliders, some of them powered by rubber bands flying for up to 20 minutes, others unsuccessfully attempting to use small gasoline and electric engines. Brief mentions of these are made in the magazine *Deutsche Luftwacht Modellflug* (*German Sky Guard Aeromodeling*). He joined the Luftwaffe as a pilot but the end of the war precluded any flying service. In January 1956 he and his wife Wilhelmine arrived in Kirchheim-unter-Tech to work for Johannes Graupner's innovative model-making firm.[2]

That year, Militky produced the "Cobra" High-A2-performance glider, radio-controlled with a solenoid magnet control fitted in the nose of the fuselage. It won a gold medal, with a 28-minute flight soaring to 600 meters (2,000 ft.) above the Rhône. Indeed, Militky took out a German patent for his "Steerable Bondage Airplane" Model Glider model (DE 1053992). This was the first of 300 models that the prolific Militky would design for Graupner during the next 20 years.

In February 1959, Dr. Ing. Fritz Faulhaber walked into the offices of the German magazine *Modell* and inquired if a motor which had an ironless rotor coil with self-supporting helical coil that his company, Schönaich near Stuttgart in Baden-Württemberg (Germany), had developed and patented for use in remote-controlled camera shutters, might be of use in modeling. The Faulhaber Type 030 was a coreless motor with integral gearbox. Militky immediately saw its advantages, and very soon after, the "Micromax" motor was powering Graupner aircraft. Its energy came from a battery made for use in electric cigarette lighters, the Rulag RL4, as developed by Artur Rudolf before the war. Encased in plastic and weighing just 1¾ ounces (50 g), the RL4 could not be recharged, but lasted long enough to get the glider into the sky.

On March 18, 1959, after countless tests, Militky's *FM241* (his 241st model), otherwise known as "Der ElectroFlug," glided out of sight after a five-minute climb. On October 4, 1959, *FM 248* made a 23-minute flight. Because of the overload of the motor, the flight was limited in its duration. By September 1960, Militky had progressed to *FM254*, an improved kit version known as the F/F Graupner *Silentius FM 254*.[3] This was the world's first purchasable series electric plane.

Militky and Hilmar Bentert worked together to make it even more remote-controllable. They were ahead of their time: By 1988, Graupner's *Elektro-UHU* motor glider was a world bestseller.

Militky was not the only one. In 1961, his compatriot Helmut Bruss built an electric model airplane with a twin pusher-puller propeller, powered a silver-zinc H-105 battery. On February 18, he created a remote-controlled record.

On October 4, 1959, Fred Militky of Graupner at Kirchheim-unter-Tech prepares his electric-powered Elektroflug *FM 248* for its record 23-minute flight (Fred Militky collection Giezendanner).

Over in Japan, in 1941, Sanwa Denki Keiki Seisakusho of Higashiōsaka, in the Osaka Prefecture, had established a company manufacturing and selling diagnostic tools, or multi-testers (circuit testers), for use in equipment repair. In 1962, his company marketed the *Electra* electric model aircraft, its fuselage made of expanded polystyrene.[4]

What if the electric airplane could derive its power from the ground, without the need for batteries? In 1964, William C. Brown, an expert in microwave radar working

FOUR. The Aeromodelers

In 1960, the Graupner *Silentius FM 254* was the world's first purchasable series electric plane. Photographed here next to a still camera (Fred Militky collection Giezendanner).

for Raytheon, flew a unique model helicopter that received all of the power needed for flight from a microwave beam (2.45 GHz microwaves). This demonstration was seen by millions of viewers during Walter Cronkite's *CBS News* program. The longest flight of this microwave helicopter lasted 10 hours. Brown's microwave-to-DC converter was patented in March of 1969 (3434678).

The big breakthrough came on September 4, 1971, when Fred Militky and Wolfgang Schwarze flew their radio-controlled twin-engined electric glider model *Silencer* to a height of 150 meters (500 ft.) in front of an amazed audience at the Seventh F3A World Championships for 22 nations in Doylestown, Bucks County, Pennsylvania. The following year Militky came up with his twin Jumbo 2000F-engined *Hi-Fly* with its 2.3 m (7.5 ft.) wingspan. The first electric flight competitions were held as early 1973, including the Militky Cup in Pfäffikon, Switzerland, a contest that it still held annually.

For Fred Militky, the next step seemed a full-scale airplane. About this time the first energy crisis hit the world. Militky met up with his old school friend from Jablonec nad Nisou, Heinrich Brditschka, now head of a sailplane-building company in Haid, Linz, Upper Austria. The Brditschka jewelry firm had been destroyed during the war. In the 1960s Heinrich Brditschka had teamed up with Franz Raab to design and build *die Krähe* (The Crow), a two-seater 36 hp gas-engined motor glider, the HB-21. In 1971, persuaded by Militky, the Brditschkas decided to take on the challenge of a full-scale gas-free electric aircraft. For the technology of this they could count on Heinrich Brditschka Jr., known as Heino, who had studied aeronautics in Vienna. Heino had been co-designer, engineer, and builder at his family's firm of their HB-3 and HB-21 motor gliders, then powered by Steyer Puch 650 or ROTAX.i/c combustion engines. Heino's son would later recall:

> My father Heinrich's childhood friend, Fred Militky, was a very calm, thoughtful guy, but he also had a sense of humor and was often joking. He was very precise in his work and was always ready to welcome the latest technology. He had no children. He and his wife Wilma would travel everywhere in one of the first BMW 2800 Coupés, in the boot of which he almost always had one or two remote-control model gliders which Wilma would carry around for him.
>
> One day, Militky came to us with the interesting project: to equip one of our sailplanes with batteries and motor and make it fly. At first we thought he was really mad! But he insisted and before long we were meeting up with technicians from Bosch motors and Varta batteries. After careful study of their documentation, we saw a real possibility to install them in the second prototype of our HB-3 OE-9023, which I had already test flown for many hours and knew well. With a wingspan of 12 m [40 ft] and fuselage length of 7m [23 ft], a wing area of 14,22m² [153 ft²] indicated that stretching came on 10.11. HB-3 380 kg resulted in a wing loading of 26.72 kg / m², so the best glide ratio was about 20. The HB-3 was not therefore an optimum performance motor-glider.
>
> We called our project *MBE-1* (Militky / Brditschka / Electric-1). Our team was made up of me and my father, Fred Militky, my pilot friend Manfred Bleimschein and our mechanic Wolfgang Weigel.
>
> We began by taking everything unnecessary out of the sailplane: gas engine, fuel tank, fire wall, various instruments, interior trim, etc. We designed and built an engine mount for the delivered Bosch unit and a battery frame. We then used a fuselage of the series HB-21 for ground tests of the provisionally installed batteries and motor. The brilliant Ernst Voss of Varta provided 100 volt batteries normally used as back-ups for conventional airplanes and helicopters. They consisted of 120 standard NiCad cells with sintered plates in four special containers of 30 cells each with a rated capacity of 25 Ah. and an overall weight about 125 kg [276 lbs.]. Bosch supplied a 10 kW Focus DC belt-driving the prop at 2400 rpm. This increased the weight of the *MBE-1* to 440 kg [970 lbs.], an additional 60 kg [132 lbs.].
>
> These run-ups with static thrust measurements, behavior of the batteries and the motor confirmed our decision to fly it. We overcame the problem of the large current circuit by using a streetcar emergency switch in the handle. The Austrian aviation authority had to agree to grant a license for the test flights. At first they thought this was a joke but after studying the documents we submitted to them, they issued a permit (0E-9023). Anyway everyone thought that we would not get beyond a few hovers in a few meters above the runway.
>
> Came the day when *MBE-1* was transported to Wels. The airfield there is relatively large and has a good infrastructure. During the week, there is very little air traffic and we almost had the aerodrome to ourselves. We estimated that flights of 12 minutes duration at up to an altitude of 380 m [1,247 ft] were just within our NiCad battery's capacity.
>
> After assembly of the aircraft and the necessary checks, it became serious. The cockpit had been simplified for weight reasons. There was an airspeed indicator and an on-off switch for the motor—just flat out or nothing. We had two sets of batteries. On my first trial I made some run-ups, rolling trials and acceleration runs. The acceleration felt much the same as with the gas engine. On the 1200m long track I could take off shortly. The second time I was able to rise to an altitude of 50m [160 ft]. After these tests, the batteries were measured and they still had sufficient capacity. The tests lasted approximately 5 minutes, but the Varta batteries still had about 5 minutes "left." So we could make a flight of ten minutes.
>
> Having replaced the batteries, everything was ready for the first flight. The date was 21 October 1973. There were only a few spectators. The manager of the aerodrome had agreed to act as official witness for the first flight. The weather was not particularly good but calm and no precipitation with sufficient visibility. My old flight instructor Hans Dorant would follow me in a Cessna 150, observing the whole thing from the air and measuring my altitude.
>
> With fully charged batteries, I was towed by a car driven by my technicians Bleimschein and Weigel to start site and placed in position.
>
> In the cockpit, I made my last checks, locked hood, belted, etc. I went over my flight plans once more in my mind: Take off and climb to about 50m [160 ft] to the runway center, then decide whether

to follow a traffic lane or whether I have to pull back and then before the end of the runway still can come to a standstill. That I was about to make aviation history did not enter my mind.

The Cessna behind me had already taken up position, ready to follow me inconspicuously.

Next followed the GO signal from the team. Motor switch ON go—good acceleration to about 55 kph [35 mph]—the plane then allowed to stand well at about 60 kph [37 mph]. The climb up to about 50m [160 ft] proceeded relatively quickly and continued unabated. This was followed by a straight climb to about 150m [500 ft]. I was flying electrically, but the prop noise the same, there did not appear to be any difference from the gas combustion engine. I banked in the traffic pattern and made the downwind leg in a few circles, decelerating. Then I switched off the engine and made about 1 minute a glide, and then I turned it on again. It also came out a little climbing performance, but not for long. I flew into the base leg and final approach to landing. I made a smooth landing and stopped. My flight had flight lasted 9 minutes 5 seconds. I opened the hood, the team and the few spectators were already there, cheering and congratulating us. It had been a very special flight.

After cleaning up and talking shop a little, the day passed like any other. For our next experiment, we would invite the Press as we wanted to introduce our novel aircraft to a broader public. We contacted the local press, television, radio and the APA (Austrian Press Agency) and the international press.

What we had not considered was that 1973 was the year of the first major oil crisis / gasoline crisis. In many countries, vehicle traffic was restricted. In Austria, you had to stick a sign with a week into the windshield, where you could not go. So for the Press, an aircraft which could fly without fuel was of enormous interest. We gave some demonstrations of "the world's first battery airplane" for over a hundred radio and television stations from around the world. In fact we made a further 8 test flights, one on October 23, the best lasting 14 minutes at an altitude of 360 meters [1,100 ft], with a climbing speed of 1.5 to 2m / s.—a performance unequalled for the next ten years.

Our *MBE-1* team got white overalls and for a few days we were the center of attention, about which we were very proud. Wels Airport was completely occupied only by us and the press photographers and reporters. Those reports continued for weeks and even months. In 1974, we proudly exhibited our *MBE-1* at the Hannover Air Show.

But, having demonstrated we could fly with electric power, we returned to our daily lives. We had stored the energy of about 1.5 l petrol in only 140kg [308 lbs.] of batteries. Now forty years later, the worldwide arrival of the regular electric airplane and its everyday acceptance is taking a little longer. I feel sad that my good friend, Manfred Bleimschein, who died young, did not live to see and fly the extended range of the gliders which we at Brditschka make today.[5]

In October 2017, the restored *MBE-1* was unveiled as a static exhibit for permanent exhibition at the Austrian Aviation Museum in Graz.

In 1974 Militky took out a patent for the engine (DE 2414704 A1). Today Brditschka electric motor sailplanes can fly for 500 km.[6] In November of that year VFW-Fokker (the Technical Aeronautical Bureau of Fokker) in Bremen applied for a patent (U.S. 3937424 A) for an electric airplane. The applicants were Hans Justus Meier, Herbert Sadowski, and Ulrich Stampa. During World War II, Stampa had worked with Kurt Waldemar Tank, chief designer of the Focke-Wulf fighters for the Luftwaffe. In their design, the interiors of central wing cases of an electric-powered airplane would serve as battery cells. Load-bearing transverse walls would take up bending stresses integrated into such walls by tops and bottoms of such a central wing case. These tops and bottoms would constitute the electrodes of the battery. Leading and trailing edge-profile completing wing cases would be release-secured to the transverse walls of the central wing case. The airplane would have two pusher propellers mounted on the wings. This patent was published in February 1976. While Ulrich Stampa came to be regarded as the design-father of the VFW-Fokker 614 jetliner with its engines mounted in pods on pylons above the wing, the electric airplane was never built.

Across the pond, the twin Boucher brothers, Bob and Roland, were also pioneering electric model aircraft. From a Quebecois family in Canada, they had become interested

Above and opposite: MB-E1 was the world's first full-scale electric airplane, equipped with a 10 kW Bosch Focus motor and Varta NiCad batteries. In October 1973, piloted by Heino Brditschka, *MB-E1* made eight flights from Wels airport, the longest for 15 minutes at an altitude of 360 meters (1,100 ft.), with a climbing speed of 1.5 to 2m/s, a performance unequaled for the next ten years (Brditschka Collection).

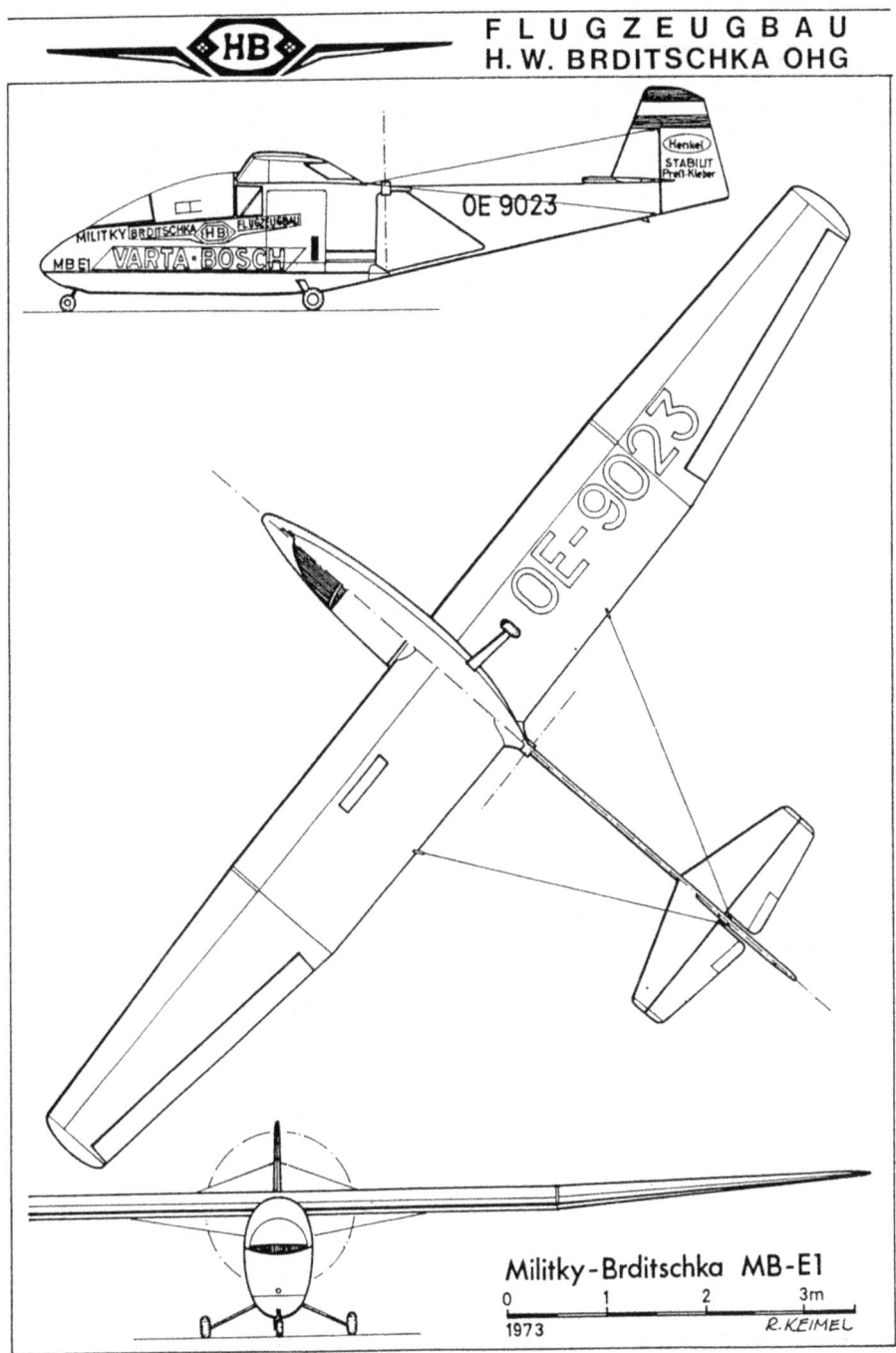

in airplanes in 1939, when their father took them on a plane ride in a Gull Wing SR-7 Stinson Reliant. At once, they were both hooked on aviation. They first started with simple rubber-powered models. After graduating from Yale with master's degrees in engineering, the Boucher brothers both moved to Los Angeles to work as engineers for the Hughes Aircraft Company on military programs. Roland was a project manager in the space division and Bob was a department manager in the missile division. Bob bought ⅐ interest in a Cessna 180 taildragger and spent many weekends traveling around California or visiting Los Vegas. One summer Bob flew his Cessna on a round trip across the USA from Los Angeles to Windham, Connecticut, and back to Los Angeles.

Again, it wasn't long before they began to design and manufacture high-performance radio-controlled sailplanes for hobbyists to use in AMA radio-controlled slope racing and soaring contests. To that end, in 1969 they founded their own company, Astro Flight, in their two garages in Los Angeles. Their first sailplane design, the *Malibu*, finished third in its very first trial at a slope contest in San Jose in the summer of that year. In order to promote the Astro Flight name and their line of radio-controlled models, they decided to try to break the existing FAI closed-course world record for RC sailplanes, at that time slightly over 100 km (60 mi). For this they would need a steady wind against a slope for at least 6 hours. When Bob visited the local federal weather bureau to find data on local wind conditions, he was told to call Paul MacCready, who was said to know where all the winds blow.

> This was the beginning of a long friendship with Paul. Paul suggested Sandberg Mountain in Southern California. We both got our Malibu sailplanes ready at dawn on the mountain. The wind was more like a gale of 40 mph (65 kph) and soon both our models were smashed. We needed slope with a steady trade wind. The next spring Bob's wife Suzanne won a raffle with tickets to Hawaii for a two-week vacation. He knew about the trade winds there and took along his Malibu for a try at the world record. That August 30, 1970, with the help of the Kapiolani RC club Bob flew my Malibu to a new FAI world record of 302 km (188 mi) on the slopes of Waimanalo Beach, Hawaii.[7]

The Bouchers now turned their attention to electric flight. The Fournier RF-4 single-seater motor glider designed by René Fournier had become a popular subject for radio control modelers on both continents. Roland had designed a ⅙ scale model of the RF-4 powered by the then popular O.S. Max 0.15 cu. in. 2-stroke model engine. It was a popular Astro Flight kit. Roland converted the model to use a rewound 12-volt ferrite motor powered by eight GE sub-C nickel-cadmium batteries and a 15-minute fast charger. Roland demonstrated his Fournier RF-4 electric R/C airplane at a trade show in Anaheim, California, in April 1971, marking the debut of America's first practical electric-powered model airplane.

Still, the demand for electric-powered models was not brisk. Perhaps another world record might help. Roland redesigned his RF-4 to accommodate a larger Astro 25 ferrite motor and a large off-the-shelf Eagle Picher silver-zinc battery. Roland flew it to three unofficial world records: distance, 19 miles (31 km); duration, 29.5 minutes; and average speed, 55 mph (89 kph). Since no official categories for electric flight existed at the time, neither the AMA nor the FAI recognized these records. But the flight was witnessed by U.S. Army Colonel H. Federin of DARPA.

After considerable negotiations, the Bouchers finally received their first military contract with the Northrop Corporation to develop a low-altitude electric surveillance drone. Bob and Roland both quit Hughes Aircraft Co. and the real Astro Flight Inc. was born. With Bob Boucher as project manager, aided by his brother Roland, their wives and their lone employee Dave Shadel, Astro Flight finished the design in only six months. The result was the Model 7212 flying wing. With a wingspan of 8 ft (2.4 m), the 7212 was powered by three of the company's Astro 40 ferrite motors each turning a three-blade, 8×8 propeller. The 7212 set another unofficial record in August 1973 by carrying a 7.5 pound (3.4 kg) lead payload over a closed course for an hour and twenty minutes at speeds reaching 75 mph (121 kph). It provided a stable platform for optical surveillance systems with no thermal or audible signature. In fact, the 7212 was completely inaudible at altitudes over 300 feet.

In July 1973 the Bouchers applied for a patent on a "Remotely Controlled Electric Airplane" and were granted United States Patent 3,957,230. Bob Boucher has recalled:

> DARPA showed little enthusiasm for either Astro Flight or our "crazy" idea for an electric powered air surveillance vehicle. John Foster was more reasonable and suggested that if we could figure out a way to increase our flight times to twelve or more hours we might have a sale. Our NiCad battery powered models could fly comfortably for fifteen minutes and with careful energy management for about thirty minutes. We experimented with one-shot lithium D cells built by Power Conversion of New York in a lightweight r/c model and demonstrated flights over three hours with no payload. But for twelve hour flights we needed a breakthrough.[8]

The DARPA contract was over by November 1973 and the model business could not pay the rent. In February 1974 Bob Boucher rented a booth at the Nuremberg Toy Fair in Germany. "There I met Militky at the Graupner booth. He was still pushing an underpowered electric sailplane. In contrast, we had determined that we would replace the internal combustion engines in RC models with electric motors. I brought two models with me, the Electro-fli and the Fournier RF-4. These could rise off ground (ROG) and perform level 1 aerobatics, loop, stall turn, etc. Simprop Electronic picked us as supplier so I came back with a $50,000 order. We were back in business."

The American and European initiative was soon followed in other parts of the world. Australian airline pilot Jack Black, having assisted at the Militky Cup, held in Pfäffikon, Switzerland, returned to Sydney. There he began to build and fly model e-sailplanes he called *Pfäffikon*, which he demonstrated in the early 1980s at the Sailplane Expo in Armidale, New South Wales.

By this time the sun had risen for electric aircraft.

FIVE

"Here Comes the Sun"

> *We have proved the commercial profit of sun power in the tropics and have more particularly proved that after our stores of oil and coal are exhausted the human race can receive unlimited power from the rays of the sun.*—Frank Shuman, inventor, writing in the *New York Times*, July 2, 1916

In 1917, Emanuel Victor Rousseau, British writer of pulp fiction, had his latest adventure, *The Messiah of the Cylinder*, serialized in *Everybody's Magazine*. Set in the year 2010, it envisages both airplanes and ray guns powered by solar energy.

Another who envisaged solar aircraft was John Wood Campbell Jr. In 1930, Campbell had his first short story, "When the Atoms Failed," published by *Science-Fiction* magazine. At that time, he was twenty years old and still a student at college. But as the title of the story indicates, he was even at that time occupied with the significance of atomic energy and nuclear physics. Born in Newark, New Jersey, to a father who was a cold, impersonal, and unaffectionate electrical engineer, the son was allowed to carry out his own experiments, going on to study at Duke University and then the Massachusetts Institute of Technology. In one of Campbell's stories, "The Black Star Passes," published by *Amazing Stories Quarterly* in 1930, then extensively edited for book publication in 1953 by Fantasy Press, the boy says to his father, "Dad, I believe that you have been trying to develop a successful solar engine. One that could be placed in the wings of a plane to generate power from the light falling on that surface...." To which the engineer replies, "[T]he one big trouble with all solar engines, eliminating the obvious restriction that they decidedly aren't dependable for night work, is the difficulty of getting an area to absorb the energy." Despite this a huge aircraft, the *Solarite*, is built and flies:

> The invisible ship darted forward. They sped past the barrier of low hills, and were again high above a broad plain. With a startled gasp, Arcot cut their speed. There, floating high in the air, above a magnificent city, was a machine such as no man had ever before seen! It was a titanic airplane—monstrous, gargantuan, and every other word that denoted immensity. Fully three-quarters of a mile the huge metal wings stretched out in the dull light of the cloudy Venerian day; a machine that seemed to dwarf even the vast city beneath it. The roar of its mighty propellers was a rumbling thunder to the men in the *Solarite*.

Although it had nothing to do with solar energy, from 1929 the Solar Aircraft Company Ltd. of San Diego, California, was registered, although it only ever built a single airplane. This was Solar MS-1, an all-metal biplane (sesqui-plane) with a 420hp Pratt & Whitney Wasp avgas engine. It carried eight passengers, with a range of about 500 miles.

FIVE. "Here Comes the Sun"

On April 25, 1954, Daryl Chapin, Calvin S. Fuller and Gerald Pearson publicly demonstrated the first practical silicon solar cell at Bell Laboratories, Murray Hill, New Jersey. They demonstrated their solar panel by using it to power a small toy Ferris wheel and a solar-powered radio transmitter. This created a sensation, and by July 4, 1955, *Time Magazine* had more to report: "When Bell Telephone Laboratories announced its silicon solar battery, it fired the imaginations of the science fictionists, and the solar system was soon abuzz with solar-powered space ships. Trimming their silicon sails to catch the sunlight, spacemen used the electricity generated by the batteries to push themselves from planet to planet."[1]

The color front cover of the October issue of *Modern Mechanix* (erroneously dated 1934, twenty years too early) showed a solar-powered aerial landing field. The article inside states:

> Recent experiments in the conversion of the sun's rays into electric power have led to an unusual idea in aerial equipment. It is a dirigible that would not only get its power from the sun but also provide space for a landing field in the air. The ordinary cigar-shaped dirigible would in effect have a slice taken from the upper half of the gas bag. This would provide a large deck on which could be mounted solar photo cells, an airplane runway, and a hangar. Planes could land on the dirigible, floating over the sea, to refuel for trans ocean passenger service. Another unusual feature of this design, in addition to the landing field, is the use of the sun's rays to power the motors of the dirigible.

Another who was inspired by the Bell breakthrough was André North, the pseudonym of Alice Mary Norton, also known as Andre Norton. Her 185-page book *Sargasso of Space* was published in 1955 in an edition of 4,000 copies by Gnome Press, a small publisher in New York City that focused on science fiction. Fresh out of Training Pool, a trade school for spaceship crews, Dane Thorson discovers that his first assignment as apprentice cargo master puts him on the free trader (basically an interstellar tramp freighter) called *Solar Queen*. During the years to come, Andre Norton would write another six novels in the *Solar Queen* series, during the last three of which solar airplanes had become a reality.

In 1973, two years after John W. Campbell's death, sci-fi became sci-fact when Roland Boucher at Astro Flight turned his attention to the creation of a high-altitude solar-powered aircraft that would have unlimited endurance:

> Heliotek in Sylmar, California, had developed a new high efficiency light weight solar cell for the Hughes Space Satellites with a power density of 100 watts per pound and a solar efficiency of 14 percent. Since this power density was comparable to the NiCads we were using in our previously successful electric powered models we knew that an airplane powered only by solar cells would fly. This was a trivial problem. But if somehow we could store enough of the sun's energy during the day, we could then use that energy to continue flying all through the night. We could theoretically fly forever. But no suitable battery was available. My brother Roland came up with the following scheme. Store the energy in the Earth's gravitational field!!! If we could build an airplane that could climb high enough during the day to be able to glide all through the night and still be at some reasonable altitude, say at 30,000 feet at dawn, we would have created perpetual flight. It all looked feasible. It would be difficult but with careful attention to detail it could succeed. We filed our patents and submitted our proposals to our prospective customer Kent Kressa of DARPA. We called our invention *Sunrise*. Our proposal to DARPA was to build a 100 foot wing span vehicle to carry the 50 pound tactical payload envisioned by the customer. ARPA counter offered to pay for a paper study to prove feasibility. Administration was by the Lockheed Aircraft Corporation of Sunnyvale, California.[2]

Astro Flight would build a ⅓ scale model of *Sunrise*, fly it on a government test range, and share the flight test data with DARPA. The Bouchers had expected to receive their contract in January 1974 so that they might make their initial flight tests in the summer months when the sun is highest in the sky and the days are longest. With the late starting date they had to work many long days from 7 a.m. to midnight to get *Sunrise I* ready for flight a few days after Labor Day.

Roland Boucher took on the task of the structural design, aerodynamics, telemetry, control and navigation. He also designed the integration of the solar panel, electric motors, gearbox and propeller. He selected an Eppler 387 airfoil for the wing. The Heliotek solar cells were 5 cm (2.0 inch) round commercial units with a 12 percent efficiency. A primary lithium booster battery capable of about one hour duration was installed to power *Sunrise* through any cloud layer and then be jettisoned. The actual airframe was constructed by a team under expert model builder Phil Bernhardt. The *Sunrise*'s wing span was 32 ft. (10 m) and the aircraft had a gross weight of 22 lb. (10 kg). The wing loading was a very low 4 ounces per square foot (0.011 kg/sq m). The aircraft structure was built from spruce, balsa and maple. Due to their roughness the solar cells were only mounted on the aft two-thirds of the wing's upper surface. The wing spars were built from spruce spar caps with maple doublers at all attachment points and two $^3/_{32}$-inch (2 mm) balsa shear webs attached to ⅛ to ⅜-inch (3 to 10 mm) balsa strips on the wing spars. The ribs were made from $^3/_{32}$-inch (2 mm) balsa. This construction resulted in a balsa spar box with tapered spruce caps. The leading edge was covered with $^1/_{32}$-inch (0.8 mm) balsa to form a leading edge D spar. The trailing edge was formed by two 2-inch (51 mm) wide $^1/_{32}$-inch (0.8 mm) sheets forming a triangle with ⅛ to ⅜ inch (3 to 10 mm) vertical spar sections in between the ribs. The covering was ½ mil Mylar. The 32-ft (10 m) span wing weighed 5 lb. (2 kg) and was capable of loads up to 100 pounds. Control was via an S&O Radio–designed and built telemetry transmitter and receiver. The standard S&O six channel radio had channels for elevator, rudder, motor on and off and solar cell operating mode. The solar cells could be set for either series or parallel operation. The telemetry functions provided gave data on motor current, motor voltage, motor RPM, airspeed and two heading references from a sun compass for navigation.

Astro Flight Model 7404–1 (*Sunrise*) made its first test flight under battery power on September 17, 1974. Shakedown flights showed that the propulsion system and the command and control systems were OK and that *Sunrise I* had an excess power ratio of seven. That meant that *Sunrise I* could maintain level flight on one-seventh of its maximum power. It could maintain altitude on 85 watts. *Sunrise I* had a wing span of 32 feet (9.8 m) and weighed 19 pounds (8.6 kg). At this weight and with this power level, the Bouchers calculated its service ceiling at 75,000 feet (23,000 m). Heliotek delivered the solar panels but they were overweight and underpowered. The panels weighed 6.5 pounds (3 kg) instead of the promised 4.5 pounds (2 kg) and the power output was only 450 watts instead of the 600 watts promised. To make matters worse, the overweight solar panels mounted on the upper surface of the wings were behind the center of gravity, so they had to add two pounds of lead ballast to the motor compartment to rebalance the airplane. *Sunrise I* now weighed 27½ pounds (12 kg) ready to fly with lead ballast, telemetry downlink and radar beacon. They recalculated the probable service ceiling to be 25,000 feet (7,600 m) in June and 10,000 feet (3,000 m) in December. It was by now late in October and the rainy season in California had begun. A very wet and windy winter was forecast that year. Astro Flight argued with DARPA that the prudent course would be to put the solar flight tests on hold until late spring of 1975 or to move the flight tests

FIVE. "Here Comes the Sun"

November 4, 1974: Bicycle Lake, a dry lakebed on the Fort Irwin Military Reservation, California: the Boucher twin brothers, Bob and Roland, prepare to launch their Astro-Flight solar glider, *Sunrise I*. It was powered entirely by 4,096 Heliotech solar cells on its wings (Bob Boucher collection).

to Australia. But they were instructed to fly at the earliest possible date at Bicycle Lake, California, a dry lakebed on the Fort Irwin Military Reservation.

The first flights would be conducted on battery power, using a bungee cord launch to 20 ft (6 m). Then at 10:00 a.m. on November 4, 1974, *Sunrise I* rose slowly and silently from the dry lake bed at Camp Irwin. It was powered entirely by the 4,096 solar cells on its wings.

The age of Solar Flight had arrived. Roland Boucher reported:

The first flight test was on battery power prior to the attachment of the solar cells. *Sunrise* was launched by a bungee cord to about a 20 foot altitude, and then the electric motor was activated. The plane climbed to about 500 feet (150 m) by the end of the runway. It then glided in a rectangular pattern turning left to fly crosswind, then left again to fly downwind, then left again on base leg and finally left again to final approach. On the first pass sunrise still had over 50 feet (15 m) altitude when passing the operators located about 300 feet (28 m) down the runway. The power-on portion was reduced gradually until an accurate measure of the average power required was established. This flight test on battery power was a complete success.

The solar panels which had been under construction at Heliotech in San Fernando California were mounted, and the electrical power verified with the aircraft in our parking lot. The aircraft was returned to Bicycle Lake, a final full power check of Solar panel and electric motor was performed, and Sunrise made its first flight powered solely by incident sunlight on the flying surfaces. The weather was extremely cloudy that year even in the desert, and for some weeks we would wait in vain for clear skies and low wind. In all, 28 flights were made on solar power alone. Take off was sluggish, but once an altitude of a few thousand feet was achieved, the cells cooled down, power increased and Sunrise maintained a respectable rate of climb. The Telemetry and control system worked flawlessly and navigation by means of the sun compass was demonstrated. The final flight was made with cumulus clouds covering about 15% of the sky. The pilot flew too close to a cloud at about 8,000 feet (2,400 m) and *Sunrise* was destroyed in severe turbulence. We were disappointed that an altitude of 78,000 feet had not been achieved. However, there was no longer any doubt that Project Sunrise had demonstrated the feasibility of solar powered flight to extreme altitudes.[3]

DARPA was sufficiently encouraged to authorize construction of an improved version, and Astro Flight received a contract on June 10, 1975, to proceed with construction of *Sunrise II*. The Boucher brothers worked 16 hour days all that summer. Roland Boucher had become physically exhausted from his work on the initial *Sunrise* and he suffered from congestive heart failure. He was admitted to intensive care at Santa Monica Hospital. While in the hospital, he resigned from Astro Flight and sold his interests in the company to his brother Bob Boucher, who continued work on the second *Sunrise* aircraft. A new digital telemetry system was built and tested and a single new samarium cobalt Astro 40 motor replaced the two Astro 40 ferrite motors used on *Sunrise I*. Heliotek was now owned by Hughes Aircraft and had a new name, Spectrolab. Bob Oliver and his Spectrolab team had developed a new high-efficiency solar cell that measured 2 × 4 cm by 8 mm thick and delivered 14 percent efficiency. The new solar panel weighed 4½ pounds and delivered a full 600 watts. This was much better than the original 100 watts per pound that was originally promised.

Sunrise II made its maiden solar power flight on a dry lake at Mercury, Nevada, that was controlled by Nellis AFB on September 12, 1975, just three months and two days after contract award! Flight tests indicated that *Sunrise II* should have a service ceiling of 75,000 feet (2,290 m) in summer and 25,000 feet (7,600 m) in winter at the 30 degrees north latitude of the test range. At this latitude 24-hour flights would be possible during the months of May, June and July. After many months of flight testing, *Sunrise II* was damaged when a failure in the command and control system caused an airframe structural failure. After recuperating, Roland Boucher returned to work at Hughes Aircraft on classified military programs.

Meanwhile, in Germany, Helmut Bruss was working on a solar model airplane in summer 1975 without having heard anything about Boucher's project. It operated without a storage battery but could only climb in a cloudless sky. Unluckily, due to overheating of the solar cells on his model, he didn't achieve level flight. Bruss was a major pioneer and sponsor of solar aeromodeling. As a physics teacher, he was able to inspire numerous youngsters to become enthusiastic about solar flying—both inside and outside school. He also wrote several successful technical books and reports on the subject of electric and solar-powered model aircraft.

But it was Bruss's friend, the veteran Fred Militky of Graupner, who produced his version, the *Solaris*, its solar cells running along the fuselage. On August 18, 1976, he completed three radio-controlled flights of 150 seconds and reaching an altitude of 50 meters (160 ft). This was probably Militky's final model as he died of leukemia the following September, aged only 55. This pioneer's name is still revered with the Militky Cup at the International Electric Flight Meeting held annually at Pfäffikon in Switzerland. Soon after, in 1978, Prof. V. Kupciks, Roland Stuck and Helmut Schenk were experimenting with solar model airplanes. At the 5th International Militky Cup in Switzerland in 1978, Helmut Schenk demonstrated his radio-controlled solar-powered model. It operated without a storage battery but could only climb in a cloudless sky.

The world's first official flight in a solar-powered, man-carrying aircraft took place

FIVE. "Here Comes the Sun"

August 18, 1976: Fred Militky (right), with an unidentified friend, holding his final prototype, which he radio-controlled while completing three flights of 150 seconds and reaching an altitude of 50 meters (160 ft.). Militky died of leukemia the following September, aged only 55 (Fred Militky collection Giezendanner).

on April 29, 1979. The *Mauro Solar Riser* was built by Larry Mauro, president of Ultralight Flying Machines, and was based on his UFM Easy Riser as a swept-wing, tailless biplane. It had been flown both as a hang glider and as a gasoline-powered ultralight. Normally foot-launched, the *Solar Riser* had wheeled landing gear added. Power was supplied by a Bosch electric starter motor of 3.5 hp (2.6 kW) connected to a 30-volt DC nickel-cadmium battery pack taken from a Hughes 500 helicopter, powering a 41-inch (104.1 cm) propeller through a reduction drive made from a timing belt and two pulleys. The battery was charged by a series of photovoltaic solar panels mounted in the top wing that provided 350 watts of power. The solar cells were not sufficient to provide power in flight. All flights were made by recharging the battery on the ground from the solar cells and then using energy stored in the battery to fly. A charge in bright sunshine for an hour and a half yielded a flight of 3 to 5 minutes.

Following a successful model test in 1974, it was Mauro's thought to demonstrate the airplane to the large crowd at the EAA fly-in, but the FAA nixed the idea. Mauro took his airplane to nearby Flabob Airport, in Riverside, California, where owner Flavio Madariaga encouraged innovation, and let aviators do pretty much as they pleased, so long as they did not get him into too much trouble. So it was that the world's first official flight in a solar-powered, man-carrying aircraft took place on April 29, 1979, at Flabob Airport. The aircraft reached a maximum height of about 40 ft (12 m) and flew 0.5 mi (0.8 km). The *Mauro Solar Riser* used photovoltaic cells to deliver 350 watts at 30 volts. These charged a small battery, which in turn powered the motor. The battery alone was capable of powering the motor for 3 to 5 minutes, following a 1.5-hour charge, enabling it to reach a gliding altitude.

Larry Mauro was quick and prophetic to point out that, while his *Solar Riser* really and truly flew entirely on electricity made from sunshine, he considered his April 29 flights to be only

> ... previews of coming attractions. The solar cells we're using are not really as efficient as some others currently on the market, and we can install at least twice as many of them on our machine as we've been using. We're also looking for a better battery. As things now stand, we have to charge for at least an hour and a half just to get a three-to-five minute flight. That's not as bad as it sounds, though, since it's theoretically possible for the rig as it is to take off, fly for three or four minutes until it catches a thermal, and then soar—shut down—for an indefinite period of time while the solar panels recharge the battery in preparation for a power assist to another updraft. Still, that's not the kind of flying I want to do. I want to improve this set-up until we can regularly take off and fly around all day on nothing but the solar-generated electricity that we're producing as we buzz along.[4]

A number of other flights of the *Solar Riser* of similar height and duration were flown, including demonstration flights at EAA AirVenture Oshkosh, before the aircraft was retired to a museum.

The next breakthrough for solar-electric aircraft came indirectly from a £50,000 prize offered back in 1959 by an industrialist called Henry Kremer for almost impossible human-powered flights across the English Channel. Hanoch Kremer, born in Latvia, had immigrated to England after World War I and was educated in Switzerland. He became an inventor of wood products and in 1941 he developed a process for making a plywood substitute from sawdust, wood shavings and resin. Structural molded boards replaced natural timber, which was then unobtainable, and were used in the war effort and later commercially. This was the first product of its type in Britain and it grew into the particleboard industry. The De Havilland DH98 Mosquito, an amazing World War II fighter aircraft with a service ceiling of over 10,000 meters (33,000 ft.), was built from Kremer's special laminated plywood; over 7,000 were built. Postwar, since working with De Havilland, Kremer had maintained his interest in aviation, and he was also very interested in physical fitness. Intrigued by the challenge of human-powered flight, he set up the monetary prizes that carried his name.

Freddie To, an architect and a member of Royal Aeronautical Society Kremer prize committee, was one of those who asked David Williams to start a project to produce a pedal-powered aircraft to compete for the prize. The resulting aircraft, at 230 lb. (104 kg), proved too heavy for human-powered flight and so was converted to solar power instead. A nose-mounted pod powerplant was installed consisting of four 1 hp (1 kW) permanent magnet 36-volt DC, 12 amp Bosch electric motors, powered by 750 solar cells of 3 inch diameter and a 65-lb (29 kg) nickel-cadmium battery pack of 24 cells with a 25 amp hour capacity, connected in series. The motors were connected by a 3:1 bicycle chain reduction drive to a 63-inch (160 cm) wooden two-bladed propeller, which turns at a maximum of 1,100 rpm, decreasing with battery discharge. The engines are controlled with a simple on/off switch. For flight the aircraft used its on-board solar cells to recharge the battery array on the ground, and then the batteries provided power for flight as the aircraft had insufficient solar cells for sustained flight. This shortcoming was not a design feature, but a problem of the cost of the solar cells as the limited project budget of £16,000 did not allow the purchase of sufficient cells. The 750 installed solar cells cost £6,000 and were the most expensive part of the aircraft.

Solar One's first flight was a short hop that occurred at Lasham Airfield, Hampshire, United Kingdom, on December 19, 1978. Freddie To was not present when this occurred, and the pitch of the propeller was found to be incorrectly set, which was why it was a short hop. Subsequent flights occurred in 1979, and those are often mistakenly taken as the first flights of the aircraft, as confirmed by Barry Jacobson, a member of the *Solar One* team. The 1979 flight took place on June 13 and covered just under 0.75 mi (1.2 km). The pilot was Ken Stewart and the aircraft lifted off at 18 to 20 kn (33 to 37 km/h) and reached 35 kn (65 km/h) and 80 ft (24 m) in height. A second flight on the same day by Bill Maidment achieved a speed of 42 kn (78 km/h). All flights were made on battery power that had been recharged on the ground from the installed solar cells. An intended flight across the English Channel was abandoned when the aircraft did not reach intended endurance targets.

Enter Dr. Paul MacCready: "The perfect combination of pilot-engineer who dared to build lighter than anyone else."[5]

Paul Beattie MacCready was born on September 29, 1925, in New Haven, Connecticut, where his father was a physician and his mother a nurse. He was dyslexic and had trouble concentrating, but showed passion for things that interested him, in particular insects. He collected them on the Connecticut shore and pored over the exquisite studies of John Henry and Anna Botsford Comstock, two 19th-century naturalists, to explore the evolution and the vein structure of the wings of lepidoptera. Nerdy already, small and unsporty, he then buried himself in making and flying model aircraft: fixed-wing and flapping-wing, out of a kit or out of his head, propelled with rubber bands or with tiny gasoline engines. He had little use for commercial kits. By the age of 15, McCready won a national contest for building a model flying machine.

Graduating at Hopkins School in 1943, McCready studied mechanical engineering at Yale, then was training as a U.S. Navy pilot when World War II ended. He returned to Yale, where he obtained a B.S. in physics in 1947, an M.S. in physics from Caltech in 1948, and a Ph.D. in aeronautics from Caltech in 1952.

During this time he became enraptured with the sport of soaring. He set an altitude record, and was a three-time winner (1948, 1949, and 1953) of the Richard C. du Pont Memorial Trophy, awarded annually to the U.S. National Open Class Soaring Champion. In 1956 he became the first American pilot to become the World Soaring Champion with a borrowed sailplane he had not flown before. He did this by taking extreme risks; in one leg of the competition he flew his sailplane far past any airport in southern France and landed after dark on the beach just to set the greatest distance flown in one day.

In 1951 MacCready founded his first company, Meteorology Research Inc., in the new field of weather modification. He was the first to use small instrumental aircraft to study storm interior atmospheric research. In 1971, McCready guaranteed a loan for a friend who wanted to start a business building fiberglass catamarans. When the company failed, MacCready found himself $100,000 in debt. Casting around for a way to deal with that problem, he recalled the Kremer cash prize for anyone who built a human-powered plane capable of sustained, controlled flight. On August 23, 1977, McCready and his colleague, South African–born Dr. Peter B.S. Lissaman of Caltech, who had recently established AeroVironment, began to recuperate that debt when *Gossamer Condor* (its name chosen by MacCready), piloted by amateur cyclist and hang-glider pilot Bryan Allen, won the first Kremer prize by completing a figure-eight course specified by the Royal Aeronautical Society, at Minter Field in Shafter, California.

Subsequent to the successful flight of the *Gossamer Condor*, winning Kremer's cash prize that had gone without a winner for 17 years (in fact, doctoral theses had been written explaining in detail why the prize was not achievable), Henry Kremer issued a second prize, this time £100,000 (approximately $200,000) for the first successful human-powered flight across the English Channel. It was assumed that, given the shortest distance across the channel was 22 miles (35 km), and the weather was usually unfavorable, it would probably be another 20 years before that prize was won. However, MacCready was undaunted. Cashing in on his publicity from the *Condor* (with the help of a publicist he hired, Tom Horton), he sought and received sponsorship from DuPont, maker of the plastic covering used on the *Condor*, to cover much of the development cost for the *Albatross*. The *Albatross*, similar to the bird, was to be a much sleeker version of the *Condor*, and used primarily carbon fiber reinforced plastic resin for its structure, allowing for light weight as well as a much improved lift-to-drag ratio. On June 12, 1979, Bryan Allen pedaled her across the English Channel from Folkestone to Cap Griz-Nez, a distance of 22.2 mi (35.7 km).

MacCready's team had actually been building three aircraft for the cross-Channel flight. The first (*Gossamer Albatross I*) was used for most of the flight testing and pilot training, where the airframe was optimized by repairing things that broke during a flight and making lighter things that did not, along with some aerodynamic refinements. The second was the "good" *Gossamer Albatross II*, meant only to be used on the Channel crossing. The third was a 75 percent sized version of the *Albatross*, named the *Gossamer Penguin*. This one, at approximately twice the wing loading, would fly faster than the *Albatross* (albeit requiring about 35 percent more power), in case the headwinds were too great an obstacle for the bigger, slower *Albatross* to make it across. (Some of the team members joked it was made so Marshall, the youngest of Paul's three boys, would have his own plane to play with after the channel had been crossed. In fact, this supposition is very close to what later occurred.) Most of the structural frame of the *Penguin* had been completed, but it was not finished at the time of the *Albatross*'s success. The actual cross-Channel flight was accomplished with the old, much repaired, first *Albatross*, because the first attempt was on a marginal weather day, and the team did not want to risk losing the "good one."

The Saga of Solar Challenger

An engineer at Arco Solar (later purchased by Siemens) in Camarillo, California, had been following MacCready's human-powered flights. He contacted Paul shortly after the Channel flight, suggesting Paul develop a solar-powered aircraft to fly the Channel using Arco Solar cells (developed for the expected terrestrial market since the oil crisis of the 1970s). He had pitched the idea internally, and received a favorable response from his management as a possible publicity showcase for the solar cells they produced. He had even picked a name for the aircraft—*Archaeopteryx* (considered an evolutionary link between dinosaurs and birds). Paul became excited by the idea. It would not only be fun to do, he felt it would show the world the potential for solar energy, steering the energy demands toward more benign, non-polluting, and renewable sources.

However, due to concerns about the financial impact on Paul's environmental consulting company, AeroVironment, as a result of becoming involved in outside projects, Paul decided to bring this project "in house." Since he had neither the staff nor facilities to run such a project inside the company, he began searching for a project manager to run this project, which included finding a building to rent, hiring the people to develop the aircraft and running the development program. He sought recommendations from several of his friends from the sailplane world, which included a gregarious, talkative fellow named John Lake, who also had become heavily involved in the budding hang-gliding industry. John had a friend named Ray Morgan, an aerospace engineer (then working in the Lockheed "Skunk Works"), whom he had met through Joe Greblo and Rich Grigsby, because Ray was teaching ground school for hang-gliding at Joe and Rich's Southern California Hang Gliding school. Morgan recalls:

> MacCready's children were learning to fly at my club. He wanted me to start a group within his company, and then find a place to design and build the new airplane. There were two unique qualities Paul possessed that all who worked with him could appreciate. Persistence, he was relentless in pursuit of a goal, once he'd made up his mind to do something. Optimism, he considered any failure a better learning experience than a success. Of course, the other side of optimism—that others did not always appreciate, was a tendency to under-estimate the time, cost, and difficulty of his pursuits ... some said by factors of 3 to as much as 7. He was good at coming up with what sounded like crazy ideas and convincing somebody to pay for them. He was intellectually fearless; he welcomed ridicule, not bothered at all about people thinking he was crazy. He was small in stature and was not athletic. He rarely smiled and spoke in a monotone, continuously sliding one sentence to the next until he was through with what he wanted to tell you and would then walk away.

A series of phone calls and negotiations led to Ray's taking a leave from Lockheed California (for 8 months, predicted by Paul to be the time it would take to achieve the cross-Channel flight under solar power) to manage this project for Paul. Unfortunately, in the interim, ARCO Solar decided against sponsorship of the *Archaeopteryx*. Fortunately, Paul and Tom approached DuPont, which had gotten a good deal of favorable publicity from the *Albatross* flight, and convinced them to fund AeroVironment the $500,000 cost of developing a solar-powered, piloted aircraft that would showcase much of DuPont's engineering plastics and fibers, and fly not just across the Channel, but from Paris to London (approximately 150 miles/240 km).

The plan was to finish the *Gossamer Penguin*, but converting it to solar-electric flight, as a testbed to learn the nuances of solar-powered flight before beginning the design of the Paris-to-London aircraft, now renamed *Solar Challenger*.

Ray's first task was to find a location that could be used to complete and modify the *Gossamer Penguin* as well as provide office and shop space for the design, construction, and ground tests of the intercontinentally capable *Solar Challenger* (as yet unnamed), and hire some folks to do that. Fortunately, there was a new, light industrial complex just being completed less than two miles from his house.

Their *Gossamer Penguin* was a ¾-scale version of the *Gossamer Albatross II*, and had a 71 ft. (21.64 meter) wingspan and a weight, without pilot, of 68 lb. (31 kg). The powerplant was an Astro Flight Astro-40 Ferrite electric motor, supplied by a 450-watt solar panel consisting of 3,920 Spectrolab solar cells loaned by Astro Flight and formerly used on their *Sunrise II* and repaired by DuPont.

Because the *Penguin*, like its human-powered predecessors, was unstable and fragile, only suited to fly in the calm air of the early morning, a solar panel would be attached to the "kingpost" above the wing, which could be tilted to the sun, low on the horizon to maximize solar power collected (the team referred to "the power in the shadow" when adjusting the panel angle). The plan for continuous, racetrack flights was to climb up to about 25 feet (7 m) above the runway, then turn and glide while flipping the panel to the other direction with a "halyard" pulled by the pilot, and then resuming solar-powered flight in the other direction. They actually flew with an empty panel initially to explore the aerodynamic effects of this large panel.

Initial test flights were performed using a 28-cell NiCad battery pack instead of a panel. The test pilot for these flights was MacCready's 13-year-old son Marshall, who weighed 80 lbs. (36 kg). The *Penguin*, like a high-aspect sailplane, had some nasty stall characteristics, and MacCready Junior got into a tip stall while practicing the "climb, turn and flip the panel" maneuver. He had quite a bad crash, escaping with nothing more than shock and bruises, but pretty much demolishing the *Penguin* wing. Fortunately, the solar cells had not been installed yet, and the crew was able to repair the *Penguin* in a few weeks, including replacing the NiCad battery pack with 350 watts of silicon solar cells gotten from Boucher. At 8 a.m. on May 18, 1980 (20 minutes before Mount Saint Helens blew up, killing any news release), the *Gossamer Penguin* was hand-towed to an altitude of about two feet and then released. Marshall MacCready guided the *Penguin* straight down the runway while climbing to an altitude of about five feet. He held this altitude for a distance of about 500 feet (150 m) before Paul asked him to land. The era of true, manned, solar-powered flight had arrived.

The official demonstration pilot for the project was Janice Brown, a schoolteacher and Piper Cherokee charter pilot with commercial, instrument, and glider ratings who weighed slightly less than 100 lb. (45 kg). She flew the *Penguin* approximately 40 times before a 1.95 mi (3.14 km) public demonstration at NASA's Dryden Flight Research Center on August 7, 1980.

After about a month spent by MacCready, Morgan and Henry Jex considering different conceptual design approaches, including both *Gossamer*'s canard design and being foot-launched, in June 1980 a second powered aircraft concept, the *Solar Challenger*, was finalized. It was over three times as heavy (without pilot) as the *Penguin* and had a shorter wingspan, but was proportionately more powerful, equipped with what appeared to be two wings, the main lift-producing wing of normal proportions in the front, and also an oversized horizontal tailplane covered with solar cells. To find the required cells, as DuPont did not make solar cells, Ray Morgan visited all the nearby Southern California manufacturers:

> None of the commercial guys had anything that was efficient enough to do what we needed to do. I went to Hughes Spectrolab in Sylmar, Calif., a maker of space grade solar cells for USAF and commercial satellites. I was told that 1) we were asking a quantity that would be 4% of their entire United States production of solar cells the previous year; 2) they could not start building them for a year due to backlog of orders; (Paul wanted this project over in eight months.); 3) that the amount of 16,000 solar cells we needed would cost us one quarter of a million dollars of, which, unfortunately, Paul had not accounted for when making his estimate to DuPont for the project cost. I felt as if I had been punched in the stomach, and the project and airplane of a lifetime was just not going to happen.
>
> However, as I was leaving, the Spectrolab head of marketing told me about some solar cells they had built for an Air Force satellite that had been rejected for low efficiency (by satellite requirements of 14%), and that were, in fact, in the Hughes Government stores facility by L.A. Airport ... suggesting

that we might get them loaned or scrapped to us. I then found out that they, the Air Force, could not scrap them to a private concern like AeroVironment, but they could loan them to another Government agency. Dale Reade, a fan of Paul's, as well as solar powered aircraft that could potentially fly into the stratosphere, was working as an engineer at NASA Dryden Flight Research Center. He had arranged funding for AeroVironment to perform a series of flight tests of the *Gossamer Albatross*' stability and control, and write a whitepaper defining its unique characteristics associated with the very low mass of the airframe relative to the mass of the air influenced by the giant wing. It turned out that the flights had been completed and the data had been taken, but Paul and Henry Jex had never quite gotten the paper complete, and had asked for and received a no-cost contract extension to do so, leaving that contract still open. Consequently, if NASA were to ask to take possession of the Air Force rejected cells, Dale could sign the paper that would "lend" those cells to AeroVironment for the *Solar Challenger*. He told Paul that he was willing to support getting those rejected cells loaned to us for the *Solar Challenger* provided, after we did the flight across the Channel, we would let Dale fit the airplane with a remote control system to see how high it could fly after we did the flight across the Channel. Paul agreed verbally. Soon after, I got in my old '65 Ford van and drove down to LAX, where I signed for 250,000 dollars' worth of space-grade photovoltaic cells (over 16,000) then and took them back to our Simi Valley shop in my van. In fact, this quantity of cells was more than we first calculated we needed when sizing the aircraft, so we enlarged the horizontal tail (easiest and quickest part to change at that point) by a factor of about 3 beyond what we needed for stability, just so we had a place to put more cells, which meant more power!

The design incorporated advanced synthetic materials with very high strength-to-weight ratios, including Kevlar®, Nomex®, Delrin®, Teflon®, and Mylar®, all supplied by the aircraft's sponsor, DuPont. The bulk of the load-carrying structure was carbon-fiber,

1981: *Solar Challenger*'s two 2.75 hp Astro Flight samarium-cobalt permanent magnet motors operated on a 70-volt system run in tandem on a common shaft to drive a single propeller (Don Monroe).

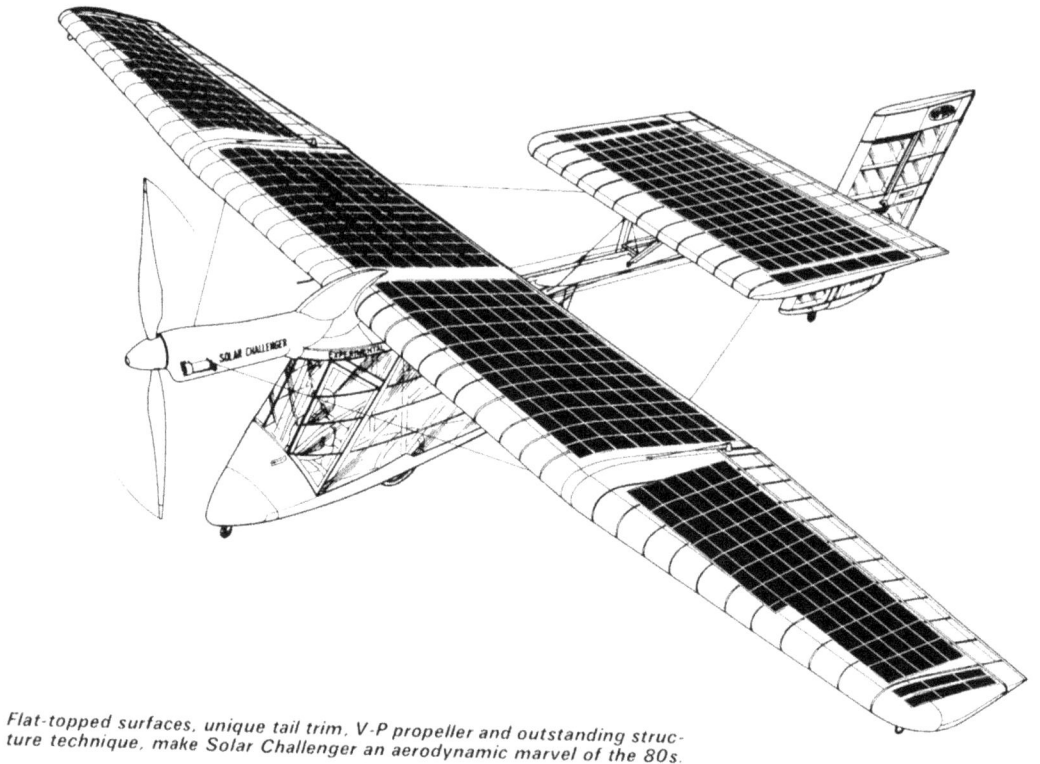

Flat-topped surfaces, unique tail trim, V-P propeller and outstanding structure technique, make Solar Challenger an aerodynamic marvel of the 80s.

This 1981 technical drawing of *Solar Challenger* shows the use of her 16,000 NASA solar cells (author's collection).

stabilized with Nomex® honeycomb and Kevlar® fabric overwrap. "We also used Lucite® (Dupont's proprietary formulation of acrylic, with excellent transparency, flexibility, and scratch resistance qualities) for the wind screen."

Boucher and Astro Flight supplied the two 2.75 hp motors operating on a 70-volt system. Each measured 3 inches (7 cm) wide and 17 inches (43 cm) long and incorporating samarium-cobalt permanent magnets, run in tandem on a common shaft to drive a single, controllable-pitch propeller. *Solar Challenger* in its ultimate form would have no batteries; it collected sufficient energy from sunlight to take off, climb to 14,300 feet (4,360 m), and cruise at 40 mph (64 kph).

Flight tests of the *Solar Challenger* began in October 1980. The airplane was first flown under battery power (36 lbs./16 kg of NiCad cells) because the team didn't want to risk losing any of the 16,128 solar cells in an inadvertent mishap while testing and learning to fly the airplane. The cells also had to be tested, matched to others in a string, and soldered together before installation, a process that overlapped the first flights on battery power. Starting with short hops down the runway, the team quickly developed enough confidence to attempt higher and longer flights.

> Morgan: I don't believe we ever measured more than 2400[1] watts in steady state conditions. But takeoffs in morning were usually made on about 1800 watts, about 1500 Watts appeared to be the absolute minimum to take off. This power increased with increased solar intensity, reduced temperature, and improvement in angle of array to sun. First two factors were improved as we climbed higher, and, of

course the sun rising higher through the morning increased the angle. Early on we were hoping for 3200 watts, but didn't get expected power from cells. After all, they were rejects!

This phase of testing, at Minter Field near Shafter California, was concluded with a 1½-hour flight at heights reaching 1,650 ft. (500 m) above the ground. The motor was used for only 15 minutes; the remainder of the flight was conducted using indirect solar power, in the form of rising air (thermals). Two days before Thanksgiving Day, on November 30, 1980, at El Mirage field in Southern California's Mojave Desert, Janice Brown made the first serious solar-powered flight of 2 minutes of the *Solar Challenger*.

Brown recalls:

Each time we flew the solar powered aircraft, we often had a group of enthusiasts that came to witness the event. I sometimes wished that we were able to work out all the bugs before the flight but this was never the case. The stress to succeed was always present. The flight plan for the day was to conduct a couple of flights down the runway and then fly over the desert and return to El Mirage airport. On that November morning, the air was crystal-clear and cool. It was perfect weather for a solar powered flight. For the first flight, the airplane was placed at the end of the runway and the wing-walkers, who are necessary to keep the wings level for takeoff, were in place. I toggled on the five electrical switches which activated five solar panels and the Solar Challenger took off with only a 20 foot ground roll. This was its first take off and the shortest take off that it ever accomplished. The airplane flew down the runway flawlessly and landed. The aircraft was then repositioned to the end of the runway again for another take off. Once again, the aircraft took off with ease, climbed sunward and began a cross-country flight over the desert. This turned out to be a short flight because the electric motor overheated and failed. As safety precaution, we always made sure that we had an acceptable landing spot below us. Down came Solar Challenger to the dry lake bed. The aircraft turned out to be an excellent glider which was fortunate because it saw many unscheduled, off-field landings during its test flights.[6]

There followed a week of 3- to 7-minute flights using the low winter sun (which provided only a fraction of the energy that would be available in late spring and summer, with the midday

1981: Paul MacCready holds a prop blade, while pilot Janice Brown prepares for the next test flight in *Solar Challenger* (Don Monroe).

sun more directly overhead). MacCready and team then took the airplane to Marana Air Park, near Tucson, Arizona. There, they hoped to make a 60-mile (100-km) flight attempt from Marana to Phoenix. Despite unusually cloudy weather, many flights were made nonetheless, and Janice Brown was able to leave the airport twice for short distances, gaining valuable off-field landing experience.

Ptacek recalls:

> Although we should not have risked it, we just started putting anyone in the cockpit to have a go at flying the airplane. When Paul got in, he couldn't get the airplane off the ground. He over-controlled it terribly and the *Solar Challenger* was not an airplane that responded well to anything other than a very light touch. It's my understanding on a successive flight when I was in the cockpit, that Paul said to one of the crew members (I think Martyn Cowley), "It's not as easy as it looks," Martyn relayed this bit of information to me later that day. That's as close as I ever got to a compliment from Paul MacCready. He spoke in the third person a lot. On one occasion time, during a (casual) meeting, that he had called, with Janice, Ray, Bob (Boucher), and me, while he was talking, Paul paused momentarily, and fell asleep! We sat scattered around Janice's hotel room in chairs and on the bed. We sort of looked at each other with incredulity and then one by one left poor Janice with Paul fast asleep in a chair.

After a week in Arizona, they returned to their rented hangar at Minter Field, California, to test new motors and other components, as well as expand the flight envelope up to approximately 14,000 feet (4,200 m), and durations of up to 8 hours. In its last flight, lasting 1 hour and 32 minutes, *Solar Challenger* made it about halfway to Phoenix, the intended destination, when it encountered a thunderstorm near Picacho Peak (about ¼ the way to Phoenix), and Janice Brown was forced to land in a clear spot among the saguaro cacti.

Solar Challenger airborne in 1981. DuPont provided advanced synthetic materials with very high strength-to-weight ratios, including Kevlar®, Nomex®, Delrin®, Teflon®, and Mylar®, all supplied by the aircraft's sponsor, DuPont. The bulk of the load-carrying structure was carbon-fiber, stabilized with Nomex® honeycomb and Kevlar® fabric overwrap (Don Monroe).

Due to funding limitations and concern over motor reliability, flights were terminated and the crew returned to Simi Valley. Bob Boucher developed a more reliable samarium-cobalt tandem motor system. Improvements were made by the AeroVironment team in power tracking and the manually controlled, variable-pitch propeller system (critical to drawing "peak" power from the solar array in varying conditions of speed, sun angle/intensity, and temperature—a major workload on the pilot), and provisions for cross-Channel flight, such as life jacket, wet suit, improved radio (with its own small solar array charging batteries for avionics), oxygen system, and basic instrumentation to permit emergency flight in IFR conditions.

After funding go-ahead, MacCready and team went back to Minter Field for a series of flight tests, trading off between Janice Brown and 28-year-old Steven R. Ptacek. The latter already had 4,600 hours of pilot-in-command time. He had flown a wide variety of aircraft, including some that were compromised with regard to flying at maximum gross weights and in marginal conditions, strong weather events, high density altitude airports, short and unimproved runways, strong and gusty crosswinds, etc.: "My experience up that point had prepared me to step in and fly *Solar Challenger* at a high level. Just dumb luck for me, and it made me appear to be better than I was when it really was just fortunate timing."[7]

During this period Janice made a flight to a reported 14,000 feet (4,200 m) (the team's Cessna 150 chase plane could not climb above 10,000 feet), and Steve Ptacek made a flight at around 11,000 feet of 8 hours' duration. Both these flights set informal records for solar (and electric) powered aircraft. Sixty-nine flights with a total of 6 hours and 4 minutes engine time had been accumulated during these tests. Ptacek actually lost about 40 pounds—slimming down to 120—that spring to reduce the power required for him to fly the *Challenger*.

That summer the *Challenger* was air-freighted across the Atlantic by Flying Tigers Air Cargo.

Ray Morgan:

> We had one day out of thirty that was good and we had to take that chance. We tried to make it that either Janice or Steve could fly the *SC*, so one would be reserve pilot. Janice had logged up 500 hours in the Piper Cherokee, but Steve had accumulated almost 5,000 hours in nearly everything that flew, including gliders. Not only would the pilot have to fly that airplane perfectly, but at the same time they must constantly look at a meter which told them how much power they were getting from the solar cells and work an overhead mechanical handle to adjust the feather angle of the propeller blades. Out of our concern for safety, Steve was chosen in the end. The 163 mile (261 km) trajectory was from Cormeilles-en-Vexin Airport near Paris across the English Channel to RAF Manston about 75 miles east of London; Air Traffic Controllers at Paris and Heathrow had rejected the idea of capital to capital.
>
> Multiple attempts were made to fly from France to England. We found difficulty from two sources. We discovered that Cormeilles-en-Vexin was actually at the bottom of a very shallow valley, not really discernible at first, and also we had a very strong north wind blowing. First time we tried the flight, Ptacek took off to the north and flew straight. However, in the early morning haze, his climb rate was only about 25 feet (7 m) per minute. Consequently, he stayed near the ground, even while climbing, until about 1 mile north of the runway, he was faced with going either over or under a set of power lines. Steve reported later that he watched a tuft of dandelion, and saw it rising above the lines, so he knew he could stay above the lines without a downdraft pushing the *Challenger* down. However, after clearing the lines, he saw a forest that continued another mile in front of him, so he wisely set the *Challenger* down in a farmer's freshly planted field, rather than risk hitting the trees, if he encountered a downdraft or a power failure. Subsequently, we reverted to the fly plan developed in Arizona, where

Steve climbed up to 2,000 feet (600 m) in a thermal before departing the airport (if he didn't catch a thermal before the end of the runway, he landed again).

Second attempt, the winds were blowing so hard that, after 4 hours, and climbing to 10,000 feet, he'd barely gotten a mile north of Cergy-Pontoise, so he returned to the airport, abandoning the attempt. Considering the predicted persistent north winds, and having been in France now almost three weeks, with our budget running out, Paul decided we should travel by land and hovercraft up to RAF Manston, figuring, if we could just get into the air at Manston, the *Challenger* could fly a zig-zag flight, keeping the solar array tilted mostly to the sun at the south, and let the winds blow us back to France. Unfortunately, the clouds socked in after arriving at Manston, and the winds became so strong that we wore out a tire from side forces scrubbing the *Challenger* sideways just rolling it out on the runway for flight attempts and back. After about a week and a half in the UK, we received a forecast for two days of calm wind, so we rapidly packed up and rushed back to Cergy-Pontoise, reading for flight the next day, south to north.

On July 7, 1981, pilot Ptacek took off in the *Challenger* and made multiple attempts at capturing a thermal for climb out, finally hooking into one just to the right of the runway, over a cloud of onlookers. He climbed to 2,000 feet (600 m), and radioed the ground crew and chase plane that he was departing for jolly old England.

Ptacek recalls:

It was the day we had been waiting for, clear and a relatively dry air mass. And most importantly light winds aloft for the entire route from Pontoise, our airport just Northwest of Paris, and R.A.F. Manston, Kent, U.K., the intended landing site. We had decided to start early, earlier than we really thought the solar intensity would provide enough direct sunlight to the solar cells on top of the wing and tail. So, at the end of the runway with Ray [Morgan] on the wingtip and Paul [MacCready] in the Cessna chase aircraft we began our take-off attempts. There were 7 altogether where I'd get airborne but only enough to land straight ahead on the remaining runway (during another attempt a few days earlier we took off and headed out at low altitude counting on the *Solar Challenger*'s ability to climb as solar intensity increased but landed but a few miles away in a bean field because I was unable to climb enough to clear a forest ahead on my flight path). Paul's son, Parker, was positioned at the far end of the runway to assist me turning around to taxi back for another takeoff attempt. In this way, there was no interruption so we were assured to get off at the first opportunity. Then, on 8th takeoff, I sensed a weak but lifting air mass about mid-field just over the small group of onlookers gathered off the right side of the runway. Coincidentally, or because of them, it was over their heads that I found lift. I stayed in this thermal until about 2,600 feet above the field and then pointed the nose at a 90-degree angle to the rising sun and headed West.

The *Solar Challenger* has a light wing loading and slow speed, 23 mph [37 kph] climb speed that day. For this reason and that I felt what winds aloft there were would increase as the day progressed, I thought it important to fly as far West as possible early and then later I could take advantage of the winds aloft and keep the sun normal to the solar cells on the northbound leg to Manston. This route put me longer over the English Channel both in distance and time but I had a strong sense that getting West was not only desirable but of paramount importance to the success of the flight. Once away from Pontoise, I settled down to trimming the *Solar Challenger* to achieve maximum solar intensity on the solar cells. I fly mostly by feel and then crosscheck the instruments. As I've found with most machines, there was a sweet spot where you could feel and hear that everything was where it needed to be. I found a small envelope where the airplane was most efficient and stayed there and not squander altitude. Fly too slow and the airplane would descend, fly too fast and it wasn't efficient and maybe descend. While I was never told how to fly the airplane, it was decided by common agreement that I should climb as high as possible and store energy in the form of altitude.

Because of concerns flying over the English Channel it was decided by some that I should wear a wetsuit just in case.... We didn't have a real good ventilation system in the cockpit of the *Solar Challenger* and the seat was hard and not at a very comfortable angle (for me) but to fly the aircraft, I never mentioned it to anyone. Apparently, I was working harder than I realized to get airborne initially and climbing through 10,000 feet [3000 m] encountered cold temperatures. It was enough that I had at least very cold water if not ice running down my back for the first few hours of the flight. Having the 16,128 solar cells positioned correctly in the sunlight meant that the pilot was always in the shade.

Just prior to the French Coast I climbed up the side of a towering cumulus cloud in blinding-refracting-light. That was the most fascinating part of the flight, simultaneously unnerving and spectacular. Unfortunately, I was unable to derive any meaningful power settings as everything was changing so rapidly. At the same time I was dodging at least one helicopter and more than one airplane presumably with media people onboard. It had been agreed upon that media representatives would ride in the MacCready aircraft and then share photos with other interested media organizations. These aircraft either didn't get the word or chose to disregard the agreement.

As things became more interesting, I purposefully flew into the cloud. "Cloud flying" as it's called in Europe is legal and I was at the time very familiar with flying without visual cues using basic instruments, something all instrument pilots practice in training. Later in the flight I had one more encounter that would be the only other trial of the day. A four-engine Lockheed Electra had descended lower than their assigned altitude and its flight path was directly ahead of mine. I waited for what I thought was a reasonable amount of time and climbed to what I thought was a safe altitude before crossing its path. What I didn't know and subsequently learned is that aircraft wakes do not always descend but can climb. This one did. I thought it was possible that the airplane might break up. But it did a great job and while the ride was exciting it survived intact and we continued on.

The remainder of flight over the Channel, the portion of the flight that I had prepared for the most, was actually very easy. I turned north, ran with the wind as I had earlier planned and I could really enjoy the beauty of where I was and the magnificent coasts of France and England. Witnessing the White Cliffs of Dover rising out of the haze is an image I will never forget. I circled Manston for about 45 minutes before landing while the chase plane landed and cleared customs, so that Paul would be available to catch the wing tip after touchdown, so the *Challenger* could be brought to a safe stop.

My only regret is for the *Solar Challenger* crew. Not only were they not there for the landing but most were unable to come to Europe at all because of the financial constraints. I was probably the least deserving person on the team as my contribution to the project was so small. I felt at the time and still contend that this was a flight that was going to happen one time ever and to that end while there was a certain degree of danger during the cumulus cloud event, I survived and experienced something that I will never forget. Granted to not experience this part would have been okay with me but I have no ill feelings to those others involved as I recognize that the media aircraft and helicopters were 1) just doing what they were told, and 2) didn't realize the impact of what they were doing had on my aircraft. I have loved aviation and flying my entire life and to experience this flight and all my flights in the *Solar Challenger* was the chance of a lifetime. I am very thankful, grateful, and honored to have been part of the *Solar Challenger* Team.[8]

Ray Morgan adds:

By common consent, for safety's sake, only one plane had been cleared to film the flight, the FIA chase plane with Paul on board communicating to Steve. But during the flight, which took 5 hours and 23 minutes, a helicopter chartered by the Press was hovering over *Solar Challenger* creating extreme turbulence that was lethal for our low wing-loaded airplane. Despite Paul's telling the Press to keep their distance it was impossible. Steve described it as like trying to order sharks in the ocean not to bite you!... We had a parachute in there above the pilot's head so that if he got in trouble he could pull it down and throw it out the side. When he hit the Electra's vortices, Steve had his hands on the parachute ready to abort when somehow the *SC* managed to right itself and pull out! As for the solar energy, he had reported a momentary peak power approaching 3000 watts when over the English Channel at 11,000 feet, and adjacent to a very bright cumulus cloud which added reflected light to the direct sunlight on the array.

Solar Challenger never flew again. She is now in storage at the Smithsonian Air and Space Museum, Washington, D.C.[9] But she has become a legend worldwide. *Solar Challenger*'s flight was reported around the world.

In 1980, a French-Japanese animation studio, Pierrot, made a television series called

Estaban, Child of the Sun. Set in 1532, it follows the adventures of a young Spanish boy named Esteban who joins a voyage to the New World in search of the lost Cities of Gold and his father. Among their discoveries of the technological wonders of the Mu Empire are *Solaris*, a solar-powered ship, and the *Golden Condor*, a huge solar-powered ornithopter (mechanical bird), capable of traveling considerable distances under the sun's power alone. *Taiyō no ko Esuteban* was premiered on NHK in 1982 and has become one of the best-loved animated series of all time.

Sun Rays Elsewhere

There were those who began to use the sun differently for aerial flight. In 1972, Dominic Michaelis, a British architect and the inventor of many solar utilities and projects, invented and built the first lighter-than-air solar balloon (1 cubic meter) with a double skin Melinex polyester film envelope with a clear external surface and dark, heat-catching internal walls. The temperature difference taken inside and outside was about 27°C, which corresponds to a lift of approximately 100 grams per m^3. Then, after building a few small balloons, Michaelis built a large one (diameter 10 meters) that was able to lift his son Stéphane (30 kg), whose captive flight saw him become the first solar balloon human pilot! But one day the balloon broke loose from its moorings and disappeared into the Oxfordshire skies. In 1981 Julian Nott piloted Michaelis's 3rd solar balloon across the English Channel. No electric motor or batteries were used.

Günter Rochelt

Over in Solln, a suburb of Munich in Germany, a long-haired, bearded, almost Biblical-looking industrial designer called Günter Rochelt was preparing his plane *Solair I*. Born in 1939 in Česká Kamenice, Bohemia, when his parents moved to Memmingen in Bavaria, the young Rochelt obtained blueprints for model airplanes but then modified them to fly better. After two terms at the engineering faculty of Munich Technical High School and two years studying industrial design at the Ulm Design High School, he began to invent. He designed a new clothes hanger, quick-change picture frame, and an energy-efficient electric shaver, which he sold to a Japanese firm.

Meanwhile, in his enthusiasm for flying, he was frustrated by the short life of the batteries, and so started to consider solar power. On June 24, 1979, Günter flew his model airplane *Silver Fox* for 3 hours and 41

Günther Rochelt (1939–1998), "hippie" inventor of Munich, helped by his wife and friends, designed and built his own solar airplane in the garage in 1980 (Carolin Scheuermann).

FIVE. "Here Comes the Sun"

minutes—the longest described solar flight yet known. Then, with his own finance and just a little help from his wife and his friends, Günter Rochelt decided to build a full-scale solar airplane in a Munich garage. He called it *Solair I*. He took a canard design by the Swiss Hans Ulrich Farner, but enlarged it from 13 to 16 meters (43 to 53 ft), extending its area by 70 percent, and overcoming the difficulty of applying 2,499 monocrystalline rigid solar panels on curved wings, to give an output of 1.8 kilowatts (kW). He used a somarium-cobalt motor built by Karl Friedel, with an output of 2.2 kW (3.0 hp). Rochelt approximated the airfoil with a number of straight-line segments, then filled in the remaining space with clear silicone rubber. It took him and his wife three months to sand the silicone to a smooth aerodynamic shape. He then covered the silicone with a layer of ½-mil Mylar. The result was a glass-smooth finish, but Rochelt estimated that the process added about 35 pounds to the airplane. Also, after being installed in the wing, the panels delivered only 1,800 watts instead of the 2,200 watts measured before covering.

Bob Boucher recalls:

> Günter took *Solair I* to Biggin Hill airport about 60 miles from Manston RAF Base where we were with our *Solar Challenger*. Günter was attempting a channel crossing at the same time that we were! We had heard some vague rumors about his airplane but we had no solid information. Then a few days after we got settled in at Manston we received a phone call from Günter inviting us to come to Biggin Hill to see his airplane. Martin Cowley and I drove over to Biggin Hill that afternoon. Günter was waiting for us with Champagne and Cognac to brighten our spirits and get us into the proper mood to witness the unveiling of his creation. The airplane was absolutely superb, the workmanship was flawless, and Günter had even installed a digital instrument panel.

On August 21, 1983, Rochelt piloted his *Solair I* for a record 5 hours and 41 minutes, at Unterwössen, Germany. By making it and its pilot lighter, *Solair I* was eventually able to clock up a total of 100 hours in the air. The Oskar Ursinus Association awarded Rochelt

1980: Günther Rochelt carefully checks his *Solair 1*, its 16-m (53-ft.) Farner canard covered with 2499 monocrystalline solar cells to provide an output of 1.8 kilowatts (kW) to the 2.2 kW (3.0 hp) somarium cobalt Friedel motor (Carolin Scheuermann).

its prize for the most advanced design. (*Solair I* is now at the German Museum in Munich.)[10]

Ever innovative, Rochelt next turned his mind to a pedal-powered airplane which he named *Musculair* and built using ultra-light carbon fiber. It was flown by Rochelt's 17-year-old son Holger to win two Kremer prizes for the flight over "the eight" in four minutes and 25 seconds in 1984. In the same year, Holger set a world speed record at 35.7 kph (22.2 mph) to receive a second Kremer prize. Later that year, Holger Rochelt and his sister Katrin, at that time still a child, flew the first passenger flight in a human-powered aircraft. In 1985, the *Musculair II* set a new speed world record of 44.26 kilometers per hour (27.50 mph). Günter's *Flair*, for which the inventor finally proposed a bridge between gliding and hang-gliding, was first conceived in 1984, but its airframe in fiberglass, carbon and Kevlar was not finished until spring 1987. After a few test flights, in which the performance was confirmed but the rudders were found to be somewhat ineffectual, Günter decided not to put the *Flair* into production, concluding that he could improve performance further with some changes to the design. This led to an ultralight glider (36 kg/80 lb.), *Flair 30*, whose design goal was to reach a 30:1 glide ratio with the pilot launching by foot, then lying prone in a special harness, and finally landing on a skid. It first flew in 1990. For his innovative aircraft concepts Rochelt was awarded the Philip Morris "Future Challenge" Research Prize. With his 90 kg (200 lbs.) *MinAir* of 1992 there was a solar-electric option, with twin motors and Heck folding propellers.

Finally the *Solair II* was built to compete for the Berblinger Prize. Modeled on glider construction, the airplane had a V-tail with electric motors mounted at the tips and trailing fins on each propeller. It was manufactured in half-shells sandwich construction with honeycomb cores. With charged batteries, it required an input of 755w power for the straight flight. Despite two 2kW electric motors and a heavy battery, *Solair II* weighed only 140 kg (300 lbs.).

Günther Rochelt in *Solair II* in 1996, waiting for some sunshine. Notice the V-tail, at the tips of which are the propellers. Rochelt's premature death precluded any further development (Carolin Scheuermann).

Its test flight was on July 7, 1996, at the Laupheim Army Airfield near the city of Ulm. But there were teething troubles; the propulsion system was overheating. Sadly, development stopped in September when Rochelt, aged only 59, died suddenly from pancreatic cancer. Eric Raymond recalls: "Gunther had a very pure design philosophy that bordered on the religious. He instinctively knew how to optimize every part, large or small, and combine them into beautiful, efficient aircraft. All his aircraft were thoroughbreds, pure Rochelt."[11]

Some continued to believe in a "pedalectric" approach, without solar panels. The Massachusetts Institute of Technology *Monarch* aircraft project was a series of two aircraft designed to win the Kremer prize for human-powered aircraft speed record, administered by the Royal Aeronautical Society. The aircraft used an electrical motor along with batteries that were charged by the pedaling action of an athlete piloting the aircraft. From the 1970s until the early 1990s, MIT had a succession of student-led projects that designed, built, and flew human-powered aircraft (HPA), starting with BURD and BURD-II, and evolving into the flight of the *Chrysalis* in 1979, the first of the MIT HPA to fly successfully. *Chrysalis* went on to have over 44 pilots, including the first female pilots of an HPA. The *Monarch B* was a human-powered aircraft built by a student team in 1983 which won a Kremer Prize of £20,000 for sustaining a speed of over 30 kph (20 mph) over a 1.5 km (1 mi) triangular course. It was a precursor to the Daedalus effort, which flew a human-powered aircraft from Crete to the island of Santorini off the Greek mainland in 1988.

Another contender, AeroVironment's *Bionic Bat*, was an aircraft built to compete for the Kremer Speed Challenge. It incorporated an electric motor that doubled as a generator while on the ground, with pilot Bryan Allen's pedaling action recharging nickel-cadmium batteries. The stored energy was used to supplement pedal power from the pilot during record attempts. In 1984, *Bionic Bat* won two segments of the Kremer Speed Challenge.

At the time of the *Solar Challenger*'s success, it was reported that its solar cells, valued at one quarter of a million dollars, had been on loan from the USAF. Unfortunately, in the team's haste to fly the Channel, the NASA plan had apparently not been coordinated with DuPont, and consequently, the planned unmanned high-altitude flights of the *Challenger* never took place. (The story of AeroVironment's entry into UAVs is told in Chapter Nine of this book.)

In July 1985, Bob Boucher presented a paper, "Starduster, a Solar-powered High Altitude Airplane," at the 21st joint propulsion conference of the AIAA/SAE/ASME at Monterey, California. The proposed *Starduster*, an improved version of *Sunrise II*, was designed to reach extreme altitudes of 200,000 feet, and although flight would be limited to daylight hours, long-distance flights over thousands of miles would be possible. In his paper, the visionary Boucher concluded: "Perhaps in this second decade of solar flight we will see solar airplanes circle the globe nonstop!" Without funds, *Starduster*, ahead of its time, was never built.

Eric Raymond

Alongside AeroVironment, others were busy. Like Paul MacCready, Eric Raymond, born in 1956 in Tacoma, Washington, had flown model airplanes at an early age. In his teenage years he started flying sailplanes, but switched to hang-gliding. While studying photography at the Rochester Institute of Technology and aeronautical engineering at University of California San Diego, Raymond won the 1979 U.S. Hang Gliding Championship. He also set world hang-gliding records and in 1983 and 1984 became world aerobatic champion. He went to work for Paul MacCready on his unmanned aircraft, meeting Günther Rochelt and flying the *Musculair II* human-powered aircraft.

From this experience, and with help from Rochelt, Raymond determined to design a solar-powered aircraft. He founded Solar Flight and in 1986 began construction of his design in his 2-car garage, in Lake Elsinore, California. It was built using usual carbon pre-preg, Nomex honeycomb, Rohacell foam, and thin film solar cells. Günther Rochelt sent Raymond the airfoils he used, and taught him the layups. *Sunseeker I* was partly assembled out on the front lawn. The solar sailplane was never fully assembled until Raymond and his team took it to the abandoned runway in Desert Center, where it made its first flight as a glider in 1989. In early 1990 solar-powered flights were made with two brush motors driving a variable-pitch propeller, and then with a brushless motor driving a folding propeller. In 1990 Raymond flew *Sunseeker I* with only 400 Wh of Sanyo battery energy across the USA from the Southern California desert, and after 21 flights ended in North Carolina, all the time flying just with natural and renewable energy sources. Two flights, covering 247 miles (398 km) and 249 miles (400 km), involved 8.5 hours in the air. The solar airplane often climbed to 15,000 ft (4,570 m) at about 35 mph (56 kph), but sometimes reached 85 mph (140 kph) on final glide when Raymond was too high. The *Sunseeker I* went on to break all previous records for solar-powered aircraft.

In 1990, Eric Raymond flew his *Sunseeker I* with only 400 Wh of Sanyo battery energy across the USA from the Southern California desert. After 21 flights, he ended in North Carolina, all the time flying just with natural and renewable energy sources. Two flights, covering 247 miles and 249 miles, involved 8.5 hours in the air (Eric Raymond).

Raymond recalls: "Since the sun was stronger in the West, as I got further east, it became harder to get up and make progress. In the Appalachians, with only 2.4 kW of climbing power, I was sometimes corralled by the trees around the airfield, flying circuits inside the tree line until I gained some height. I often flew with birds, and the day I crossed the Appalachian ridge, I had 3 birds in tight formation with me when I climbed into the clouds, and they tucked in tight and flew with me the whole time I was IFR."

Sunseeker II, built in 2002, was updated in 2005–2006 with a more powerful motor, larger wing, 1,953 Wh lithium battery packs and updated control electronics. *Sunseeker II*'s use of the lithium battery announced a new chapter in the development of electric airplanes. As of December 2008 it was the only manned solar-powered airplane in flying condition and was flown regularly by Solar Flight, using 2,841 Wh of Lipo cells. In 2009 Raymond flew *Sunseeker II* all over southern Europe, becoming the first solar-powered aircraft to cross the Alps, from Butwill, near Zurich, to Turin, Italy, 99 years after the first crossing of the Alps by an aircraft. In June, at the World Air Games in Turin, Raymond flew the *Sunseeker* for demonstrations and won a Gold Medal for the best experimental aircraft. He also set two world records for solar-powered aircraft during the event: one for time in the air, the other for absolute altitude, about 20,380 feet (6,211 m).

During this time, China had also been testing its first unmanned solar aircraft. Danny H.Y. Li had had a great longing for flying since childhood. He became a test pilot, piloting different kinds of airplanes, helicopters and gyroplanes. He had twice escaped from airplane crashes during his test flight career. In 1989, he and his friends flew across China with ultralight planes that set the world record for long-distance formation flight. Li obtained his Ph.D. in aeronautics and astronautics at Beijing University. In 1991, Li and Zhao Yong teamed up with Matsushita Electric Industrial Co. to create a pilotless solar aircraft. The body and wings were hand-built predominantly of carbon fiber, Kevlar and wood. The design used winglets to increase the effective wingspan and reduce induced drag. On August 19, 冲天 (*Soaring*) successfully completed its first flight in northern China. Li and Zhao were granted a patent license from the Chinese State Patent Office to further develop the aircraft.[12]

Rudolf Voit-Nitschmann, born in 1950 in Eisenach, studied aerospace engineering at Stuttgart University. He then became senior engineer at aerospace companies such as Gyroflug and Dornier, then in 1995 became professor of the Aircraft Institute in Vaihingen. Two years later, a project was launched at the University of Stuttgart to build a practical solar aircraft. Voit-Nitschmann joined the Akaflieg (Academic Flying Group), a team of 45 students, for the project they called *Icaré*, a combination of the name of Icarus, the hero of the Greek mythology, and Re, the Egypt sun god. With the help of sponsorship of the State of Baden-Württemberg, *Icaré II* was built. The 12 kW electric motor was mounted on the tail fin so as to increase the efficiency. For the self-launch, a 915 Wh NiCad battery was used, charged by the photovoltaic panels on the wing, enabling the Stuttgart sailplane to rise up to 400 meters (1,312 ft). After the first solar long-distance flight on July 7, 1996, over more than 350 km (218 mi) from Alen to Jena with Rudolf Voit-Nitschmann at the controls, *Icaré II* won the Berblinger Prize in Ulm for the most efficient solar plane, followed by the EAA Special Achievement Award in Oshkosh, the Golden Daedalus Medal of the German Aeroclub, and the OSTIV-Prize in France in 1997. Increased altitude and range became possible by the installation of lithium ion batteries.

Danny H.Y. Li continued to develop his solar-powered aircraft. During 2000 to 2002, directing New Concept Aircraft (Zhuhai), the China Aviation Industry Development Research Center, and the China Academy of Space Technology, Li developed the *Green Pioneer*, an integrated-wing solar powered aircraft with optimized aerodynamic design. *Green Pioneer* made its test flight in 2004.

Another approach was adopted by Alan Cocconi of San Dimas, California, a Caltech-trained engineer. Cocconi had served as an engineering consultant and developed the drive and solar tracking systems for the General Motors SunRaycer car, which won the 1987 World Solar Challenge, a cross-country race for solar-powered vehicles held in Australia. He designed and built the controller for the original GM Impact that was introduced at the 1990 LA Auto Show, and which evolved into GM's EV-1. In 1992 Cocconi and Wally Rippel founded AC Propulsion specializing in AC-based drive train systems for electric vehicles.

In 2005, to prove the sustainability of solar-powered flight, Cocconi built an unmanned solar glider he called *SoLong* with the goal of making a 48-hour nonstop flight around-the-clock. *SoLong* had a wingspan of 4.75 m (15 ft.) and weighed 12.6 kg. (28 lbs.). Power from 76 SunPower Corp. (Sunnyvale, California) solar cells supplied the plane's energy. Power distribution among the onboard systems was controlled by management software developed by Cocconi. Twelve PIC18 microcontrollers from Microchip Technology Inc., of Chandler, Arizona, controlled and monitored all vehicle systems. Systems under control of the PICs included the autopilot, motor drive, power tracker, six servomotors, the battery monitor, and a tracking downlink antenna. For example, the autopilot controller decoded 13 PWM control signals from the uplink receiver, input serial data from the GPS module, and monitored 23 analog sensor channels.

During daylight flight the nominal 225-W solar array powered all systems and recharged 120 Li-ion cells from Sanyo Corp. The Li-ion cells fulfilled the craft's energy demand at night. Propulsion came from a high-efficiency electric motor driven by a split-phase power controller developed by AC Propulsion. A variable-pitch propeller fine-tuned thrust for different rpm and power settings using a load cell for in-flight thrust measurements.

An earlier 24-hour test flight showed the original battery reserve couldn't keep the craft airborne. "We split the first test flight's night in two, flying midnight to midnight," said Cocconi. "We were getting enough solar energy during the day but we didn't have quite enough battery to take us through the night." The Sanyo cells pack 220 W-hr/kg and have a charge-discharge efficiency of over 95 percent. "That made the difference," Cocconi stated, allowing the *SoLong* to pass the 48-hour mark.

SoLong took off at 4:08 p.m. on Wednesday, June 1, 2005, from the sun-baked runway at Desert Center Airport just east of Eagle Mountain in California's Colorado Desert. It remained aloft until Friday, when it skidded to a stop at 4:24 p.m. after 48 hours and 16 minutes in the air. From the 5 ft × 8 ft (1.5 m × 2.44 m) *SoLong* trailer serving as ground station, Cocconi and the team of seven crack radio-control and hang-glider pilots took turns monitoring flight conditions from the twenty-three channels of telemetry plus GPS navigation and video downlink data available in the ground station. Nothing, save the flagging energy of its pilots on the ground, kept the *SoLong* from flying for another two days, or ten, or a whole month! Equipped with a video downlink, *SoLong* was the first solar electric airplane to take film images from the sky.[13]

Lockheed Martin's 240 ft × 70 ft (70 m × 21 m) HALE-D (High Altitude Long

Endurance-Demonstrator) was built and tested as a remotely controlled solar-powered UAV airship designed to float above the jet stream at 60,000 ft (18,000 m). Its 15 kW thin-film solar array supplied energy to 40 kWh Li-Po for two 2kW motors. On its maiden flight on July 27, 2011, from Akron, Ohio, the Lockheed HALE-D reached 32,000 ft before a problem with the helium levels cut the test short, forcing an emergency landing in the deep woods of southwestern Pennsylvania. Two days after the landing, before the vehicle could be recovered from the crash site, it was destroyed by fire.

Meanwhile, the Institute of Aircraft Design (Institut für Flugzeugbau) continued to pursue its R&D. *e-Genius* was built for the CAFÉ 3rd Green Flight Challenge sponsored by Google to be held in late September 2011 at the CAFE Foundation Flight Test Center at Charles M. Schulz Sonoma County Airport in Santa Rosa, California. The design was a converted motorglider using a tailfin-mounted 80 hp (60 kW) electric motor. The *e-Genius* performed its first 20-minute flight on May 25, 2011. In July 2011 the aircraft flew for over two hours between two points near Mindelheim, Germany, at an average speed of more than 100 mph (161 km/h). During the Challenge, held July 10–17, three aircraft competed and two met the challenge requirements to fly 200 miles (320 km) in less than 2 hours and use less than one gallon of fuel (or energy equivalent) per passenger. The first place prize of $1,350,000 was won by the Pipistrel USA.com team led by Langelaan LLC of State College, Pennsylvania. Second place prize of $120,000 went to the *e-Genius* team led by Eric Raymond of Ramona, California.

On June 20, 2013, Eric Raymond test-flew his new two-seater *Sunseeker Duo* from Milan, but with a takeoff tow from his car. On December 17, 2013, he used the *Duo's* motor to take off. The *Duo* has 1,510 solar cells on its 72-ft span (22 m) wing and on its empennage surfaces, plus 8,158 Wh of Li-Po batteries, driving a tail-mounted 16 kW electric motor. During 2014 Eric Raymond and his wife Irena, a fully qualified pilot, took a number of passengers for flights in the *Duo*. But to go one further, the Raymonds decided to repeat the 2009 adventure of the *Sunseeker II*, a crossing of the Alps in both directions, this time with the *Sunseeker Duo*. The mission was to stop and show the airplane at different airports. For the final destination, Munster-Geschinen airfield in Switzerland was chosen: "This is a perfect starting point to explore the Swiss Alps and visit the highest peaks and glaciers. We decided to fly to Torino Aeritalia airport on 2 August 2015 and then continue to Switzerland. On August 3rd, we climbed up the foothills, with low clouds at the base of the Italian Alps. We crossed the high mountains near Zermatt and then detoured to the Aletsch Glacier, Jungfrau, Eiger and Mönch. The distance flown was 384 km and the maximum altitude reached was 4090 m. The battery pack was completely charged after we landed."[14]

The weather on August 5 was very promising, so Raymond and his hang-gliding friend Stefan made a flight over the Aletsch Glacier. On August 7, 2015, Eric and Irena Raymond took off for the flight to their home base.

> We headed first to the Matterhorn and then direction south, toward Genoa. The flight was very easy in comparison to the crossing of the Alps a few days earlier, in the opposite direction. The weather in the Alps was more than perfect. The Matterhorn just got covered with a cloud cap when we approached it, however, the view on this giant, steep mountain was spectacular. The maximum altitude we climbed was 4,545 m. We could easily go much higher, unfortunately we do not have a dual oxygen system yet. Descending toward the Po Valley the air became hotter and the visibility was poor. After 230 km and less than 4 hours our *Sunseeker Duo* reached her home base.

The *Duo* has since been used to train a number of pilots.

Irena Raymond braves the cold on a winter 2015 flight in *Sunseeker II* over the Apennine Mountains in Northern Italy. To date she has flown 110 hours in solar-powered airplanes (courtesy: Irena Raymond).

At the Sustainable Aviation Symposium held in San Francisco in May 2016, Eric Raymond stated that Solar Flight was ready to build and fly a solar-electric six-seater, called the *Observer*, based on an Italian design, the Partenavia. The wing is optimized for best aerodynamic efficiency with a large camber changing flap for short takeoff and landing capability. The main power source is a lithium battery pack, but an optional range extender can be fitted into part of the baggage compartment. It consists of a generator running on unleaded auto gas. A ten-seater would follow.

Then there is the *Sun Flyer*, under development by George Bye of Aero Electric Aircraft Corporation (AEAC) of Denver, Colorado. An engineering graduate of the University of Washington and an indefatigable entrepreneur, Bye has immersed himself in aviation for decades. After earning a bachelor of science degree in civil engineering from the University of Washington, from 1981 to 1993 he was a U.S. Air Force instructor pilot flying Northrup T-38 and C-141B strategic transport, accumulating over 4,000 flying hours in aircraft. Once in mufti, Bye began designing aircraft and developing businesses. In 2005 Bye designed and developed the Javelin Jet, a lightweight category twin turbofan, tandem-seat jet aircraft. In 2007, he created Bye Energy, an engineering services organization, with a special concentration on various unmanned aerial vehicle projects. In July 2010, Cessna announced it was developing an electrically powered 172 as a proof-of-concept in partnership with Bye Energy. In July 2011, Bye Energy, whose name had been changed to Beyond Aviation, announced the prototype had commenced taxi tests on July 22, 2011, and a first flight would follow soon. In 2012, the prototype, using Panacis batteries, engaged in multiple successful test flights.

To further exploit recent advances in battery and solar technology, George Bye founded AEAC in 2014 with the goal of putting an all-electric trainer aircraft into production. Joining him was Charlie Johnson, the former president of Cessna. He then obtained a license agreement and engineering contract with Calin Gologan's PC-Aero to incorporate the work done in Germany on the *Elektra One*, which AEAC then modified with new landing gear, prop and instrumentation. The two-seat prototype was installed with the EnstrojEmrax 268 high voltage electric motor, rated at 100 kW and 400 volts nominal supplied by four Panasonic lithium-ion battery packs and solar panels on the wings and the horizontal tail and behind the canopy. A regenerative propeller is also used. The solar panels energize the batteries whenever the sun is shining, whether the plane is in the air or on the ground. This brings over three hours of endurance and a 30-minute recharging time. To enable this swift recharging, *Sun Flyer* partnered with the Bloomington Corp. of Orlando, Florida, now working on a national network of battery charging stations. Alternately, for more advanced flight schools, it is possible to replace depleted batteries with fully charged ones in a matter of minutes. When a line-up of *Sun Flyers* is parked on the airfield, they can use their solar panels to supply electrical energy to the local grid. *Sun Flyer* will use broadband and iPad connectivity as part of a high-tech flight training system to enhance the student pilot-instructor experience. The Redbird Flight Simulations Sidekick will keep track of motor parameters, as well as flight time, airplane position, attitude, and landings, and wirelessly transmit the data to the flight school or ground station where the operator can track it via Redbird's customizable Sidekick software.

Flight schools seriously need this product. In May 2016, AEAC hosted a rollout event of its proof of concept (POC) *Sun Flyer*, built by Arion Aircraft of Nashville, Tennessee, at Centennial Airport. Performance data from the POC prototype airplane would be used to help finalize the design for the FAA-certified production version. During late 2016 the POC was put through extensive ground testing. Following 10 deposits received for the electric airplane unveiled at EAA AirVenture Oshkosh, in September 2017 AEAC gave a ground test to their POC, registered N502. Spartan College of Aeronautics and Technology and a local flight school signed for the first 35 *Sun Flyers*. In addition, a Centennial Airport–based school, Independence Aviation, also just recently signed on. Bloomington Corp. ordered 30 *Sun Flyers*. An agreement was also signed with the Aero Touring Club de France, based at Toussus le Noble, to purchase *Sun Flyers* for their club.

Bye referred to Boeing's estimate that 533,000 commercial pilots will be needed by 2033, and that with the current training fleet being 40 years old, something has to change. Bye noted the high cost of flying, the high demand for new pilots, and the need for an answer to these problems, and then asked, "What is the solution and how do we get it?" Cessna delivered the first 172s in 1956. As of 2015, Cessna and its partners had built more than 43,000, a large number for pilot training.

For those pilots training to fly the new generation of electric airplanes, all-electric flight simulators, more energy efficient and easier to maintain, are already the standard. Recall how in 1929 J.P. Buckley patented an electrically driven aeronautical instructing device (or flight simulator) for pilots of biplanes. The first manufacturer to tap improved electric technology and apply it to flight simulation was Moog Inc. of East Aurora, New

York. Moog presented the concept of all-electric motion systems for full-flight simulators with several FFS manufacturers that already used Moog's high-performance servo valves in their hydraulic motion systems. In 1994, Moog engineers had designed and tested their first 4,500 kg (10,000 lb.) electric platform. But going from hydraulics to electric actuation posed a number of engineering challenges—among them, handling heavier payloads, providing smooth motion, ensuring safety, and preventing unwanted noise. Moog teamed up with Ron Jantzen and Niddal Samur of Flight Safety International of Broken Arrow, Oklahoma, the world's premier professional aviation training company and supplier of flight simulators. This involved the development of 36- and 60-inch actuators, a 12-pole, brushless servo-motor with custom rotor and stator design to deliver 5,600 lb.-in (633 Nm), and motion control software. The work paid off, and in May 2006, the FAA granted first Level D Certification to the first all-electric high-payload flight simulator. Since then adoption of all-electric FFSs has grown quickly. *Civil Aviation Training* magazine recently noted there are 1,150 FFSs in use around the world built by companies such as FSI and CAE in Montreal, Canada. Of the motion-control solutions for flight simulators now being built and sold, industry experts say more than 85 percent are electric.

A second AEAC airplane envisaged by Bye is the graphite composite solar-electric StratoAirNet, with solar cells on its 15-m (49-ft.) wing, supplied by SolAeroTechnologies. It will be used as a prototype for a medium-altitude, multi-day persistent unmanned aircraft. In August 2017 SolAero Technologies Corp (SolAero) delivered the first solar wing for the StratoAirNet. The initial wing-solar cell configuration will deliver sufficient power, approximately 2,000 watts, under suitable daylight conditions at altitude. The Bye Aerospace StratoAirNet family of "atmospheric satellites" are intended to provide support for commercial and government security requirements. Final assembly and integration of the wings and power systems began in November 2017, with ground and flight tests to follow from Bye's new facility at the Northern Colorado Regional Airport.

On July 23, 2017, at EAA AirVenture, AEAC announced plans for a four-seater pure-electric airplane, Sun Flyer 4, with an autonomy of four hours. In addition, Spartan College of Aeronautics and Technology was the first flight school to hold a deposit for a Sun Flyer 4. The two-seat Sun Flyer, Sun Flyer 2, will be the first FAA-certified all-electric trainer aircraft under FAR Part-23. The new four-seater will closely follow the certification of the two-seat version. Features of the Sun Flyer 4 included a 46-inch cabin width, 38-foot wing span, ballistic parachute recovery system and a gross weight of 2,700 lbs., with a full 800 lbs. of payload for pilot and passengers.

Not related in any way to solar power, in March 2017, Bye Aerospace also allied itself to XTI of Denver to develop a hybrid/electric prototype of XTI's revolutionary TriFan vertical takeoff airplane. Transmission, gears, two large heavier engines, and other components would be replaced with electric motors, batteries, generators, and a single smaller turboshaft engine. The *TriFan 600* is designed to travel at over 300 mph (480 kph), with a range of over 1,200 miles (2,000 km). Using three ducted fans, the *TriFan* would take off vertically, and then its two wing fans would rotate forward for a seamless transition to cruise speed and its initial climb. It would reach 35,000 feet (10,000 m) in just ten minutes and cruise to the destination as a highly efficient business aircraft.

In 2012, Anne Réale, a painter and writer, with Thierry Vigoroux, journalist and pilot, published *De Glace et de Lumière* (*Of Ice and Light*) with Editions Pascal Galodé, an adventure novel about a solar-powered amphibian that takes off from the French Alps, crosses the Atlantic, and lands on the Greenland glacier. The authors may have been inspired by the *Hy-Bird*, a solar-powered hydrogen fuel-cell amphibian, with foldable wings as projected by Erick Herzberger of the LISA Aeroplane Company in Savoy, France, at the side of Lake Bourget. The plan was to fly it around the world, landing on land, water and ice. Such an electric aircraft is still to be built.

Another e-airplane, which on June 10, 2016, was test-flown for ten minutes in the skies above Calverton Air Base, New York, was the solar-wing-powered Luminati Aerospace VO-Substrata prototype, with Robert Lutz at the controls. After the flight, Lutz described the flight as "very birdlike. You feel like you're in the environment up there with the creatures. Hawks will be circling around, and they kind of flock to you. It's the only aircraft I've ever flown where I can hear a helicopter next to me. It's a little spooky but pretty cool."

The Luminati project was to manufacture a fleet of manned and unmanned aircraft for perpetual, solar-powered electric flight in the stratosphere, as a platform for commercial Internet and government ISR (Intelligence, Surveillance, and Reconnaissance) applications, also to provide aerial Internet service for an estimated four billion people worldwide. The founder of Luminati Aerospace, Daniel Preston, brought together an "enlightened" team of engineers, professors and advisors from MIT, UIUC, and Georgia Tech, leading industry professionals, and visionary technological advisors such as Dr. Anthony Calise, former professor of aerospace engineering at Georgia Tech. Their prototype airplane was based on a modified "Elektra One Solar," bought from the German company PC-Aero, with its four-hour flight autonomy, but with its aileron span increased, wing flaps excluded, and rudder chord slightly increased. In 2015, Luminati Aerospace paid $3.4 million to acquire the old Calverton Air Base, 16.3 acres (6.6 hectares) of land on Long Island. In June 2017, it was announced that Luminati was searching for new sponsors after ties with its earlier backer, Facebook, were severed. To buy the sprawling Enterprise Park at Calverton, an approach was made to John A. Catsimatidis, Greek-American billionaire. The departure of key members of what Preston called his team, and revelations about the startup founder's past and current legal troubles, have cast doubt on his ability to fulfill his promise to bring the defense aerospace industry back to Calverton.[15]

Although still in the embryonic phase, the solar helicopter now exists. After successfully completing the longest duration flight for a human-powered helicopter in fall of 2013, the University of Maryland Gamera Team, a student team originally inspired in 2012 by the American Helicopter Society's Sikorsky Prize, has continued raising the bar. In 2014, a new group of undergraduate students took over Team Gamera, reinventing itself as Solar Gamera to test the feasibility of applying solar power in achieving human helicopter flight. Solar Gamera is powered solely by four banks of solar panels, with lift provided by four sets of rotor blades. It measures 100 ft (30.5 m) square. On August 26, 2016, it successfully carried a passenger, the lightweight Michelle Mahon, over one foot (0.3 m) into the air, staying airborne for nine seconds. According to the team, once its electronic control system is better able to compensate for drift, that duration figure should rise significantly.[16]

If ongoing progress with solar cells is observed, solar airplanes and UAVs have a promising future. From 2009, the University of Washington's Multidisciplinary University Research Initiative (MURI) project team, with lead researcher Dr. Minoru Taya, has been working on the airborne solar cells and found that dye-sensitized solar cells made from organic materials, which use dyes and moth-eye film, are able to catch photons and convert them into synthesized electrons that can harvest high photon energy. As an optimum energy-harvesting source, DSSC may lead to longer flight times without refueling.

Alta Devices of Sunnyvale, California, made up of scientists from Caltech and the University of California, Berkeley, have developed the flexible AnyLight™ cells to give as much as 5 times more daytime endurance, and at one gram per watt of power, with virtually no impact on aerodynamics. Based on gallium arsenide (GaAs) which is a III-V semiconductor with a zinc blende crystal structure, Alta's units hold single- and dual-junction solar efficiency records at 28.8 percent and 31.6 percent respectively. AnyLight has a solar cell thickness of 110 mm, a mass of only 170 g/m^2, and the ability to bend to cover curved surfaces. Equipped with AnyLight cells, in March 2008, an AeroVironment RQ-20 Puma UAV, which normally has an endurance of 2 to 3 hours, flew for 9 hours 11 minutes and 45 seconds.

More experimentally, Martin Kaltenbrunner, Niyazi Serdar Sariciftci, and Siegfried Bauer of the Department of Soft Matter Physics, Johannes Kepler University, Linz, Austria, have shown perovskite solar cells just 3μm thick can power miniature model aircraft for several hours. Plastic foil substrates and chromium oxide interlayers are used in a novel technology that combines high efficiency, low weight, and extreme flexibility in a single platform. For these tests, they powered a 4.8 g model airplane with a 58-cm (23-in.) wingspan with a 3μm-thick, 5.2g/cm, light solar panel (with 64 individual cells) on the tailplane and flew it on a sunny winter afternoon above the campus of Johannes Kepler University. A lighter-than-air model blimp and a "solar leaf" were also tested. These three aeronautic models were still fully functional more than six months after their initial flights. In their future research, the Koblenz team plans to focus on realizing perovskites with improved efficiency and moisture resistance, by exploring electrode transport materials, alternative metals, and superhydrophobic coatings. They suggest their design could initially find applications in robotic insects and drones.[17]

SIX

The Lithium Advantage

Alongside those who were equipping airplanes with solar panels, while batteries still lacked the required energy density, the idea of an electrically propelled airship was resurrected. English aeronautical engineer Graham E. Dorrington, Ph.D. at Queen Mary and Westfield College, University of London, wanting to bring researchers and scientific instruments close to the crowns of forest trees, developed what he called *dendronautics* (from the Greek words *dendron*, tree, and *nautica*, navigation). In 1993 Dorrington built a pedal-electric airship with Yuasa lead-acid batteries to enable Dieter Plage, a celebrated German cinematographer of nature documentaries, to film above the canopy of the Sumatran rain forest. During one flight, a gust of wind caused the airship to buck and become entangled in a treetop. Plage fell to his death, having unfastened his safety belt in order to try to reach and save his camera.

Dorrington, determined to improve stability, embarked on "Project Hornbill," designing and building *Dirigible-4* with a helium capacity of 380 cubic meters, perhaps the first electric airship since *La France* took to the skies above Paris in 1884, over a century before. Again with energy from gas-free recombination Yuasa lead-acid batteries to provide nominal 24V DC power for about 1 or 2 hours duration, *D-4*'s two main motors, 24-volt 650W EMDs, were used to drive two side-mounted 0.6m diameter propellers that could be vectored through 360 degrees, to provide about 70 N thrust (combined). As a backup, for forward propulsion, *D-4* was also fitted with a large 2.6m-diameter propeller that could either be electrically or human-powered, although this was rarely used.

Between January and April 1995, Dorrington flew *D-4* over a reserve of pristine tropical rain forest near the Danum Valley Field Centre (DVFC) in southern Sabah, Malaysia, collecting pollinating insects as well as counting orangutan nests in a forest area of about one square kilometer. These pioneering efforts proved that controlled movement close to the forest canopy is in fact viable. A total of thirty successful and silent flights were made, ranging up to 3 km from the takeoff point, with safe return in all cases; one flight reached 450 ft (137 m) at 20 kph (12 mph), while another lasted 114 minutes. On its penultimate flight, the *D-4* was soft-landed atop a flowering Merbau tree and rested there for twenty minutes, further demonstrating the effectiveness of similar platforms for canopy research. After operations began, *D-4* was also fitted with two small, laterally facing Astro Cobalt electric motors (200W each) driving propellers in the tail, and two electric motors (250W each) driving propellers slung on the keel structure, to act as bow and tail thrusters. These thrusters proved to be essential for maneuvers at low speed or during hover.

If electric airplanes were to progress, they needed men of vision. One of these was Brien Seeley. After graduating from UC Berkeley, Seeley obtained his M.D. degree from UCSF in just three years, specializing in eye surgery. While a medical student, he designed and hand-built his own 70 mph, street-licensed electric car and drove it to the hospital each day as an intern. During his residency in eye surgery at UCSF, he devoted his two-week vacation to earning his pilot's license, and this began a life-long passion for aviation. He studied aeronautical engineering and helped build two experimental homebuilt aircraft. In 1981, Seeley founded the Comparative Aircraft Flight Efficiency (CAFE) Foundation to host the CAFE 400 flight efficiency aircraft races. From 1981 to 1990, each summer's CAFE race at Oshkosh attracted the premier aircraft designers in the U.S. to bring their sleekest aircraft to compete. Oshkosh Airshow became the focus of an annual aircraft race that emphasized fuel efficiency; the winner had to complete the 500 mile triangular race course while consuming no more than an allotted amount of fuel. Many felt that the fuel allotment was too arbitrary and constrained entries in the event. The race's minimum qualifying speed also limited participation.

As well as vision, electric airplane builders also needed incentives.

In 1927, at the age of 25, Charles Lindbergh had emerged from the virtual obscurity of a U.S. Air Mail pilot to instantaneous world fame as the result of his Orteig Prize–winning solo nonstop flight from Roosevelt Field on Long Island to Le Bourget Field in Paris. He flew the distance of nearly 3,600 statute miles (5,800 km) in a single-seat, single-engine, purpose-built Ryan monoplane, *Spirit of St. Louis*. In 1997, Lindbergh's grandson Erik, a commercial rated pilot, artist, and entrepreneur, keen to push the boundaries of electric aviation, founded the X Prize Foundation and followed up with the LEAP (Lindbergh Electric Aviation Prize), including an electric flight program and the Electric Aircraft Development Alliance. On September 29, 2011, Lindbergh stated, "The first company to produce a certified two seat electric aircraft with a 1.5 hour range will dominate the aviation training market."

In addition, there was also need for a technical breakthrough in energy storage. Although, as we have seen, solar energy had been the mainstay of electric flights, the age-long quest continued for a lightweight battery with high energy density. The "Grail discovery" came in the form of the lithium battery, which would make the myths of aviation a reality. In the fall of 1972, M. Stanley Whittingham, an English-born chemist working with a team at the Exxon Research & Engineering Company, announced that they had come up with a new battery, and patents were filed within a year. Within a couple of years the parent company Exxon Enterprises wheeled out a 3W 45Ah prototype lithium cell and, linking it to a diesel engine, started work on hybrid vehicles. When solid-state physicist Professor John Goodenough became head of inorganic chemistry at Oxford University in 1976, his research group included assistant Dr. Phil Wiseman, Dr. Koichi Mizushima and Dr. Phil Jones. They set themselves the task of looking at the potential of rechargeable batteries, which began by simply "kicking around ideas on a blackboard." "We looked at it in a different way using lithium cobalt oxide at the positive terminal and pulling the lithium out; this produced a huge cell voltage, twice that of the Exxon battery," Dr. Wiseman explained. It was this spare voltage that allowed alternatives at the other terminal where Exxon had been forced to use lithium metal, which was fraught with problems. Instead lithium-ion material could compose both electrodes. The group's research was published in the *Materials Research Bulletin* in 1980. In 1977, Whittingham teamed up with John B. Goodenough to publish a book, *Solid State Chemistry of Energy Conversion and Storage*.[1]

In 1979, a paper on lithium-thionyl chloride batteries with energy of 200 kw/h was presented at the Near-Surface Ocean Experimental Technology NORDA Workshop, National Space Technology Laboratories, Mississippi, by three technicians from the Naval Ocean Systems Center, San Diego, California.

Development continued. By the late 1990s, Japanese companies, in particular SONY, had made great strides in the commercialization of lithium rechargeable batteries. In 1997, Tsuyonobu Hatazawa, R&D manager at the Sony Corporation in Kanagawa, invented the polymer gel electrolyte and lithium-ion polymer battery. In 1997 Nissan Motors produced its Altra, an electric car, equipped with a neodymium magnet 62 kW electric motor and run on lithium-ion batteries manufactured by Sony. The following year, Ji Joon Kong in South Korea founded Kokam to manufacture polymer processing equipment, including polyester film and polarized film manufacturing systems, and breathable (porous) film casting systems. In the late '90s, Kokam expanded its business to designing and manufacturing lithium-ion/polymer secondary batteries and succeeded in developing the world's first high-capacity Li-Po batteries.

It was about this time that the lithium-ion battery went afloat and took to the air. Its first planned use was for a top-secret defense application. In 2003, Northrop Grumman had been commissioned to design and build the Advanced SEAL Delivery System (ASDS), a 65-foot (20 m) midget electric submarine for piggyback operation by the United States Navy. With power from a 67 hp (50 kW) electric motor, when it was found that the silver-zinc batteries were depleted more quickly than planned, the pioneer decision was made to develop a lithium-ion pack. Yardney Technical Products of Pawcatuck, Connecticut, was awarded a $44 million contract modification to provide four lithium-ion batteries for the ASDS program by May 2009. But in April 2006, while this was progressing, the program for new submarines was canceled and Northrop Grumman was notified of termination. The current submarine was still in development and use from Pearl Harbor when on November 9, 2008, while being recharged, the lithium batteries caught fire, and burned for six hours. The specter of inflammability would continue to haunt the Li-Po battery.

During the next decade, the commercial availability of these lightweight, energy-dense batteries heralded the long-awaited arrival of electric airplanes built, as we shall see, by individuals and groups all over the world. But as we have already noted with the pioneer Fred Militky, the Li-Po adventure began with aero *modelers* in 2004. These were very low-current models due to the small size and low current ability of the early cells as manufactured by Kokam and E-Tec in Korea, although the range was soon extended.[2]

The second adaptation of auxiliary electric propulsion for full-scale sailplanes was a German-Italian project (the first being the *MBE-1* described in Chapter Four). In 1992, Stefan Gehrmann of AirEnergy of Aachen, Germany, teamed up with sailplane builder Mario Beretta of the Alisport company in Cremella (Lecco), located in northern Italy near Milan. Beretta had set up a program to make gliding less expensive and more accessible to young people. As part of this, AirEnergy fitted out an Alisport *Silent Targa* sailplane with an electric motor and battery to combine with the front electric motor and folding propeller on the nose, enabling self-launching without the need for a tow. The *Silent Club AE1* was fitted with a single-blade propeller belt-driven by a 13 kW (17 hp) DC electric motor running on 40 kg (88 lb.) made up by AirEnergy using Sanyo cells that provided 1.4 kWh of power. The *Silent Club AE1* was certified in Germany under that country's light aircraft category after numerous tests published in August 1997, including the controlled destructive testing of two airframes. It first flew in 1997 and reached a height of 600 meters. Following this, Stefan Gehrmann did a tour through Germany with the NiCad *Silent*, introducing it to almost twenty different airfields. "We also flew it in Italy at the field near in Alzate (near Cremella the home of Alisport). We built 4 AE-1s with NiCad batteries."[3]

Another who used batteries to power his sailplane was Axel Lange, based in the Rhineland Palatinate, Germany. A passionate glider pilot since his youth, Axel Lange had previously worked as an engineer at sailplane manufacturer Glaser Dirks. Here he had, amongst others, designed the highly successful motor-glider DG 800. While working at Glaser, he conceived the idea of the electric motor-glider as a way to solve many of the reliability and safety problems bothering traditional motor-gliders. Finding no backing for the idea at Glaser, he quit the job and started his own company in 1996. Lange worked in silence, and his flying testbed and technology demonstrator, the *LF 20*, caused a sensation when it was unveiled in 1999. The *LF 20* was a heavily modified DG800 powered by the same 42 kW outrunner motor as would later power series production aircraft by Lange. The battery system in use was an experimental NiMh battery provided by Panasonic. The aircraft could self-launch and then climb up to 1900 m (6200 ft) above the airfield. This was more than enough for the mission of a self-launching high-performance sailplane.

In 2003, AirEnergy entered into the Li-ion technology market with the development of the first battery management system, and the following year they delivered the first

Axel Lange flies the *LF20* motor sailplane with its experimental Panasonic NiMh battery (1999) (courtesy Lange Aviation GmbH).

Korean Kokam Li-Po battery system for an electric boat (220V / 40kWh). It therefore seemed natural that for the first AE-1 sailplane, they should use Li-ion polymer batteries, weighing 10 kg less than the NiCads. The examination certificate for this sailplane, dated June 15, 2005, mentions lithium batteries as a special feature. According to his logbook for AE-1 (Registration D-MFLP), Stefan Gehrmann made the first 40-minute flight with the Li-ion battery from Aachen/Merzbrück (EDKA), near AirEnergy's base, on June 30, 2005. Two flights followed, then Gehrmann took it on a holiday trip to Fayence, France, and made twenty-two flights over the Alps, the additional energy density enabling the e-sailplane to climb even higher and achieve three times the range. Gehrmann also demonstrated it to Alisport at Alzate Airport in Italy. A total of 38 hours of flying was sure proof of the Li-Po's efficiency.

Meanwhile, using funds from selling his *LF 20* motor glider, Axel Lange had gone ahead with his *Antares 20E*, a single-seat high-performance motor-glider designed from the start for electric propulsion. (Antares is the brightest star in the constellation Scorpio.) With a wingspan of 20 m (65 ft.) and a MTOW of 660 kg (1440 lb.), it first flew in 2003, and achieved full EASA certification in 2006. It is the world's first series production electric aircraft. Rather than using NiMh batteries, the *Antares 20E* used new and revolutionary Li-ion batteries from Saft-Batteries in France. These batteries, which were far ahead of their time, allow the *Antares 20E* to self-launch and climb up to 3,500 m (16,400 ft.) on a single charge. The same cells have since then been used in many aerospace applications, including the F35 and the A350. In the years that followed, Lange Aviation GmbH continued to produce and upgrade the *Antares 20E*, as well as its younger sibling, the *Antares 23E*, which has a wingspan of 23 m (75 ft.).

During the summer of 2005, Stefan Gehrmann of AirEnergy of Aachen used the Li-Po battery to extend the range of his electric assisted sailplane *Silent AE1*. He logged twenty-two flights over the French Alps, the additional energy density enabling the e-sailplane to climb even higher and achieve three times the range (Stefan Gehrmann).

Axel Lange in 2006 in the series-produced e-sailplane, the Antares 20E, named after the brightest star in the constellation Scorpio (courtesy Lange Aviation GmbH).

Proving that electric propulsion need not result in a performance disadvantage, Antares sailplanes have been used to perform numerous noteworthy soaring flights. On April 8, 2007, John Williams of the UK spent some 10.5 hours flying over the Scottish highlands. When he landed that evening, he had flown 1243 km (772 mi.) over terrain not usually associated with soaring. The flight resulted in four new national records. Williams has continued to make spectacular flights, and at the time of writing, he holds 16 UK national records. For a while, he also held the world record for fastest 1500 km out and return, with 180.3 km/h. This flight was flown over the Andes.

Another example of the transition from NiCads to Li was with the RC model helicopter. Although there had been one-offs in the late 1970s, it was in 1980 that Ishimasa Company of Japan made the first production electric model helicopter, the *Skylark EH-1*. It had an onboard 9.6 volt NiCad battery, but power could also be supplied via a silicone/silver cable to hook up to a 12-volt auto battery to power the Mabuchi 540 S motor. Although a technological success, because the *Skylark* had a limited flying time, a very basic control system, and a high price, it was not bought in any great numbers. A second model electric helicopter was developed in 1989 by Hiroyuki Oki of Kalt Sangyo Co. Ltd. in Gotenba, Japan. With energy from 20 NiCad cells and an AstroFlight 40 cobalt sport motor, the *Kalt 30 Baron Whisper* was an immediate success. Then there was Kyosho of Tokyo, who had already introduced their gas-engined *Concept 30* in 1988 then developed the *Concept EP*; a powerful AP36L with an 8.4V NiCad gave moderate performance for the tyro, while increases to a 9.6V NiCad battery gave it high speed and an aerobatic performance. However, the 540-sized brushed-motors were on the limit of current draw, often 20–25 amps on the more powerful motors, hence brush and commutator problems were common. With the arrival of the Li-Po battery, sales of RC model helicopters took off—such as Horizon Hobby's *Blade CX RTF* Electric Coaxial Micro Helicopter, which has sold by the thousands.

Another to adapt the Li-Po battery was airship designer and pilot Graham Dorrington. Desiring even greater autonomy, the Englishman planned a more streamlined envelope with twice the capacity, to be built by Cameron Balloons, while searching for more efficient engines and batteries. For the engine he found the radial-armature permanent magnet pancake motor developed by Cedric Lynch, while for batteries he began to examine a revolutionary chemical couple—rechargeable Thunder Sky lithium-ion phosphate batteries. Dorrington first flew *The White Diamond* in December 2003 inside Cardington Airship Hangar in Bedfordshire, England. The operation then shifted to Guyana, where in July 2004, flights were made over the forest canopy. A Nexa hydrogen fuel cell was also briefly tested although a 28V rechargeable lithium battery with a mass of 40kg and a nominal capacity of 200Ah was used to provide most of the propulsive power. The maximum measured discharge rate of this battery was about 165A. Additional power for the bow motors and aft gondola motor was provided by either one or two pairs of 12V, 26Ah sealed lead-acid batteries (wired in series to give 24V), each with a combined mass of about 21kg. Power control was achieved using off-the-shelf MOSFET controllers. With Dorrington was filmmaker Werner Herzog, who edited his footage into a documentary film called *The White Diamond*.

Arguably the first to adapt the advantage of lithium batteries to an aircraft which could fly under power for over an hour was Randall B. Fishman of Cliffside Park, New Jersey. Even at his elementary school in Lynbrook, Long Island, Fishman would assemble little gadgets. His 4th-grade science project was a working hydroelectric powerplant that used the water faucet pressure to turn the turbine wheel. Some of the kids good-naturedly nicknamed him "Doctor Gizmo."

During the early 1970s, Fishman was flying hang-gliders along the beach cliffs on the north shore of Long Island. During the 1990s, on nice days, he was using an electric bicycle to commute to work at a jeweler's shop in Fort Lee, New Jersey. His bike could use either just electric power, electric and pedaling, or just pedaling. In the late 1990s he bought a Trikke 8 sports scooter. Although he could get it to go fast on a smooth surface, it did not work very well on rough pavement. In the early 2000s Fishman took the motor from his electric bike and bought some lithium-ion laptop computer batteries off eBay: "I was able to open the plastic boxes and take out the cylindrical cells. I rearranged the cells from several laptop packs to provide the voltage and capacity I was looking for. I set the whole thing up on the Trikke 8 so now I had an electric scooter. Wow! That thing was really fast and would run for a solid half hour or more. I had a lot of fun with that and took one or two serious headers when I made less than the best decisions. This was my first experience with using lithium polymer batteries for transport. I was impressed by the amount of power these batteries could hold."[4]

His experience with the Li-Po batteries from the laptop computer battery packs gave Fishman confidence to wonder whether he might use the same technology on an ultralight aircraft. He wanted to take off quietly and without vibration, and also reduce his ground maintenance. He consulted Erwin Rodger, an old hang-glider friend, designer of the *Cloud Dancer* motor-glider. Erwin confirmed that the power available from the motor was sufficient and the Li-Po battery pack Fishman was proposing had enough capacity and discharge rate to do the job. The 53-inch folding propeller was correct to optimize the thrust for the motor power available and the speed range of the aircraft. After running the formulas, it was clear that the trike could take off and fly fully loaded with batteries and Fishman's 200 lb. (90 kg) weight. This was in the summer of 2006. During the fall and winter of 2006–7 Fishman built his prototype in Florida, first in a boatyard, then a hang-gliding resort, and finally in a rented apartment: "I vacuumed up the aluminum drillings, filings, etc. every day from the carpet. I assembled the battery pack there as well. I finished up the trike and did a static test run with the front wheel up against the apartment building and the prop blast angled out into the field behind the place. At full power the motor was getting a little warm but the batteries were almost completely cool. Cool!"

In April 2007, Fishman took his prototype, which he had called the *ElectraFlyer*, to Sun-N-Fun, an event run by the Lakeland, Florida, chapter of the Experimental Aircraft Association (EAA). It was a static exhibit.

> People were very interested. The motor system had run but it had not yet flown except for a little hop at the South Lakeland airport a few days before. Tom Peghiny of Flight Design USA came over to check out the trike. A couple of years later, Tom built an electric version of his old flight design aircraft in a deal with Yuneec. I also met Mike Theeke from Fly Hard Trikes of Wildwood, Georgia. He invited me to bring the trike to his location in Jasper, Tennessee. He loaned me a North Wing Stratus high-performance wing and from May 5, I made some test flights there. First I flew straight down the runway at about 10 feet off the surface. The *ElectraFlyer* flew as expected so the next time I made a circuit. Mike took some video of the first circuit flight and the short, 1 minute video clip on the website is from that flight. I was really excited (and probably a little bit lucky) that everything worked so well! The next day I got a little fancier and made longer flights and more maneuvers. There was a gentleman there with his son. He took a lot of photos that day. I was full of enthusiasm since the flights went well the evening before. He must have taken about 50 pics of me flying around the airport. I was doing 360s and the normal maneuvering I used to do in my gasoline powered trike. On the third day I took off and flew down the valley. I saw Mike take off in his two-seater gas powered trike with a student. I flew for over an hour at cruise speed of 25 to 30 mph (40 to 50 kph). Words won't describe the feeling I had! Mike landed with the student, then I came in. He was amazed at the length of time I had stayed aloft. According to my analog volt meter I still had quite a bit of capacity left in the battery packs!

Having wowed the ultralight world at the AirVenture Show in Oshkosh in 2007, and winning the Ultralight Grand Champion award there, Fishman traveled back to New York, where he enjoyed silent flying around his old haunts in the Ellenville, Ulster County area. Searching for the right batteries took time, involving Fishman's obtaining sample cells from several manufacturers. He would try five other suppliers over three years until he finally found cells that had the build quality for this demanding use, staying with the same battery cell supplier from then on.

Next came his *ElectraFlyer-C*. The basis for this was a gas-powered Monnett Moni motor-glider that he had built with a friend in the '90s:

> It was a nice little airplane but the gas engine with small prop was not a good match. When I converted it to electric, the power was reduced from the 25HP gas engine to the 18HP electric motor but I made some mods on the aircraft to raise it high enough to install a much larger prop turning at a much slower rpm. The static thrust increased 60% and this also changed the airplane from loud and vibrating to a smooth quiet flyer. For this project, I rented a room in the back of a machine shop in New Jersey. Much of the work was done there to convert the aircraft in the fall of 2007. I moved down to Sebring, Florida, and rented a hangar at the airport. I also rented a house nearby and built the 2 battery packs for it on the kitchen table there. Final assembly was completed just before the 2008 Sebring Light Sport Aircraft Show. I took a booth and showed the new C model and the completed trike there. I

Randall B. Fishman of Cliffside Park, New Jersey, was the first American to adapt the advantage of lithium batteries to an airplane which could fly under power for over an hour. Here he is in May 2007, flying his *ElectraFlyer* trike over Jasper, Tennessee (Randall B. Fishman).

was flying the trike at the show but the C was not ready for flight yet. I moved over to the south Lakeland airport and the DAR there issued the airworthiness certificate under the experimental category.

In May 2008, the aluminum *ElectraFlyer Model C* flew for the first time at the Ellenville airport with Joe Bennis at the controls. Then it was demonstrated before the crowds on August 2 at AirVenture in 2008. Fishman and his Electric Aircraft Corporation successfully applied for the Lindbergh Aviation Award and received $10,800, which is how much *The Spirit of St. Louis*, Lindbergh's airplane, cost in 1927. It was their only aviation award given that year. In 2008 Fishman won the August Raspet Memorial Award for his electric aircraft work. The award recognizes the "person who has made an outstanding contribution to the advancement of light aircraft design" each year. Fishman now used the Lindbergh grant to develop the two-seater, all-composite *ElectraFlyer-X*, powered by a new 40 hp (30 kW) brushless electric motor, to eventually qualify for light-sport aircraft status. Although *ElectraFlyer-X* was scheduled to fly in the summer of 2010, the FAA objected to its operating under the light-sport rule. He therefore sold the project to an interested builder, then turned his inventive mind to a project he had started in 2007–8 with a Slovakian aeronautical engineer, to build a ready-to-fly electric 245-lb. ULS for sale to the public under the FAA Part 103 ultralight rule. With a 20 hp direct-drive motor and 53-inch propeller, the carbon fiber/foam airplane is available with a fixed or folding prop. Fishman would eventually log 140 flying hours on his trike, and over 350 flying hours on his ULS, securing his position as one of the true pioneers of electric flight.

Other aviators would also make use of the lithium advantage.

By eventually logging up 140 flying hours on his trike, and over 300 flying hours on his ULS, Randall Fishman secured his position as one of the true pioneers of electric aircraft (courtesy James Lawrence).

The Pipistrel powered hang-glider had been built during the late 1980s by Ivo Boscarol of Ajdovščina, who then set up the first private aircraft company in Yugoslavia. Boscarol was born on April 15, 1956, in Postojna, Slovenia, at the time part of Yugoslavia. His father August Boscarol, a machine engineer, spent several young years as a test pilot at Aermacchi, an Italian aircraft manufacturer. The family lived in Ajdovščina, a town in western Slovenia near a small military airfield. After elementary and high school in Ajdovščina, Boscarol studied economics at the University of Ljubljana. From 1976 to 1986, Ivo was involved in publishing, owning a photography studio, part of the ŠOLT group, holding expositions in several countries, cooperating with several magazines. During this time he was the official photographer at the Šentjakobsko Gledališče Theatre in Ljubljana, an advertising manager of the student radio station in Ljubljana, and a manager of several musicians and rock bands. He also organized several art photo exhibitions, including nudes.

Due to legal restrictions imposed by the Yugoslavian government during the 1980s on flying alternative and ultralight aircraft, Boscarol and his friends had to fly their first hang-gliders secretly in the evening, between dusk and dark and using lights at the front of the aircraft. Since powered hang-gliders had triangular shaped wings, the local people started to call them "bats"—and "Pipistrellus" is a Latin word for a bat. Boscarol decided to name his company Pipistrel, and during the first 10 years they produced over 500 hang-gliders and exported them all over Europe. In 1991 he also organized the first national championship for ultralight aircraft.

In the mid–'90s the market started to change and new composite materials appeared, resulting in the 503 Rotax engined Pipistrel Sinus ultralight motor-glider, of which some 600 were built and sold. Sinus is still one of Pipistrel's best-sellers. In 1999 the closed-cabin Virus was added, having shorter wings and built for higher speeds, not so much for gliding.

Then the Taurus self-launched microlight glider made its first flight in 2003, again Rotax-engined. As the world's first microlight class two-seater with parallel seats, the Taurus was much wider than conventional two-seat gliders; with its wider nose and its wingtips turned up like horns, it seemed appropriate to call the glider Taurus (Latin for bull). In 2007 Boscarol and his team decided to electrify the Taurus, installing a 30 kw Sineton motor with Li-Po batteries. On December 21, 2007, Ivo Boscarol took off in the modified airplane, silently climbed to 1,000 m (3,250 ft), and flew for 20 minutes at speeds of up to 150 kph (90 mph). They were delighted that their electric bat behaved exactly as predicted and that they could now go into serial production. On their website, they posted: "First flights of the Taurus Electro, the first two (2) seat self launching glider with electric engine in the World are behind us. We are happy to announce that the bird behaves exactly as predicted and that we will publish additional information shortly. For now, let the pictures do the talking! Merry ELECTRO Christmas and an ELECTRO New Year everybody!"

In 2008, *Popular Science* magazine named the Taurus Electro one of the ten best innovations of the year. In 2010 the Pipistrel team won both a gold medal at the Biennial of Industrial Design and the European Business Award for Innovation from among 15,000 competitors.

One of the team was Tine Tomažič. He learned to fly at age 15, became an instructor at age 19, and in 2001 became test pilot at Pipistrel. A deep believer in electric flight, Tomažič dedicated his Ph.D. thesis at the Faculty of Electrical Engineering, University

On December 23, 2007, Ivo Boscarol took off in the electrified Pipistrel Taurus airplane, silently climbed to 1,000 m, and flew for 20 minutes at speeds up to 150 km/h (Pipistrel).

of Ljubljana, Slovenia, to "the development of control strategies and systems to optimize flying with hybrid-powered aeroplanes," researching technologies to identify the future of electric flight, including distributed electric propulsion architectures, power train modeling, and flight mission optimization. One of his first projects was to electrify the Taurus G2 as a two-seat side-by-side self-launching sailplane. Its motor peaks at 40 kW for takeoff and allows continuous climbing at 30 kW power. It is controlled by a specially developed power inverter/controller and governed by the cockpit ESYS-MAN instrument. All components are networked via CAN-bus, feature proprietary multilayer protection logic, and produce a true throttle-by-wire experience. Alongside this Pipistrel had developed their Solar Trailer, which offers 1 kW of usable energy; this means charging time of 10 hours for 30-Ah batteries and 12 hours for 40-Ah batteries. Powered endurance was 17 minutes, intending to allow for self-launching to an altitude of 2000 m (6500 ft), after which the engine is retracted and the aircraft then soars as a sailplane.

In 2011, for the NASA Challenge, Pipistrel now linked two G2 hulls into a four-seater one-off, the Taurus Electro G4. All components of the aircraft were developed and made by Pipistrel. Aerodynamic studies were carried out by to Dr. Gregor Veble, head of research at Pipistrel. The construction of the aircraft was accomplished by Vid Plevnik in cooperation with structural specialist Rado Kikelj. The development of composite technology parts and systems was done by Sašo Kolar and Franci Popit with their teams. The most challenging parts, namely the development of electronic systems and their regulation as well as the system for charging the batteries, were developed by Jure Tomažič. With a 75-ft. (22.9 m) wing, the G4 had a gross takeoff weight of 3,300 lb. (1,497 kg),

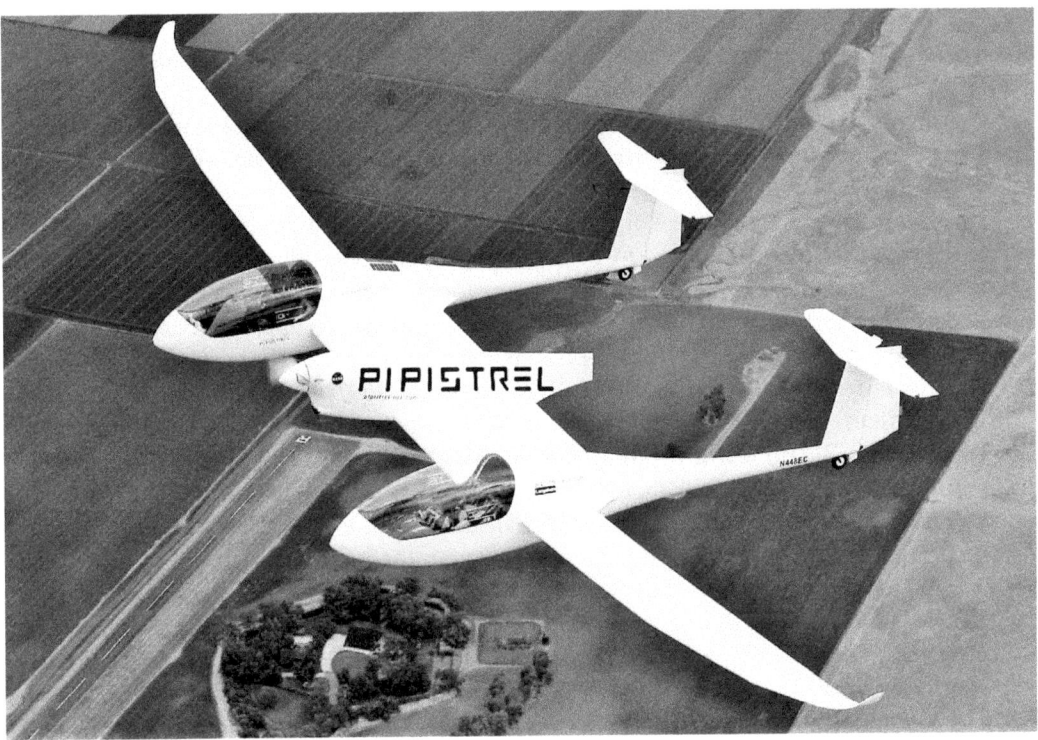

In 2011, Pipistrel developed a unique four-seater, the *Taurus Electro G4*. It made its first flight at 7 a.m. on August 12, at AirVenture, Oshkosh, Wisconsin, winning the $1.35 million CAFE Green Flight Challenge prize. It would later be used as a testbed for fuel cell propulsion (Pipistrel).

making it the heaviest manned electric-powered aircraft built to date. It had a glide ratio of over 30:1 at 100 mph (161 km/h). The 145 kW (54 hp) engine powered the aircraft from internally-mounted Li-Po batteries, weighing 1,100 lb. (499 kg) for a 17-minute climb.

In April 2011, the Taurus G2 won the 2011 Lindbergh Electric Aircraft Prize for "best electric aircraft" at the aero show held in Friedrichshafen, Germany. It was first presented to the American public at the Oshkosh AirVenture in July. According to the informal vote by the visitors, it was amongst the 10 most attractive aircraft on display among 15,000 exhibited airplanes; each year, the EAA awarded a virtual "Dead Grass Award" prize (for the most viewed exhibit) to the ten aircraft with most votes.

Taurus G4's first flight was at 7 a.m. on August 12, 2011, at AirVenture, Oshkosh, Wisconsin. NASA had put up for the CAFE Green Flight Challenge (200 miles at 100 mph) the largest aviation prize of all time, $1.35 million, which was also won by the Taurus G4, with the Pipistrel-usa.com team led by Jacob W. Langelaan, covering 403.5 passenger miles per gallon (649.4 km) gasoline equivalent. During the challenge, at the Santa Rosa airport, there was no power supply to charge the G4 and the e-Genius overnight, so Google sponsored a basic charging point.

Because of all these achievements, the president of the Republic of Slovenia, Dr. Danilo Türk, decorated the Pipistrel team in 2011 with the highest civil award in the country, the Golden Order, for services to the Republic of Slovenia! The Design Museum in London chose the Taurus G4 as one of the nominees in the transport category for the "Designs of the Year 2012" exhibition. Taurus G4 was nominated for the prestigious Colli

Trophy, the "Greatest Award in Aviation." Because of all the achievements of this aircraft, a scale model of Pipistrel's Taurus G4 is displayed hanging in the main hall of ICAO's headquarters building in Montreal.

By an extraordinary coincidence, the first flight of a French electric airplane had also taken place on Christmas Eve 2007. In the Hautes Alpes, one start-up company had also realized the potential of lithium-ion polymer batteries. It was set up by Anne Lavrand, a business executive and keen amateur pilot. Beginning with an SME manufacturer of ULMs, Lavrand heard about the Li-Po batteries produced by Kokam. In 2007, she applied for a grant to modify a wooden ultralight *Souricette*, a Michael Barry design, and retrofitted it with an 18-kW (24 hp) disk-brushed Lynch electric engine, a golf cart–type controller, and a 47 kg (104 lb.) Kokam Li-Po battery power pack. There would be no solar panels. *Electra* made its first test flight on December 23, 2007, at Aspres sur Buëch airfield, Hautes Alpes. Test pilot Christian Vandamme flew the strut-equipped aircraft (officially registered as BL1E Electra F-PMDJ) for 48 minutes, covering 50 km (31 mi) cruising around the southern Alps at about 90 kph (55 mph). It was the first time since 1887 that a French battery-electric flying machine had taken to the skies. The French team was unaware that the Slovenian team had also flown on the same day.

Encouraged, Ann Lavrand created a start-up she called Electravia, based at Sisteron airfield, in France. Further developing their prototype, by 2008 they had developed the *Electravia Electro Trike*, a single-seater delta trike with an electric propulsion system from Electravia. With a 26 hp GMPE 102 electric motor supplied by a 3 kWh pack of Li-Po batteries, the *ElectroTrike* could fly for an hour. One ultralight sailplane which benefited from the Electravia motor was the *Alatus-ME*, built by Aerola of Kiev in Ukraine. The electric motor gave the *Alatus* a half-hour endurance. Another spin-off from Electravia's initiative was the *E-Fenix*, the first electric two-seater paratrike in the world. Developed by Planète Sports & Loisirs, headquartered on Re Island, off the coast of La Rochelle, France, the trike would carry visitors on discovery flights over the scenic island. Equipped with a 35-horsepower GMPE 104 motor, an E-Props QD2 four-blade propeller, and a 6 kWh Kokam Li-Po battery, the *E-Fenix* made its maiden flight on May 12, 2011, with Michaël Morin at the controls, circling around Electravia's Sisteron airfield. With its Bulldog wing, the *E-Fenix* can fly for 35 minutes with two people on board.

With no further funds for his electric airships, Graham Dorrington turned to electric airplanes with Project Orion. A group of ten master of engineering 4th-year students at Queen Mary and Westfield College, University of London, were electrifying a 335 kg (739 lb.) "Optimist" glider. To do this, they added eight under-wing pylons attached by moveable straps that carried 16 Graupner/Ripmax Li-Po battery-powered wireless link activated electric motors that drove sixteen 280mm (11-in.) diameter propellers. Such a configuration is known as displaced electric propulsion. On September 20, 2008, Derek Piggot piloted the *EA9* for its maiden test flight at RAF Tibenham Airfield, Norfolk, England. It was towed to 2,500 ft (760m) and then released; it made two further flights. This was the first time that a manned e-airplane had taken to English skies. Further development work was halted through lack of sponsorship, and the British Isles are still waiting for a second electric aircraft to take to the skies.

Since the 1980s, ICARO, located in Sangiano, North Italy, had been making Atos and Swift hang-gliders, and by 2008 sales totaled almost 9,000. This is largely due to the innovative Manfred Ruhmer, holder of the absolute hang glider distance world record at 701 km and 4 times rigid world champion with a Swift. Ruhmer decided to equip the

September 20, 2008: Derek Piggot pilots the *EA9*, designed by a University of London Queen Mary and Westfield College team led by Graham Dorrington for its maiden test flight at RAF Tibenham airfield, Norfolk, England (courtesy Graham Dorrington).

rigid-wing Swift with an electric assist. In April 2008, a 10kw Werner Eck-design paramotor weighing just 3.7 kg, powered by a small Li-ion phosphate battery, enabled Ruhmer to reach a height of 600 meters. Building a slightly larger battery enabled him to fast-charge in 15 minutes between 20-minute flights. In 2011, series production of the Electric Swift, with its 1.4 m folding-blade carbon fiber propeller, began with both new eSwifts while existing ones were retrofitted. Using battery swap systems, to date almost 40 Electric Swifts are in use. Ruhmer also experimented with an electric trike fitted with two rear-facing Eck motors.

During this time, Sweden had been developing its own electric airplane. In 2001, the Swedish Defense Administration asked Saab, who had been building aircraft since 1937, to produce a demonstrator for practical education in More Electric Aircraft technology. The project, led by Lars Austrin, was to buy and modify a flying platform, the 12m wingspan Windex 1200C motor-glider designed by Sven Olof Ridder and built by Windex Air AB in Åtvidaberg. The task was to retrofit the glider with a 20 kW, 11 kg custom-designed Stridsberg PM electric motor in the tail plane. The plane modifications were by Kåre Ljung at Windex, including a custom carbon fiber propeller. SAFT provided 44 lithium-ion MR39 cells weighing 1.1 kg each, with 22 fitted into each wing. The total electric power was 6kWh @ 176VDC. The central electric distribution system was designed by Jonas Larsson and Lars Austrin at Saab. In March 2005 the Saab *MERA 01* was delivered to Saab, and in 2008 the airplane was ready for taxiing trials at Saab/Linköping airport.

Following trials being towed into the air, it made its first powered flight of 45 minutes in June 2009, piloted by Fredrik Müchler, a Saab Gripen fighter test pilot, reaching a speed of 150 kph (93 mph) and an altitude of 1,100 m (3,600 ft). Throughout its short career, the *MERA 01* clocked up a total of 5 hours. By 2012, when it was put on static display at the 2012 Malmen Airshow, Sweden, its flight career was over. During the project

Top: In 2009, the Swedish team stand proudly behind their e-airplane; among them, their positions unidentified, are Jonas Larssen, Lars Austrin, Sven Olof Ridder, Kåre Ljung and pilot Fredrick Müchler (Saab). *Bottom:* The Saab 20 kW, 11 kg custom-designed Stridsberg PM electric motor derived its energy from 44 lithium-ion SAFT MR39 cells weighing 1.1 kg each with 22 fitted into each wing. Its custom carbon fiber propeller was made at Air AB in Åtvidaberg (Saab).

a study of electric actuators for aileron control was initiated. This was continued with a demonstrator program including a prototype actuator based on Stridsberg's PM-motor and a test rig. As a spinoff from these programs, a business concept was proposed for the Boeing 787, the "Dreamliner." As a result, the high lift system in the Dreamliner is based on the SAAB electric actuators.[5]

For Brazilians, the Father of Aviation is not the Wright Brothers, but Brazilian-born Alberto Santos-Dumont. In 1906, Santos took off in his 14-bis and made the first officially observed flight of more than 25 meters (27 yd), certified by the Aéro-Club de France, and won the Deutsch-Archdeacon Prize. Remaining in Paris, he continued to develop his airplanes, which he called Demoiselle. A century later, it seemed natural for the Brazilians to produce an electric airplane. The *SORA-e* (Portuguese for "sister") is a two-seater developed and built by Alexandre Zaramella and a team at ACS (Advanced Composites Solutions) from an original design created by Professor Claudio Barros of Universidade Federal de Minas Gerais (UFMG). With support from the state financier of studies and projects (Finep) in 2010, an electrical system was developed by the Electric Vehicle Research & Development Center (CPDM-VE) at Itaipu; two Emrax motors of 35 kW each, produced by the Slovenian company Enstroj, were powered by six sets of Li-Po polymer batteries, which together provide 400 volts. Following taxiing trials from ACS's runway, the *SORA-e* made its first public flight on May 18, 2015, at Professor Urbano Ernesto Stumpf Airport, São José dos Campos, in the interior of the state of São Paulo. Its rate of climb was 1,500 feet (460 m) per minute. Its flight range was from 90 minutes to 190 kph (120 mph) and it was designed for a top speed of 340 kph (200 mph). In April, the airplane went to Iguaçu Falls, in the state of Paraná, where it was put through a second and final series of tests, performed by ACS-Aviation. The next challenge will be to develop a commercial version of the *SORA-e*.

Canada's first gas airplane, the *Silver Dart* (or *Aerodrome #4*) had been flown off the ice of Baddeck Bay, Nova Scotia, in February 1909. Now, one hundred years later, on December 8, 2012, Canada's first electric airplane, *Green 1*, took to the skies above the Pitt Meadows Airport in British Columbia. Behind the project were Randy Rauk and John McClintock of eUP Aviation in Lumby, British Columbia. Rauk, of Freedom Flight School, Inc., piloted the glider trike, with its complete Electravia electric solution: a 19 kW (25 hp) electric motor and a 3.9 kWh lithium polymer battery. Bench and ground testing had been completed in the Vancouver and Okanagan areas during the previous months. In 2013 *Green 1* was flown across western Canada, then up and down the Pacific Coast, demonstrating its self-launching and soaring capabilities over California deserts, at Torrey Pines, and in the mountains of Oregon.

Equally the Czech Republic developed an e-airplane using the SportStar EPOS. Financed by the Technology Agency of the Czech Republic under the guidance of the company Evektor, based in Kunovice in the Czech Republic. Evektor is one of the biggest design companies in Central Europe, with activities in aerospace and automotive industries. In 2012, Evektor developed its *SportStar EPOS* ("Electric Powered Small Aircraft") powered by a Rotex Electric RE X90–7 50-kW electric motor, which directly drives the three-blade composite propeller manufactured by VZLÚ Prague. Engine performance was controlled by an electronic control unit, developed by MGM COMPRO, manufacturer of the motor control unit, and the Faculty of Information Technology of Brno University of Technology, supplier of the display unit for motor parameters. The system was divided into "airborne" and "ground" sections to reach the lowest possible weight of its

Kokam 90S, 330-volt, 40Ah battery. The first test flight, in cooperation with the Czech Light Aircraft Association, was made by factory pilot Radek Surý on March 28, 2013, from Kunovice Airport. After taxiing tests at 8:25 a.m., the aircraft took off for its first test flight. After a short, approximately ten-minute flight, it landed successfully, and a second flight, lasting 30 minutes, followed immediately afterwards. The Czech *SportStar EPOS* was presented to the public for the first time at Europe's largest aviation exhibition, Aero Friedrichshafen in April 2013. By 2014, power output had been stepped up to 75 kW, more than the most powerful Rotex 914.

In July 2010, with the success of his Antares e-motor-gliders, Axel Lange was honored with the prestigious Lindbergh Award, presented by Lindbergh's grandson, Erik Lindbergh. In 2011 Lange and the *Antares 20E* won the Berblinger competition for electric aircraft, which was held in Ulm, Germany, and in 2017, Lange received the OSTIV award for his pioneering work on electric propulsion in sailplanes.

Lange Aviation's *Antares 20E* and *23E* propulsion system was also adapted by the Holighaus family of the Schempp-Hirth sailplane firm based in Kirchheim-unter-Teck, to produce their *Arcus-E* self-launch two-place glider. After Lange had installed the propulsion, following a first flight in 2010, the *Arcus-E* entered series production shortly afterwards, with several now delivered. Kirchheim-unter-Teck, it may be recalled, was the home of Fred Militky, who had pioneered electric model airplanes half a century before.

What if powered sailplane pilots could use an electric motor installed just behind the nose propeller to launch themselves into the sky? Once the plane was airborne, the propeller blades would fold back flush with the nacelle. In 2008, Luka Znidarsic and his father Matija Znidarsic, both experienced sailplane pilots and mechanical engineers in Logatec, Slovenia, developed their FES (Front Electric Sustainer). This a power system that provides a climb of 1.5 meters (5 ft.) per second (198 feet per minute) for their prototype configuration on an 18-meter (60-ft.) *LAK 17* sailplane. The Znidarsics developed their own brushless DC synchronous permanent magnet electric motor and controller for the application, which is light, small and mounted on top of the main well box. The motor is 7.3 kg (16.1 pounds), and puts out 22 kW (30 horsepower) at 116 volts. Total weight for the power package is 45–50 kg (99–110 pounds), with Kokam Li-Po batteries in two packs (each behind the rear spar, balancing the weight of the motor and propeller in the nose). If batteries are discharged during flight using the motor, they are easily accessible from the top of the fuselage so that you can take them out for charging. To date more than 70 sailplanes of 7 different types are already equipped with the FES propulsion system. The maximum endurance in powered flight is almost 1 hour or 100 km (60 mi).[6]

In 2010, another American e-airplane took to the skies. The Sonex Waiex could trace its origins back to 1994, when John Monnett, a builder of aircraft kits based in Oshkosh, Wisconsin, teamed up with Pete Buck, who had spent two semesters of his engineering degree analyzing and building the battery/power system for a hybrid electric vehicle (HEV) sponsored by Ford Motor Company. Together they came up with a feasibility study for a project dubbed *Flash Flight* for a small, electrically powered, and manned aircraft that would be capable of a short duration flight in order to set or establish speed records for this new class of aircraft. Built of many off-the-shelf components at relatively little risk, it would only have a 10-minute autonomy. Sonex Aircraft was founded in 1997 with John Monnett building a succession of Moni sailplanes and gasoline-motor gliders, and the electric was put on hold. In 2007 Monnett and Buck got round to building the *e-Flight Waiex*, complete with a 54 kW brushless DC-Cobalt motor. The motor was designed be very lightweight, at only 50 lb. (23 kg), to operate at 90 percent efficiency, and to use a 14.5kw-hr Li-Po battery system. It was displayed at AirVenture in 2009 and made its maiden flight at Wittman Field, Oshkosh, on December 3, 2010. With John Monnett as pilot, the yellow N270DC made a short hop on runway 27, intended to be a conservative non-pattern flight to break ground-effect and analyze in-flight system performance as the next step in testing. On March 30, 2011, the *e-Flight Waiex* made its second flight, achieving a full circuit of the Wittman Airport traffic pattern, as planned. A third flight was also conducted on April 13, 2011, in which a full pattern was again flown. The electrical system would be 270 volts and 200 amps, and adjustable to different power outputs. At the time of writing, Sonex, like others, is waiting for the increase in battery energy density.

In 2010, Calin Gologan, a Bucharest-trained aeronautical engineer formerly at Aerostar Bacau, and his team at PC-Aero of Nesselwang, Bavaria, Germany, teamed up with composites expert Carbon Wacker GmbH of Hurlach, to develop a 660-lb. advanced glass/carbon-fiber composite single-seater aircraft for electric propulsion. Powered by a 16 kW motor and ultra-lightweight Kreisel battery units, the *Elektra One* sailplane, complete with retractable main wheel, made its first flight on March 21, 2011; it won the Lindbergh electric aircraft prize presented at the EAA AirVenture airshow in July 2011. With help from Solar Hangar and Solar World, equipped with molded solar panels, PC-Aero's *Elektra One Solar* motor-glider was projected to have an endurance of almost four hours/400 km with a top speed of 100 mph (160 kph). PC-Aero was planning a whole line of aircraft including a version of the *Elektra One* with longer wings and built-in solar panels and an aerobatic version with double *Elektra One*'s power and airframe strength. Two- and four-seaters are envisaged. On June 25, 2015, *Elektra One Solar* took off from Unterwössen (Germany) for the Alps, crossed over the Grossglockner, and landed in the sunny town of Lienz in East Tyrol (Austria). The flight took around 2.5 hours. After the successful flight on the south side of the Alps, *Elektra One Solar* started on the way back on July 2 (a few days before the *e-Genius* of the University of Stuttgart also flew over the Alps) in quite difficult weather conditions. Despite headwinds and strong gusts, the plane crossed the Alps at an altitude of more than 3,000 m (9,800 ft) and landed after a flight of about 2 hours and 190 kilometers (120 mi), as planned at the airfield in Zell am See, Austria. It used only 18 kilowatt-hours of electricity for the complete trip.

In 2011, researchers at Nottingham University, England, published a paper concerning the feasibility of equipping the main landing gears with electric motors for the aircraft traction during the taxi phase. Those electromechanical wheel actuators make possible a "Green Taxi" operation by considerably reducing the on-ground carbon emission. Moreover, this would enable important fuel saving for short-distance flights with high frequency of landing and takeoff. In this work, a direct-drive wheel actuator was considered for energy efficiency and mechanical reliability. Two possible locations of the actuator were examined and the weights of the corresponding electric machines compared. The most weight efficient location was then selected. A high torque density permanent magnet machine was then designed to fit in this envelope to satisfy peak torque, weight, and flux weakening capability requirements.

The *Cri-Cri* (English: the chirp-chirp sound made by a cricket) is the smallest twin-engined manned aircraft in the world, designed in the early 1970s by French aeronautical engineer Michel Colomban. In 2004, Didier Esteyne, designer and amateur pilot of Aero Composites Saintonge in the Charente Region of France, had thought about an electric *Cri-Cri*. The Kokam Li-Po batteries and 4 electrical model engines weighing about 7.5 kW could come from aeromodeling. But with the batteries' total weight of 40kg (88 lb.), Esteyne calculated that the airplane could only stay airborne for around 13 minutes with five as aerobatic! So he decided to wait. But Esteyne was not one used to giving up. In 1979, as the son of a pilot in the French Air Force, Esteyne had been training in fine arts, including technical drawing, in Bordeaux. Following his first flight, at age 25 he caught the flying bug and obtained his pilot's license the following year. From then on he combined his passions, designing, building and flying his first aircraft three years later. He went on to design and fly a succession of eight aircraft in France, Mexico and the USA, improving each aircraft based on his own experience as a pilot.

Five years later, Didier Esteyne felt more positive about the weight and energy of batteries. During the Le Bourget Air Show in June 2009, with his friend Gérard Feldzer, the Director of the Musée de l'Air et de l'Espace, Esteyne decided to create an association "GREEN CRICRI" and try to find money for this exciting idea. With Gérard Feldzer's contacts, they met staff from "EADS (European aerospace company) Innovation Work" in Paris and convinced them to support them technically and financially. A classic 2-stroke *Cri-Cri* was modified by Esteyne and his 9-strong team at Saint-Sulpice de Royan, including Marc Faure and Alain Bugeau.[7] Four Czech Republic-built Free-Air electric engines each developing 5.5 kW each weighing just 2.2kg (5 lb.) derived their energy from 26.5 kg (58 lb.) of Kokam, Li-Po batteries. The *Green Cri-Cri* had a flight potential of 30 minutes at 86kn (160 kph) or 10 minutes of aerobatics at speeds up to 120 kt (222 km/h) with a climb rate of 1,000 feet (300 m) per minute. As Jean Botti, EADS's chief technical officer, explained: "The Cri-Cri is a low-cost test bed for system integration of electrical technologies in support of projects like our hybrid propulsion concept for helicopters."

In September 2013, electric aviation had its first fatality. At the practically disused airport of Marville in Northern France, two recently fitted Electravia electrical motors in a Colombian MC-15E *Cri-Cri* were being given a high-speed taxi test when the aircraft became airborne unintentionally, and during the landing, it bounced several times. The aircraft left the runway, flipped over and caught fire, killing its 70-year-old Belgian amateur builder and pilot, Toon Jacobs. He had been the fourth to own an electric *Cri-Cri*, which he had recently shown at the Federation RSA (Network of Sport Air) Rally in Vichy, France.

In 2014, Chip W. Erwin of Aeromarine LSA, based in South Lakeland Airport, Florida, equipped a 220 kg (485 lb.) Zigolo MG12 ultralight kit-built motor-glider built by Aviad in Italy with an electric unit and flew it over to the Sun'n Fun Fly-In Expo. Through his PSA or Personal Sport Aircraft, Erwin's mission is to bring an affordable airplane to the market.

Now it was the turn of the amphibian aircraft, able to use both land and water for takeoff and landing, to be re-engined electrically. Amphibians are not new. In 1920 the British Air Ministry organized a Commercial Amphibian Competition at Martlesham and Felixtowe; entrants must take off from the land and touch down on the water, then vice versa. Four prototypes took part and so kick-started a category which during the 1920s and 1930s would challenge aircraft builders from France, Germany, Canada and the USA with such amphibians as the Keystone-Loening K-85 Air Yacht, an eight-seat biplane amphibian (1930).

Back in 1977, Dale Kramer, an aircraft engineering student at the University of Toronto, got his inspiration to design the 5.5 hp 100 cc Pioneer chainsaw-engined *Lazair*, a majestically slow ultralight aircraft with an average speed of only about 70 kph. The plane was sold in kit form, and between 1979 and 1984 more than 2,000 were built, making it the most produced Canadian-designed aircraft. In 2010, Kramer decided to "electrify" his plane, using twin Bevirt JM1 Jobymotors with Jeti SPIN Pro 300 controllers and dual 16-cell 4 amp-hour battery packs that produced 63 volts, mounted in the wings. He moved to Hammondsport and bought the estate of Glenn Curtiss, the father of naval aviation. In July 2011, Kramer took off from a field behind his house and made the first-ever landing on water with an electric-powered amphibian *Lazair* after a 7-minute maiden flight. Past the historic flight, after one hour of recharging, he took off again, this time from water. The aircraft won Antique Ultralight Champion and Best Ultralight Amphibian at EAA AirVenture, Oshkosh. The aircraft has remained an experimental project with no production planned.[8]

Since 2013, E. Brian Robinson and his team at Lindsay, in the Kawartha Lakes region of southeastern Ontario, have been developing the 6 PAX *Horizon X2* to operate from land, water, ice, or snow. The company envisions a hybrid with 700 hp powered from two advanced axial-flux electric motors, coupled with a range-extender high-efficiency gas engine, and designed around system redundancy. Traveling at speeds up to 200 mph (320 kph), the *X2* will be able to carry 2,000 lbs. (900 kg) of useful load over 1,000 NM (1850 km) at altitudes up to 18,000 feet (5000 m).

It was perhaps inevitable that prolific aircraft innovator Burt Rutan of Coeur d'Alene, Idaho, would direct his innovative genius to electric propulsion. In 1986 Rutan had designed the record-breaking Model 76 *Voyager*, which was the first plane to fly around the world without stopping or refueling, and the sub-orbital space plane *SpaceShipOne*, which won the Ansari X-Prize in 2004 for becoming the first privately funded spacecraft to enter the realm of space twice within a two-week period. During 2014, although technically retired, working as an individual out of his one-car lakeside garage, 73-year-old Rutan conceived a new two-seater seaplane which would fly up to 2,100 NM (3,890km) nonstop, survive a 10g impact on rolling seas, and fit inside his garage. The *SkiGull* can cruise at 200 mph (320 kph), taking off or landing in about 400 feet (120 m) on challenging surfaces including rough terrain, seas, grass, snow, or ordinary runways, fueled by ordinary automotive or marine gasoline, and having small auxiliary electric motors for 30 percent power assists or emergency landing. It can also can fly in cruise mode for about 7 NM using electric power alone. In water, the electric power system can also be used to maneuver the aircraft to the dock. It is designed to resist corrosion, even in salt water, and will have a docking system using two 12 hp electric motors driving a folding propeller. Its autonomy will enable it to travel from California to Hawaii, or from Boston to Iceland, without having to pick up fuel along the way. Rutan has said the *SkiGull* will be his last design and, having spent more of his life developing new airplanes than flying them, he wants to fly this one around the world with his wife Tonya. In October 2015, Rutan's pilot Glenn Smith made the *SkiGull*'s first flight test at Hayden Lake north of Coeur d'Alene; on November 24, further flights were made from the Coeur d'Alene Airport.[9]

The 2014 CAFE Electric Aircraft Symposium saw an incredible meeting of experts and thought-leaders in the areas of electric and sustainable aviation. Dr. Qichao Hu of MIT presented a multi-fold advance in energy storage density, Dr. Ajay Misra of NASA presented super-magnets made by nanotechnology, and Dr. Gecheng Zha of Miami

University presented high-efficiency forward-swept ePropellers. Brien Seeley wrote the seminal AIAA paper in 2015 on Regional Sky Transit.

In 2015, after Seeley's 34 years as president of the renowned CAFE Foundation, including the setting up of the Personal Aircraft Design Academy (PADA), an annual gathering of prominent aeronautical designers at Oshkosh AirVenture, this visionary left to found the nonprofit Sustainable Aviation Foundation (SAF). Its aim is to advance technologies and innovations pertinent to environmentally friendly, electrically-powered aircraft and to help bring forth their implementation into safe, quiet, useful aircraft that can benefit the public, the environment and the transportation system. In May 2016 the SAF Symposium, a global group of renowned presenters focused on the future of quiet, electrically powered aircraft, was held in San Francisco. It focused on early entry practical market opportunities in all-electric, hybrid, and autonomous flight. With its history of drawing together green aviation leaders from industry, government, and academia worldwide, EAS had now become the go-to conference for electric flight.

Erik Lindbergh continued to work towards electric flight. In 2013 with Eric Bartsch, Lindbergh founded a private company, "Powering Imagination—the Future of Flight." One of the first moves was to enter into a partnership with the Museum of Flight in Seattle to help graduate students from the Embry-Riddle Aeronautical University in Daytona Beach, Florida, design *The e-Spirit of St. Louis*, an extremely light (1,700 pounds), 52-foot (15m85), 100-horsepower two-person prototype that can soar in near silence. The design is based on the Austrian Diamond HK-36 motorized glider. There is the potential for more than 200,000 sightseeing flights over the USA's parks every year. And while the views are spectacular, the noise spoils the show. The National Park Service has been concerned about the racket, and its effects on wildlife and visitors, since the 1970s.

For a decade, experienced airplane builder Brian John Carpenter of Adventure Aircraft Inc. at Corning Municipal Airport, California, has been working on a low-cost electric motor glider *kit* using complete video instruction for the build and incorporating over one hundred 3D-printed components. The EMG-6 began development in March 2013 from the lessons learned on the EMG-5 after the FAA's response to the manufacturer's request for clarification on Li-Po weight and the maximum ultralight empty weight specified in FAR 103. Power is from twin Plettenberg Predator 37 electric motors developing 16 hp (12 kW). Carpenter gave EMG-6 its first test flight on December 20, 2013. Those wishing to build their own EMG-6 were able to visit an Internet website and blog to get guidance.

Meanwhile, Yuneec of China had also been adapting good designs. One of these, an ultralight, was by Martin Wezel of Germany, renowned for his ultralight motor gliders such as the Condor, Star, Sting and Sirius. The build-it-yourself *Apis 2* wing design, with its upper wing telescopic air brakes as well as flaps, was initially intended to be powered by a 50 hp (37 kW) Rotax 503 two-stroke or 60 hp (45 kW) HKS 700E four-stroke powerplant. The rights to this motor-glider were purchased by Yuneec in 2011 with the intention of making an electric version. The standard engine fitted is the 40 kW (54 hp) Yuneec Power Drive 40 electric motor, controlled by a Yuneec Power Block 40 400-amp power controller and powered by two Kokam Li-Po battery packs of 31 ampere-hours (Ah) each 62 Ah total. But in June 2012, Wezel announced that the production move had been delayed indefinitely. In 2013, the Yuneec eSpyder 24kw electric propulsion system received the world's first electric aircraft engine certification from DULV.

In the Netherlands, during the air combats of World War I, among the aircraft doing battle was a Dutch-built bus called the Spyker-Trompenburg V.2, a low-powered, tandem-seat gasoline biplane. Now 100 years later, Spyker linked up with the American electric-aircraft startup Volta Volaré, to produce the *SpykerAero*, powered by an electric motor weighing just 32 lb. (14 kg) but delivering upwards of 680 pound-feet of torque.[10]

Axel Lange's *Antares* sailplanes continued to satisfy their pilots. On November 10, 2015, Ludwig Starkl (Austria) flew a cross-country flight over Namibia. He flew 1104.5 km (686 miles) with a very high average speed of 165.6 km/h (89.4 kts). On December 17, 2016, Anja Kohlrausch of Eberbach (Germany), also flying in Namibia, set a new (female) world record for free distance using up to three turn points of 1141.7 km in an *Antares 20E*.

There is also a requirement for a reliable e-airplane for training pilots. On August 8, 2014, Pipistrel's WATTsUP two-seater electric trainer made its maiden flight. As part of its 25th anniversary celebrations, that month Pipistrel displayed the airplane at the Salon de Blois Airshow, France.

In November 2015, Contra Electric Propulsion (CEP) was set up by Nick Sills, formerly technical director of the Electric Lightning P1 single-seat pylon racer, to develop an electrically powered version of the Falco composite Furio kit plane designed by legendary Italian engineer Stello Frati, fitted with two electric motors driving twin contra-rotating propellers. The CEP development, manufactured by Potenza Ltd., a Coventry-based company, a leader in electric and hybrid power units, is a bolt-on self-contained twin-engine contra-rotating fixed-pitch 300hp system for existing piston engine light aircraft. The contra-rotating props will be supplied by Hercules Propellers, which has already manufactured similar items for the Bugatti 100P project. The Furio has a carbon fiber monocoque airframe light enough to allow 400kg of Li-Po battery to be added and enable an autonomy of 1 hour. The component parts, fixed-pitch contra-rotating propellers, twin YASA 750 series motors, their inner and outer coaxial shafts and splined drive rings, were next assembled in a ground test vehicle to undertake both static and mobile tests to simulate aircraft taxing and ground maneuvering at up to 60 mph. This would lead to a design using 4 motors to offer a 500kW (625 shaft horsepower) system.[11]

Following Graham Dorrington's Dirigible D4, the return of the electric lighter-than-air ship continued. Indeed in April 2013, French pilots Pierre Chabert and Gerard Feldzer of transoceans.fr flew their electric-powered lenticular blimp, *Iris Challenger 2*, across the English Channel from France to Dymchurch in Kent, covering 45 km (30 mi) in 2 hours and 23 minutes. It was built by Chabert's company, Airstar Space Lighting, which won an Academy Award for technical achievement in 2003, specifically given for "the introduction of balloons with internal light sources to provide lighting for the motion picture industry." Chabert's balloons have illuminated *Titanic, Pirates of the Caribbean*, and countless other films. The e-airship's envelope contained 568 cubic meters (20,059 cubic feet) of helium, and could carry a payload of 200 kilograms (440 pounds). Equipped with two electric motors of seven kilowatts (9.38 horsepower) each, and two counter-rotating 1.3 meter (51.1875 inch) propellers made by E-Props, the 14-meter (46-foot), six-meter-high *Iris Challenger 2* navigated at a cruising speed of 20 kph (12.4 mph). An air-filled floater gives amphibious capability to the airship. They set a world record for distance, duration and speed, which was certified by the FAI (the World Air Sports Federation) in the 400-m^3 to 800-m^3 balloon class.

The French are not alone. By 2015, researchers at the University of Lincoln (UK) School of Engineering had completed a three-year investigation into stratospheric passenger airships as part of a multinational engineering project designed to provide a future sustainable air transport network. They were members of a pan-European research team that believed airships may be the "green" answer to the future growth of aviation. The Multibody Advanced Airship for Transport (MAAT) project aims to position airships as the solution for future air transportation that is safe, efficient, cheap and environmentally friendly. The EU-funded MAAT project, made up of eight nations and led by the Universita di Modena e Reggio Emilia in Italy, envisages the design of a cruiser which can travel across the globe on a set route. Smaller feeder ships carrying people and goods would then be able to dock onto the cruiser while it is still moving. The primary energy source for the MAAT is through harvesting sunlight from photovoltaic arrays mounted on the upper airship surface to provide sufficient electric power during the day to operate the airship's systems and provide life support, propulsion and control, while also producing sufficient excess energy that can be stored to facilitate continuous MAAT operation at night.[12]

The electric airplane had come a long way thanks to the Li-Po battery. But could it be made even more efficient? In March 2017, after four years of R&D, South Korean battery manufacturer Kokam launched XPAND, a nickel-manganese-cobalt (NMC) drone battery pack for EVs including a ceramic separator and battery pack thermal containment technologies; its energy density of 265 watt-hours per kilo would increase operating time by 20 percent over the average Li-Po battery.

Elon Musk of Tesla announced that their Model S P100D automobile would be equipped with a 100kWh battery, extending range up to 300 miles. Backing this up, Musk invested in building a "Gigafactory" outside Sparks, Nevada, with the aim, by 2020, of mass-producing more lithium ion cells annually than were produced worldwide in 2013. Musk revealed having his own design for a VTOL electric plane, and says that such a system becomes possible once battery energy density reaches over 400 Wh/kg, while his Tesla vehicles are believed to be powered by battery cells with ~240 Wh/kg.

Then there is Ann Marie Sastry of Sakti3 in Ann Arbor, Michigan, who discovered that Li-ion batteries could be improved by a factor of 2 to 3 by replacing the liquid electrolyte with a thin film of solid material. Though Sakti3 batteries are not yet on the market as a stand-alone product, the company has announced that it has achieved an impressive energy density of 1,143Wh/liter.

One serious player in the quest for greater specific energy density is OXIS of Oxfordshire, a team led by Huw Hampson-Jones, with their Lithium-Sulfur (Li-S) battery. When the Li-Po battery was first presented in the 1990s, it showed 80 kw/h per kg, and this has increased to 220. The OXIS battery lighter, safer and maintenance-free, is already showing an energy density of 400Wh/kg. Sulfur represents a natural cathode partner for metallic Li and, in contrast with conventional lithium-ion cells, the chemical processes include dissolution from the anode surface during discharge and reverse lithium plating to the anode while charging. As a consequence, Li-S allows for a *theoretical* specific energy in excess of 2700Wh/kg, which is nearly 5 times higher than that of Li-ion. In view of the price of cobalt mounting, and the toxic nature of the Li-Po, the Li-S stands a greater chance for the future. As a demonstrator, OXIS has built a UAV called *Centurion* fitted with Li-S batteries. The craft is so named because during World War II the iconic *Centurion* tank was built by the MG car company at Abingdon, where OXIS is based.

Whichever comes out the winner, it can only prove beneficial. Meanwhile other solutions are being investigated and tested to give the electric airplane its much-needed autonomy range.

SEVEN

"H" Is for Hybrid, Hydrogen, Helium

Alongside energy from the sun, then from the Li-Po, there is the potential from alternative forms of power sources to extend flying range.

At the start of the 20th century, three engineers had come up with the idea of extending automobile range with a hybrid-electric drive system. Harry E. Dey of Jersey City fitted a gasoline-powered, 6.5-liter, two-cylinder engine and a dynamo flywheel connected to an onboard battery into an Armstrong Phaeton; Ferdinand Porsche at the Lohner works in Vienna and Louis Kriéger of Paris achieved a similar drive train. In 1911, London engineer Jack Delmar-Morgan equipped his cabin cruiser *Mansura* with a gasoline-electric drive and went cruising out to lands beyond the Thames estuary.[1] Yet nobody before World War I thought of extending the range of an airplane with an additional gasoline engine.

Then on April 10, 1920, an article was published in *The Electrical Experimenter* headlined "The Electrical Airplane." It reported that an experiment in aviation had drawn up plans for an airplane with electrical transmission. A power of 6,000 hp would be provided by gas engines, then transformed into electric power distributed to engines powering the propellers. "This device in regard to the advantages it will procure will not go without causing a considerable increase in weight—the author envisages it for a giant plane which would carry 75 to 100 passengers, whose wingspan would be 72 m (236 ft), length 54 m (177 ft)." Could this be the first mention of a hybrid-electric airliner? The size is interesting compared to the 60-m (197-ft.) wingspan and 76-m (249-ft.) length of the Boeing 747 jet airliner a lifetime later.

Four years later, in 1924, Alphonsus Ligouri Drum of Chicago obtained U.S. Patent 1511448 for a hybrid-electric biplane with a central gas motor generating power to three electrically driven propellers: two mounted on the upper wing gave it vertical lift, while a third gave it traditional forward motion, so achieving improved stability in flight. As far as we know, the Drum hybrid was never built.

In 1943 a patent application, titled "electrical airplane propulsion," was filed by a team working at the Westinghouse Electrical Corporation at East Pittsburgh, Pennsylvania. On the team: Frank W. Godsey Jr., Lee A. Kilgore (who had developed a variable speed motor drive), Frank B. Powers and Bennie A. Rose:

> The principal object of our invention is to convert the perennial suggestions of electric power-transmissions in airplanes from the realm of impracticability to the realm of practicability.... In our

design, we have moved the propeller driving engines of a multi-propellered plane, from nacelles or wing-fairings or bulges, to the fuselage, and we have replaced them with high speed, lightweight, small-diameter four-pole squirrel-cage *electric motors* and gear-units, which can be submerged entirely within the airfoil section of the wing, for driving the propellers. By taking advantage of the much lighter weight, per horsepower, of large-sized gasoline engines or turbines, developing more power than can be absorbed in a single propeller, and necessitating some sort of power-transmission from a single large prime mover to a plurality of propeller-shafts, we have produced a design of electrical airplane-propulsion which may actually increase the cruising-radius or the speed of the airplane, while reducing the amount of gasoline required per flight. In this manner, we have eliminated the 20% drag. In our design, we are enabled to utilize a large number of relatively small-diameter three-bladed propellers, which can be readily balanced. The use of a plurality of electric motors for driving a plurality of propellers distributed over the wing also reduces stresses and vibration in the wing-structure, which is at best quite flimsy in comparison with its size. The use of an electric power-plant makes it feasible, also, to utilize auxiliary propelling means during take-off, to supplement the propelling force of the propellers....

The British engineer Lorne Campbell has commented:

They claim a number of advantages like small nacelles for the motors in the wings, hence less drag; lighter weight in the wings so that there is less strain on wing structure; the capability of running the motors at different speeds from the engines so that both can be operated at their most efficient speed, etc. The engines are mounted in the fuselage, and they claim this, also, is an advantage because it will mean the main weights are around the Centre of Gravity and the plane will be more manoeuvrable. They also say that having fewer, large engines is more efficient than having more, smaller, engines from the power to weight point of view. The big disadvantage that I see is that they don't mention (as far as I have read) the losses in the conversion of the power from the main engines to the electric motors and the elephant in the room, in my opinion, is that mounting the engines in the fuselage takes up a great deal of passenger/cargo space—this, to my mind, is a big disadvantage. Chucking the engines out on the wings opens up all this space.[2]

It is interesting to note that the filing date for this Patent was February 2, 1943. *On exactly the same date*, one of the co-inventors, Frank W. Godsey Jr., filed another patent for an airplane engine "in which air is drawn through a tube and discharged in a jet or blast at the rear of the aircraft"; in other words, a turbojet engine. Indeed, history records that Westinghouse Electrical Corporation developed and built their J30, the first American-designed turbojet to run, and only the second axial-flow turbojet to run outside Germany. The patent for the turbojet (U.S. 2404954 A) was published in 1946, while the patent for the electric airplane design (U.S. 2462201 A), again never built, was only granted in 1949. Frank Godsey Jr. went on to accumulate 100 patents in fields that included radio aviation and undersea warfare. He later became a consultant to James E. Webb, administrator of NASA. Lee Kilgore continued to develop lesser-known but reliable electric motors for wind tunnels. In IEEE interviews about their careers recorded in the 1970s, neither Godsey nor Kilgore made mention of their involvement in the electrical airplane transmission.[3] Their ingenious idea would only be resurrected some seventy years later.

Unknown to these men, in 1945 Stanley Bizjak, a humble corporal in the U.S. Army, based at his family dairy farm in Crivitz, Wisconsin, received U.S. Pat. No. 2,368,639 for an electrically powered glider. He had applied for the patent on June 3, 1943, specifying that its electric propulsion *should be driven in part by gasoline*. We do not know whether it was ever built. Bizjak also patented a motorized sledge.

Hybrid philosophy for mass-produced gas automobiles returned some fifty years later with the release of the Toyota Prius in Japan in 1997, followed by the Honda Insight in 1999. Meanwhile, aviation was looking at a variety of options, one of them hybrid.

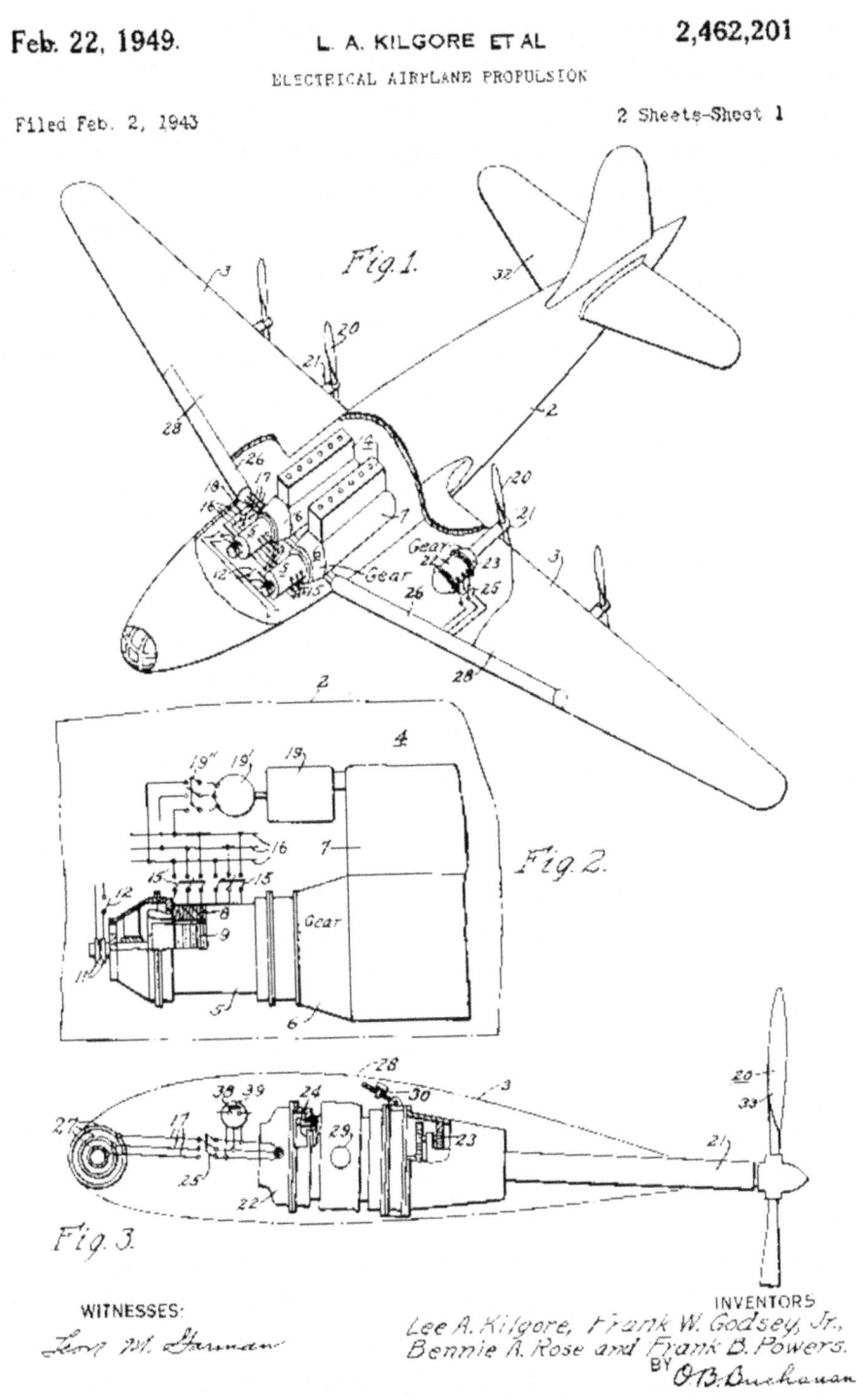

In February 1943, Frank W. Godsey Jr., Lee A. Kilgore and a team working at the Westinghouse Electrical Corporation at East Pittsburgh filed a patent for this hybrid-electric "multi-propellered" airplane. Patent U.S. 2462201 A was finally granted in 1949, and then forgotten.

In 2003, a team of researchers led by Dave Bushman at the Marshall Center, NASA's Dryden Flight Research Center in Edwards, California, and the University of Alabama in Huntsville flew an 11-ounce (312 g) model airplane with a propeller powered with solar panels illuminated by a laser. The plane with its five-foot wingspan was constructed from balsa wood and carbon fiber tubing and was covered with Mylar film, a cellophane-like material. The photovoltaic cells had been selected and tested by team participants at the University of Alabama in Huntsville. The prototype was flown indoors at Marshall to prevent wind and weather from affecting the test flights. After the craft was released from a launching platform inside the building, the laser beam was aimed at the airplane panels, causing the propeller to spin and propel the craft around the building, lap after lap. When the laser beam was turned off, the airplane glided to a landing.

Another exponent of laser-powered airplane concepts is Leik N. Myrabo with his LightCraft, a 1-kg. (2-lb.) launch vehicle, made from high-temperature ceramic materials, that could fly into space on a megawatt laser beam. Myrabo first got the idea in 1988 while working on the "Star Wars" anti-missile shield. A LightCraft channels the heat generated by a laser into its center, heating the air to about 30,000 degrees and causing it to explode, generating thrust. Small jets of pressurized nitrogen spin the LightCraft at 6,000 rpm to maintain stability. It was all just theoretical research—which the U.S. Air Force, NASA and the Strategic Defense Initiative provided $600,000 to help finance—until July 1996, when Myrabo, working with the U.S. Army at the White Sands Missile Range in New Mexico, propelled a small LightCraft prototype, measuring 6 inches long and weighing ounces, 50 feet into the air. After that, Myrabo conducted 24 test campaigns at HEL-STF using the PLVTS 10-kW laser. On October 2, 2000, sponsored under a nonprofit grant to his company LightCraft Technologies, Inc., Myrabo, using a 10-kilowatt pulsed carbon dioxide laser, saw the LightCraft climb to 233 feet (71 m) during a 12.7-second flight. This experiment established a new world altitude record for laser-boosted vehicles in free flight. On December 2, 2002, having made 140 test flights using small prototypes, LightCraft Technologies Inc. of Bennington, Vermont, was awarded U.S. Patent #6488233—"Laser Propelled Vehicle." Myrabo's challenge since then has been to find the finance to build the first full-scale craft. Once achieved, Myrabo sees laser flight carrying people around the globe and into space by 2020. Ground-based lasers called LightPorts would provide the energy needed to propel the aircraft. One is reminded of Nikola Tesla's patent of a century before.

Another option is the fuel cell.

In the early 1950s, an English engineer, Professor Thomas Bacon of Cambridge University, was making considerable progress developing the first practical hydrogen–oxygen fuel cell to present large-scale demonstrations. One of the first of these demonstrations consisted of a 1959 Allis-Chalmers farm tractor powered by a stack of 1,008 cells. With 15,000 watts of power, the tractor generated enough power to pull a weight of about 3,000 pounds. Allis-Chalmers maintained a research program for some years, building a fuel cell–powered golf cart, submersible, and fork lift, but not a boat. Interestingly, the U.S. Air Force also participated in this program.

The first experimental fuel cell airplane came thirty years later, with the Soviet Tupolev Tu-155, a Tu-154 airliner retrofitted by the Kuznetsov Engine Design Bureau. After two years of R&D, it made its first flight on April 18, 1988, by a crew under the command of Andrei Talalakin. The liquid hydrogen–fed engine was tested at altitudes up to 7,000 m (23,000 ft.) and speeds up to 900 kph (560 mph). In-flight starts and failures of the experimental engine were simulated, and the fire extinguishing system also was tried. It made another four flights before the fall of the Soviet Union and it is currently stored in the Ramenskoye Airport near Zhukovskiy. The Tu-156 was intended to fly commercially around 1997 but was canceled due to the fall of the Soviet Union. At the beginning of the new millennium, the economic situation in Russia stabilized, and interest in cryogenic fuels was reignited. Financing of a new project—the Tu-2016, a modified Tu-204K—began in 2002 with approximately 50 percent coming from the state budget.

In 1997, Klaus Graage of Ballard Power Systems and Daimler-Benz combined their vehicular PEM fuel cell and fuel cell system businesses to form dbb GmbH. The following year, they expanded to take in the Ford Motor Company and dbb became Xcellsis. Although the plan was to develop PEM fuel cells for autos and trucks, an additional patent was obtained in 1999 (DE19821952A1) for the energy supply on board an airplane.

Ever at the cutting edge, in 2001 Paul MacCready and his team at AeroVironment (see Chapter Five) applied for a patent for a fuel-celled flying wing. Their system would use liquid hydrogen as fuel, but gaseous hydrogen in the fuel cell. A fuel tank heater would be used to control the boil-rate of the fuel in the fuel tank. Although Patent U.S. 20020005454 was obtained in 2002, it was not until 2011, three years after MacCready's death, that AeroVironment's *Global Observer* took off with a hydrogen-fueled propulsion system.

In 2006, a team led by Professor Giulio Romeo at the Department of Aerospace at Turin University, funded by the European Commission, began a UAV program called the VESPAS (Very Long Endurance Solar-Powered Autonomous Stratospheric), a fleet of whose *Heliplats* (Helios Platforms), using a combination of solar and fuel cell technology, would be able to monitor the Mediterranean Sea from Turkey to Spain. By October 2007, a Super Dimona 2400 model motor-glider with a 7-m (22.9-ft.) span modified with solar lithium technology made its maiden flight. From this, the Turin Polytechnic—supported by 11 partners, including Israel Aerospace Industries, Evektor, Metec, Air Products, Enigmatex, Infocosmos, and Intelligent Energy, Brno University of Technology in the Czech Republic, and the Université Libre de Bruxelles in Belgium—adapted a Jihlaven Rapid 200 two-seater light plane into their ENFICA-FC (Environmentally Friendly Inter-City Aircraft powered by Fuel Cells) prototype. From 2010, the fuel-cell airplane was successfully tested at the Reggio Emilio Airport by POLITO during six experimental flights, lasting a total of 2 hours and accumulating 147 miles (237 km). Climbing was obtained at a combined fuel cell and battery power of 35 kW. Level flight was attained up to 100 mph (160 kph) using only a fuel cell power setting. A new world speed record of 84 mph (135 kph) and an endurance of 39 min were established for the airplane during several flights conducted in the FAI Code Category C for motorized aircraft. The next step would be for an 11–16 passenger ENFICA-FC commuter.

The same approach was taken up by the major airliner builder Boeing. From 2003, during the formative stages of the project, it became clear that a NASA-funded U.S. effort, led by Worcester, Massachusetts-based Advanced Technology Products, had come to many of the same technical conclusions as the Boeing team, developing a similar fuel cell–powered electric aircraft based upon a French DynAero airplane. Instead of working in parallel, a decision was made to combine the two teams' efforts in order to speed up progress and take advantage of mutually beneficial funding opportunities. A four-strong research team, Elena Bataller, Jonay Mosquera, Nieves Lapeña-Rey and Fortunato Ortí, engineers at Boeing Research & Technology Euro in Madrid (part of the Boeing Phantom

Works advanced research-and-development unit), began to work around an Austrian Diamond HK-36 Super Dimona motor glider. For the motor, proton membrane fuel cell, and li-ion batteries, they worked with Advanced Technology Products of Massachusetts, then Aerlyper of Spain integrated the propulsion package into the airframe. In February and March 2008, the Boeing Fuel Cell Demonstrator, piloted by Cecilio Barberan Alonso of Senasa, achieved three successful straight-level flights out of an airfield at Ocana, south of Madrid. The aircraft took off on a combination of battery power and the fuel cell, but used the fuel proton membrane cell alone to cruise at 3,300 feet (1,000 m) and about 55 knots for 20 minutes.[4]

As Boeing, so Airbus: Deutsche Aerospace AG had cooperated in a German-Russian effort started in 1990 to investigate an environmentally compatible airliner using fuel other than kerosene. A two-year feasibility study by Deutsche (later Daimler-Benz, then DaimlerChrysler) Aerospace AG, completed in September 1992, concluded that liquid hydrogen is safer than natural gas, kinder to the environment, and more readily available over the long term. Following this feasibility study, a new three-year phase was begun to develop critical technology and components such as tanks for liquid hydrogen storage at -253° C at 1.5 bars (21.75 lb./sq in), together with pumps and seals. This was supported by the German Ministry of Economics. Participants in the Cryoplane project included Tupolev, Samara, Daimler-Benz Aerospace, Munich, Linde, MAN Technologie, Messer-Griesheim, UHDE, Honeywell, Bodenseewerk, Drägerwerk, Deutsche Lufthansa, Munich Airport, Berlin Airport Corporation, and Hamburg's Max Planck Meteorological Institute.

In 2002, Koni Schafroth of Zurich, Switzerland, inspired by the shape of a tuna fish, began flights with an unconventional low aspect ratio model airplane he called *SmartFish*. It was initially powered by batteries, but following successful flights, the transition was soon made to a fuel cell plant. For this, Team SmartFish joined forces with the Institute for Technical Thermodynamics of the German Aerospace Centre (DLR) in Stuttgart with the goal of adapting the *SmartFish*'s impeller engine to run on hydrogen fuel cell power as developed by Horizon Fuel Cell Technologies of Singapore. So the name chosen was *HyFish*. In addition, Schafroch was joined by Ulrich Scheifer, a former aerodynamics expert on BMW's Formula One team. Following a successful flight in April 2007 in Bern, Switzerland, the team began to plan a 20 Pax full-scale airplane.

In 2005, financed by the European Commission, Airbus launched the CELINA (cell in airplane) to investigate the complete fuel cell system including kerosene reformer, fuel cell stack, air supply and all subsystems based on simulation models and tests. By 2008, Airbus, DLR and Michelin had progressed to performing flight evaluations on a testbed A320. The earlier fuel cell was installed on a cargo pallet and produced some 25 kW of electrical power—operating the electric motor pump for the aircraft's back-up hydraulic circuit, and controlling the spoilers, ailerons and elevator actuator. It has now partnered with the DLR German Aerospace Centre and Parker Aerospace to study usage of a Multifunctional Fuel Cell (MFFC) system on aircraft to replace today's gas turbine–based auxiliary power units. The system could provide an estimated 100 kW of electricity, acting as an independent source capable of providing power throughout an aircraft.

Axel Lange, one of the pioneers of the e-assisted sailplane, also decided to pursue the fuel-cell flight path. On July 7, 2009, the German took off from Hamburg airport in his fuel-cell powered *Antares DLR-H2* for a 10-minute maiden flight to an altitude of 837 feet (255 m). Designed as a flying test bed for fuel cell technology aimed at civil aviation

(APU replacement), the heavily modified *Antares 20E* has a top speed of 170 kph (105 mph). Commissioned by the Stuttgart-based Institute for Technical Thermodynamics of the German Aerospace Center (DLR), Lange strengthened the structure of the *Antares 20E*, added two underwing external pods and integrated an HT-PEM fuel cell system designed by Serenergy in Denmark, with additional input from BASF and Airbus. Pressurized hydrogen was used for fuel. Three years later, the aircraft was re-equipped with LT-PEM fuel cells and a larger pressure vessel for fuel storage. Located in the starboard external pod and weighing some 95 kg (210 lb.), the new tank holds some 5 kg (11 lb.) of hydrogen at 350 bar (5,076 pounds per square inch), compared to the 2 kg (4.4 lb.) capacity of the previous tank. To achieve this weight-saving, they wrapped hydrogen-trapping magnesium with an atom-thick layer of graphene. These changes yielded a maximum range of 500 km (270 miles) and an endurance in excess of 5 hours.

Next followed the *Antares H3*, whose target was to set range and endurance benchmarks beyond those accomplished by its predecessor: higher-performance and more compact fuel cell systems were incorporated in cooperation with Hydrogenics. The additional option of hybrid operation using storage batteries installed in the wings maximized the aircraft's peak power. A transatlantic crossing was even planned. It was hoped the *Antares H3* would demonstrate significantly increased performance—the developers planned to achieve a range of up to 3,700 miles (6,000 km) and endurance of more than 50 hours. For the *Antares H2*, these values had been only 430 miles (700 km) and 5 hours respectively. The aircraft would have a wingspan of 76 feet (23 m), a maximum takeoff weight of 1.25 metric tons (2,200 lb.) and it would carry payloads of up to 450 lb. (200 kg). It would use four external pods to house the fuel cells and fuel.

Fuel-cell electric airplane progress has gone on elsewhere with the *HY4* as presented by H2FLY at the German Aerospace Center, Deutsches ZentrumfürLuft und Raumfahrt

Antares H2 flying over Zweibrücken, Berlin (courtesy Lange Aviation GmbH).

(DLR), working with Hyrodgenics. The *HY4* is a four-seat hydrogen fuel cell electric air taxi, equipped with a low-temperature fuel cell with a proton exchange membrane (PEM) and an electric motor with a power of 80 kW that allow a maximum speed of about (125 mph (200 kph) and a cruising speed of 90 mph (145 kph). Depending on the speed, altitude and load, the autonomy of *HY4* can vary between 750 and 1,500 km (450 and 930 mi) away. In cruising flight, the fuel cell will only power the electric motor; it will be helped by a 21 kwh Li-Po battery to provide the extra power required during takeoff and climb in altitude. Using the airframe of the Pipistrel Taurus G4, the *HY4* eventually appeared at the 2016 Hannover Trade Fair. During the spring of 2016, the DLR research team successfully tested the drive train in the laboratory. In order to take off, the engine must reliably provide a maximum takeoff output for three minutes. This has already been successfully demonstrated for more than 10 minutes. The interaction of the fuel cell and the high-performance battery, used as a buffer and additional safety system, has also been successfully demonstrated in a simplified form in the laboratory. Hence, the road is clear for installing an initial version of this propulsion system in the four-seater *HY4* passenger aircraft. The first short 15-minute demonstration flight of the *HY4* was made on September 29 at Stuttgart Airport above the public and the media; air traffic control had all the other air traffic stopped, so spectators could appreciate the almost complete silence of the fuel cell airplane, flown by pilots Johannes Anton and Nejc Faganelj in one cockpit, with two dummy passengers in the other.[5]

In March 2017, Professor Josef Kallo, head of the Institute for Energy Conversion and Storage at Ulm University, describing this flight, announced plans to test the technological platform over the coming years before the target will be upped to six or eight seats. He explained: "Recent studies on commercial aviation show that there are indeed feasible propulsion designs for regional air travel with up to 40 seats and a range of 435 miles (700 km) or below, even though the technical challenges are significant."[6] During the summer of 2017 the *HY4* continued to make a large number of short proving flights.

The origins of the Thevenot fuel-cell E-Trike go back to 1974, when Gérard Thevenot of Fontaine Les Dijon in the Côte-d'Or Department, while still a student at the engineering school of Nancy ISIN, acquired the plans of the Seagull (an American-manufactured glider) and decided to build his own wing, based on this model, in order to fulfill his wishes to become airborne. He called the wing "La Mouette" (Seagull in French). By mistake he built the leading edges stronger than required due to the materials he had at hand, thus improving the wing's stability and performance. Enthusiasm about hang-gliding and the new wing spread rapidly among his friends, who begged him to build some for them. Thevenot built a 13 kW Flytec-engined Mouette E-Trike, which could be folded up and put in a car trunk. In July 2009, Thevenot flew his hydrogen-powered ultralight "trike" across the English Channel, with a Geiger/Eck HP-10 electric motor powered by a hydrogen fuel cell, without the onboard presence of an accumulator or battery. Its light weight allowed a consumption of only 550 grams per flight hour, and the craft's 5-liter tank allows about one hour flying time. In an environmentally friendly hint of its passing, the craft leaves behind only a mist of water vapor.

On January 8, 2017, a test flight of 350 yds (320 m) made in Shenyang, Liaoning, by an RX1E, equipped with 20 kilowatts of hydrogen fuel cell power supply, made China the third nation in the world to successfully test an airplane using hydrogen fuel. With a charging time of 90 minutes, the airplane is designed to be able to fly at a maximum altitude of 3,200 feet (3,000 m), for 45 to 60 minutes.[7]

Both photographs: Axel Lange in *Antares H2* fuel cell aircraft (courtesy Lange Aviation GmbH).

At the time of writing, Axel Lange is working on his next electric aircraft, the *Antares E2*. While not being a competitive aircraft on its own, the *Antares H2* had shown that if the fuel storage problem can be solved, then fuel cells can work in aviation. The *Antares E2* solves the problem of fuel storage by carrying a liquid hydrocarbon fuel and reforming it into hydrogen onboard the aircraft. This solution results in a higher empty weight compared to IC. However, the high fuel efficiency of a fuel cell means that the system becomes more and more weight-competitive as the endurance increases. With a planned endurance in excess of 32 hours, it goes without saying that the *E2* must be operated as an unmanned air vehicle (UAV). The aircraft targets the civilian market, with the goal of fulfilling hitherto unfulfilled sovereign and industrial requirements. As with the *Antares H2*, fuel cells and fuel are once more housed in underwing pods, leaving the fuselage for flight controls and payload. The fuel cells are of the HT-PEM variety, and once more, Serenergy is delivering the fuel cell systems. The battery pack of the *Antares 20E* is maintained, making hybrid operation possible. This allows for high peak power consumption during takeoff and in adverse weather. Contrary to the previous gliders, the *Antares E2* will be equipped with anti-icing and lightning strike protection. The *Antares E2* has a MTOW of 1650 kg (3640 lb.), of which 300 kg (660 lb.) are fuel, and 200 kg (440 lb.) are payload. The motor installation *Antares 20E* and *23E* had been optimized for climb and soaring flight. This resulted in a high-torque electric motor with a large-diameter propeller located on a pylon over the center fuselage. For soaring flight, motor and pylon are retracted into the center fuselage, yielding a pure sailplane. For the *Antares E2*, which is to operate under continuous power, this propulsive solution does not represent an optimum. The increase in power required would result in an unacceptably large propeller diameter, and the propeller wake would continuously hit the empennage, resulting in deteriorated aerodynamics. This problem was solved by distributing the power over six motors mounted on pylons above and behind the wing trailing edge. Altogether, this results in a unique aircraft that may very likely prove to be another milestone of electric flight.

The main challenge for fuel cell aircraft will be the constrained durability and instability of nanoparticles. The robustness of fuel cell systems is lower compared with the internal combustion engines, particularly in the specific temperature and humidity ranges in which an aircraft driven by a fuel cell would operate. The durability of a commonplace fuel cell stack is half the optimum durability required for its use in commercial aviation.

In Israel, aerospace engineers led by Dr. Shani Elitzur at the Technion-Israel Institute of Technology have developed and patented a process that can be used onboard aircraft while in flight to produce hydrogen from water and aluminum particles safely and cheaply. The hydrogen can then be converted into electrical energy for in-flight use. According to the Technion researchers, fuel cells can even play an energy-saving role in airline and airport ground support operations when they are also used for systems such as de-icing and runway light towers.[8] Non fuel-cell hybrid-electric research has continued in parallel.

In late 2016, a team led by Pat Anderson at Embry-Riddle Aeronautical University's Eagle Flight Research Center, at Daytona Beach, Florida, announced that it was developing a nine-seat hybrid electric turboprop aircraft.

Another hybrid-electric prototype developed by a team at University of Cambridge's Engineering Department led by Dr. Paul Robertson was given flight tests in late 2015 at Sywell Aerodrome in Northamptonshire, England. The Cambridge demonstrator used a

Honda engine in parallel with a custom lightweight electric motor while a set of 16 large lithium-polymer cells was located in special compartments built into the wings. These tests consisted of a series of "hops" along the runway, followed by longer evaluation flights at a height of over 1,500 feet (460 m).[9]

In 2013, General Electric opened the $51 million Electrical Power Integrated System Center (EpisCenter) on the campus of the University of Dayton, Ohio. The facility was sized to test electric power systems ranging in size from 500kW to 2.5MW. In the lab, university researchers and students began to work side by side with GE Aviation scientists and engineers to create new advanced electrical power technologies such as new power systems for aircraft, longer-range electric cars, and smarter utility power grids for more efficient delivery of electricity. On August 25, 2017, the GE EpisCenter published a white paper reporting how they had modified an F110 engine, a propulsion option for the Boeing F-15 and Lockheed Martin F-16, to generate 1MW of electric power. By siphoning compressed air from the core, GE had extracted 250kW from the high-pressure turbine and—an industry first for a two-shaft engine—750kW from the low-pressure turbine, according to the white paper. As a megawatt of electric power is equivalent to 1,341 hp, the F110 still has plenty of thrust to continue powering even in a single-engined aircraft. They had also produced an advanced electric motor in a separate project where the 1MW motor drove a propeller designed by Dowty, another GE subsidiary. When coupled with a gas generator, such a hybrid propulsion system could produce the same thrust as a large version of the Pratt & Whitney Canada PT6A turboshaft engine. Whereas most aviation motors are designed to achieve 90 percent efficiency, the new motor demonstrated by GE is 98 percent efficient. Such efficiency means a 1MW motor produces only 20kW of waste heat, rather than at least 100kW if a conventional aviation motor is used. GE Aviation did not reveal the size or weight of the device, but announced that they were currently engaging with several prospective companies on hybrid electric aircraft concepts.[10]

GE is also running a lightweight megawatt-class power inverter, which is a key step toward development of a viable hybrid-electric aircraft propulsion system. Tests of the liquid-cooled inverter, conducted at GE's Global Research Center (GRC) in Niskayuna, New York, form the next phase of an accelerating company-wide drive to perfect technology for hybrid-electric, turboelectric and all-electric aircraft.

Also in England, with the Griffon Hoverwork 995ED, another air cushion vehicle also known as the hovercraft, whose origins go back fifty years, finally embraced electric power. Designed by a team led by Mark Downer in Woolston, Southampton, southern England, the 28-ft (8.6 m) 8-seat craft sees two standard hybrid power modules containing a 67kW Ford Tiger diesel engine providing lift through highly efficient fans and supplying the two 45kW lightweight axial thrust electrical motors. No batteries are involved. The 995ED went on trials in August 2016 and since then shakedown trials have continued.

The success of one-off electric airplanes, be they supported by solar panels or hybrid fuel cells, is one thing. Taking them to the stage of commercialization is another.

EIGHT

Towards Commercialization

To understand how one-off prototypes can be transformed into commercially manufactured units, we should glance back at commercial aviation history. In February 1909 the three Short brothers obtained the British rights to build the American Wright Flyer. An initial order for six aircraft was taken, all of them taken up by members of the Royal Aero Club. Short Brothers thus became the first aircraft manufacturing company in the world. They built these biplanes at their workshop on unobstructed marshland at Leysdown, near Shellbeach on the Isle of Sheppey. During World War I, the Sopwith Aviation Company built more than 16,000 aircraft and employed 5,000 people. From 1913 to 1933 the Avro Company built over 8,000 AVRO 504 training aircraft at several factories. By the Armistice, the Curtiss Aeroplane and Motor Company would claim to be the largest aircraft manufacturer in the world, employing 18,000 in Buffalo and 3,000 in Hammondsport, New York. Curtiss produced 10,000 aircraft during that war, and more than 100 in a single week.

Sixty years later, in 1967, came the Boeing 737; eventually 8,800 were built. The 747 first flew in 1969; 1,520 were built.

The threats to the climate, the demands of government, and above all the needs of business have pushed on the development of electric aviation. But would these pressures have the same commercializing effect as the needs of air war?

As in the past, individuals and small operations have often led the way, from the land first. In 1967, Wilt Paulson of Lektro, which had pioneered the electric golf cart, produced a small electric aircraft tug for an Oregon FBO using a chassis originally built for an electric cart for area mink ranchers. Wilt Paulson's friend, Cy Young, owned an FBO called Flightcraft. Paulson and Young noted that towbars often caused damage to nose gear and were overall problematic. Young wondered if the nose gear could be lifted with a scoop to cradle the gear, eliminating the towbar. From this idea, Wilt Paulson produced a small electric aircraft tug by turning around the mink feeder chassis and attaching a hydraulic scoop and winch, and towbarless towing was born. By 2014 Lektro had manufactured some 4,500 electric towbarless tugs.

In July 2010, Axel Lange was honored with the prestigious Lindbergh Award, presented by Lindbergh's grandson, Erik Lindbergh. Lange had extended his range of motor gliders to 18, 20 and 23 meter models. The 20- and 23-meter (60- and 75-ft) variants can be equipped with a 42-kW electric motor and SAFT VL 41M lithium-ion batteries. Lange Aviation is currently also the most experienced serial producer of certified electric drives with more than 90 aircraft delivered since 2004. These are more than all other electric aircraft globally combined. The fleet has almost 100,000 hours of flying experience. For many years, there has been an annual Antares meeting organized by owners, where up to twenty-five Antares participate.

On February 24, 2012, the first ground works for the new Pipistrel facility for the production of the new 4-seat aircraft Panthera started in Italy, 15 miles (25km) away from the current Pipistrel headquarters in Slovenia. The new facility was built at the Duca d'Aosta airfield next to the town of Gorizia. The value of the building, together

2016: The Pipistrel factory production line at the Duca d'Aosta airfield next to the town of Gorizia in Italy takes electric airplanes such as the two-seater *Taurus Electro G2* and the hybrid-electric four-seater *Panthera* into the series production phase (Pipistrel).

with the administration and management extension, exterior and all the equipment, amounts to 5 million Euro. The building is completely energy self-sufficient, beginning with a 1.1 MW solar power plant—later increased to 1GW—on the roof of the complex, the same quantity of energy as a medium-sized city uses in one week. The combined worth of the investment was 7 million Euro. The first electric airplane in the production line is their Taurus Electro G2.

The Pipistrel Panthera is a 4-seat "General Aviation"-class aircraft with three versions of propulsion: the gasoline-engine version, powered by a Lycoming IO-540 fuel injected engine; the hybrid; and the fully electric versions, for both of which the propulsion systems will be developed entirely by Pipistrel. Panthera also features all-electric systems for component actuation. Its titanium trailing-link undercarriage, flaps and trim are all electrically operated, resulting in low weight and maximum reliability by removing the need for complex and heavy hydraulic systems. All internal and external lighting is realized using state-of-the-art LED technology, providing for better clarity, recognition and feel. *Panthera Hybrid* will have a 145 kW hybrid-electric power train, supported by the state-of-the-art battery system and range-extender generator unit. Panthera Electro will have a pure-electric 145 kW able to cover 400 km (215 NM), quietly, efficiently, with absolutely zero emissions and for a fraction of cost. The platform is open and ready to accept future generations of battery technologies, which will increase the operating range.

In November 2016, at the China International Aviation & Aerospace Exhibition in Zhu Hai, Pipistrel announced a trade deal with Sino GA Group, a general aviation (GA) company in China, whereby eventually up to 500 electric and hybrid airplanes a year would be manufactured, the Alpha Electro and Panthera Hybrid models. The value of the seven-year project, which will include building new aircraft production facilities for both models, a runway, and a maintenance and training facility for both models in Jurong, Zhejiang Province and Yinchuan, is more than 500 million euros. The deal grants exclusive rights for the sales of the two aircraft models in Chinese territory and in the neighboring countries, namely Myanmar, Laos, Thailand, Vietnam, Cambodia, Taiwan, Philippines, Indonesia, Malaysia, Singapore, Brunei, Korea and Mongolia. The contract also includes the delivery of 50 aircraft of each model, all required know-how on assembly of the two models, training of the personnel in Slovenia, implementation of the assembly process in new facilities in China, as well as supervision of newly established production to assure the same quality of the produced aircraft as the ones made by Pipistrel in Slovenia. Pipistrel will also use some of the money for the development of a new, ambitious zero-emission 19-seat airplane. It will be powered by hybrid electric technology and hydrogen fuel cells, and planned for public transport between the cities in China and all over the world.[1]

On November 15, 2015, Pipistrel CEO Ivo Boscarol, aged 60, was chosen as a member of the New Europe 100 list, a list of outstanding innovators from Central and Eastern Europe. This ranking is published annually by Res Publica together with the Visegrad Fund, Google and *Financial Times*. The founders explain that "only the people who have courage to think big, seek new ideas and show their skills have a chance to be listed." In 2015 he was listed among the "Top 28 most influential people in the European Union" by *Politico Magazine*.

In December 2016, Pipistrel purchased a new 4200-m² building for R&D. Still in Ajdovščina, the building, designed by architect Boris Podrecca, pioneer of postmodernism, has an underground garage for 40 vehicles and is energy-efficient. At the end of

2015, Pipistrel already had 89 employees, and by 2017 the number had been increased to 140, recruited from around the world, including the USA. Taja Boscarol, joint company owner with her father Ivo, confirmed: "We are also convinced that the future is electric—not just the future of aviation but the future of the entire transport on Earth."[2]

From May 2017 the second step of the Hypstair project, MAHEPA (Modular Approach to Hybrid-Electric Propulsion Architecture), with funding from the European Union's Horizon 2020 research and innovation program, was launched at Pipistrel's headquarters at Ajdovščina. Also taking part are Compact Dynamics, DLR, the University of Ulm, H2Fly, Politecnico di Milano, TU Delft and the University of Maribor. MAHEPA's aim is to tackle current limitations of electrically powered aircraft by introducing new serial hybrid-electric power trains. The project will develop new components in a modular way to power two, 4-passenger hybrid electric airplanes scheduled to fly in 2020. The first will be equipped with a hybrid power train utilizing an internal combustion engine, and the second will be a fuel cell hybrid–powered aircraft. As with the exponential increase in the number of electric automobiles, provision must be made for recharging e-airplanes. In partnership with students from three universities, Pipistrel also developed the first taxi-up-and-plug-in public electric aircraft charging station, incorporating a computer, charging protocol, AC/DC converter, communication with owner's cell phone, and WiFi. On September 30, 2017, the charging station at Pipistrel was used to charge an Alpha Electro, its battery fully charged in an hour. The station is capable of charging two aircraft simultaneously.

In December 2017, a Pipistrel Alpha Electro G2, delivered to a flight school at Pitt Meadows Airport, West Vancouver, Canada, was cleared for takeoff, while on January 2, 2018, at Perth's Jandakot Airport in Australia, a G2, carrying Recreational Aviation Australia (RAAus) light sport aircraft registration 23-0938, made its first flight of two circuits. Passengers could soon be flying to Rottnest Island protected nature reserve in a lithium battery–powered electric plane if the idea takes off. Avinor with Norges Luftsportforbund (the Norwegian Air Sport Federation), in the country determined to become the first in which electric-powered airplanes take a significant market share, also acquired a *G2*. The British Civil Aviation Authority issued a BCAR A8-21 Organisational Approval Certificate for microlight aircraft design and manufacture to Pipistrel, the first company outside the United Kingdom to receive this.

During this time, Airbus, Europe's giant airliner manufacturer, whose A320 airliner variants had sold in the thousands, had become fully involved with the French *Cri-Cri* airplane. With the official green light from given in October 2012, the demonstrator was renamed Airbus E-Fan. The new approach was for an airplane which used twin electric ducted variable pitch fans, spun by two electric motors powered by a series of 250 V Li-Po batteries (180 wh/kg). These were also linked to an electric drive for taxiing along the runway. The ducting increases the thrust while reducing noise, and when centrally mounted, the fans provide better control. The 120 assembled batteries weighed 137 kg (302 lb.); the weight of the airplane was 600 kg (1,323 lb.) at launch. Work began in earnest by an 18-strong team at ACS to prepare it to fly at the Paris Air Show at Le Bourget. Airbus Group Innovations concentrated on the e-FADEC system for managing energy and data flows within the plane. In 2013 E-FAN was presented at Le Bourget, but as a static exhibit. Its maiden and low-altitude flight, with Didier Esteyne at the controls, took place on January 30, 2014, at the Rochefort Airport. Its first higher-altitude flight took place on March 11 at Bordeaux-Mérignac International Airport, comprising a 29-

minute flight to a height of 700 ft. Esteyne landing in front of a large applauding audience. On April 25, at an E-Aircraft day, the E-Fan was appreciated by the media, VIPs and France's Minister of Industry, Arnaud Montebourg. Flight noise measurement tests were conducted in order to make the comparison with a conventional aircraft. It was then presented in flight during the Berlin and Farnborough Air Shows of that year.

Airbus Group now formed a subsidiary called VoltAir SAS in France to build a family of plug-in and hybrid-electric light airplanes, for which a factory would be built in the south of France. This would begin with a two-seat trainer called the E-Fan 2.0, slated to reach the market in 2018. A follow-on, hybrid-electric four-seater called the E-Fan 4.0, targeted primarily at buyers in the United States, would emerge soon after and is projected to go on sale in 2019.

On July 25, 1909, the world read about French pilot Louis Blériot's crossing of the English Channel in his Type XI gasoline-engined monoplane. The publicity gained by this achievement brought the company orders for large numbers of the Type XI, and several hundred were eventually made. On July 7, 1981, pilot Steve Ptacek flew *Solar Challenger* across the English Channel. Now came the opportunity for a pure electric airplane to achieve the same feat. Pipistrel's French dealer had been planning to make this crossing with an Alpha Electro when, it is reported, Siemens had abruptly warned the Slovenian team that they should not use its Dynadyn 60kW motor for overwater flights.[3] Two Czech-built motors powered the Airbus E-Fan, and Didier Esteyne did have the same nationality as Louis Blériot. He has recalled:

> At the end of May 2015, we began the test flights of *E-FAN 1.1* to prepare it for its flight demonstrations both at the Paris Air Show and in particular for the "Channel Crossing." On the afternoon of July 9th, a practice run enabled our reduced team, based at Lydd Airport in England, to "call" our device, the flight being followed by a rescue helicopter at sea, in which my friends Dominique and Paul were providing security by continuously scanning telemetry data transmitted by the E-FAN. A second helicopter would produce the video images for live transmission to the Control automobile located in Calais, in turn responsible for their broadcasting. A debriefing following this flight allowed each to confirm his role and his benchmarks. In 1909, Louis Blériot had taken off from a beach in Calais to reach England at Dover. We decided to make the crossing in the opposite direction to benefit from favorable winds and also allow both guests and spectators to be present at our arrival at Calais Airport.
>
> The following day, Friday, 10 July 2015, at 10:15 a.m., helped by Francis, I am harnessed into the narrow cockpit with all the safety equipment, life jacket, oxygen mask, anti-smoke goggles ... somewhat cluttered! At 10:00 the top start was given by a plane 15 minutes ahead of my flight with confirmation of the air traffic activity on the course. This was not a straight line between two points, but a navigation towards the Saint Inglevert field for the first leg of the course, in order to reduce the time over the sea and to land if one of the engines were to fail me. The first difficulty was to achieve a steady climb to the selected safety altitude of 3,500 ft, making the best possible use of the batteries and making sure to stay within the envelope of operating temperatures of the traction chain. A very precise point-of-no-return had been defined and confirmed permanently by the calculations of Dominique and Paul in the helicopter. My concern was not reaching France and having to turn back to Lydd. Weather conditions were almost perfect, except a little front wind at the end of the climbing time, temperature and visual conditions were great. All went well and I continued to follow the coastline. I could finally enjoy some flying and the scenery below. In Calais, Olivier and Romain were "glued" to the telemetry screen, following the flight data. The French coast is in sight at this altitude, and in the serenity of this moment, with a clear and turbulence-free sky, I was happy and surprised by the number of boats cruising under my wings! At regular intervals, Dominique, who was in radio communication with me and who was officiating as "Chef de Mission," confirmed the parameters indicated on my dashboard and informed me of any discrepancies with the data provided.
>
> At 10:47 a.m., I entered the safety zone of the Saint Inglevert airfield, and all being normal, I made

a change of course with a left turn and continued towards Calais which I already had in my sights. After checking all the data with Dominique, and in compliance with our flight plan, I reduced power to make a gentle descent towards the next turning point, "BlériotPlage." 10 minutes later, welcomed with enthusiasm by the airport control service, it was time to turn right towards the entrance of runway 06 with, as planned before the landing, a passage on the axis to greet the public. The altitude margin allowed me to accelerate up to 210 kph and the E-FAN glided obediently and without noise. Then with a tailwind, I slowed down, pulled out the train, the flaps, and almost reluctantly put down the electric bird and let it roll to the last exit lane.

With my arrival at the car park after 38 intense minutes and 74 km (46 mi) traveled, Jean Botti of Airbus welcomed me, happy and relieved. I just had time to observe that the calculated reserve of 19 percent of batteries remaining on arrival was actually 21 percent. I switched off all the contacts and jumped out of the cockpit to answer questions from the surrounding journalists. This is not the time to savor the adventure! Some people wanted to give me the title of "Hero." I clearly refused; those who preceded us at the beginning of the last century agreed to go into the unknown, with everything still to understand, to conceive, to try, and it is through these "heroes," their courage, their determination, that we are able to fly safely today. Although I was alone in the cockpit, ours was a team effort.[4]

But their glory was short-lived because it soon came to light that French pilot Hugues Duval had flown from Dover to Calais 12 hours earlier on Thursday in his *Cri-Cri E-Cristaline*, powered by two 35 bhp Electravia electric motors. By the beginning of 2014, about 70 aircraft had been equipped with French Electravia propulsion systems, including the MC30E *Firefly* in August 2011, and the electric motor-glider *ElectroLight2*.

According to the Associated Press, Duval's flight took only 17 minutes, which was shorter than E-Fan's flight because the latter aircraft circled Lydd Airport after taking off while a helicopter carried out a visual safety check. One difference in Duval's flight that could have been a point of controversy is that he did not have formal permission to take off from Dover, so his aircraft was lifted into the air by a gas-engined Broussard monoplane called *La Navette Bretonne*. This assistance may place recognizing Duval's record flight in jeopardy. Between March 2014 and February 2016, E-Fan had made a total of 117 flights including 39 in 2015.

On Friday, July 10, 2015, Didier Esteyne pilots the Airbus *E-Fan* out across the English Channel (G. Bassignac/Capa pictures–Airbus).

Top: Airbus's chief technical officer Jean Botti congratulates Didier Esteyne after his 2013 cross–Channel flight in *E-Fan* (G. Bassignac/Capa pictures–Airbus). *Bottom:* When Hugues Duval flew the Cri-Cri *Cristaline* across the Channel, he was given an airborne launch from a conventional Broussard monoplane called *La Navette Bretonne* (photograph: Jean-Marie Urlacher).

Russia has also decided to take the challenge of an electric airliner very seriously. This program is led by Alexander Inozemtsev, general designer of Aviadvigatel, and has brought together 100 enterprises of aviation, electronic and electrical industries and a number of leading academic institutes of the Russian Science Academy. A Tupolev Tu-214 (No. 64501) passenger jet will be used as a flying testbed for the "Integrated program for the creation of an all-electric aircraft." Deadline: 2022. Inozemtsev was born on April 9, 1951, in the town of Kamyshin, in the Volgograd region of the USSR. In 1973, he graduated from Perm Polytechnic Institute and started working at Tupolev's Engine Design Bureau (today OJSC Aviadvigatel) as a design engineer. The AEA program will use an experimental engine based on the PS90A turbofan developed by Aviadvigatel, which is currently used on the Ilyushin Il-96 and the Tupolev Tu-204/Tu-214 series, as well as the Ilyushin Il-76 transport aircraft. The Tu-214E will have a single, centralized power supply system that provides all the energy needs of the aircraft. The pneumatic and hydraulic systems will be replaced with electric ones. The Zhukovsky Central Aero-hydrodynamic Institute (TsAGI), will be actively involved in the development project of the electric chassis that ensures taxiing airplanes without turning on the engine and the use of special trucks. The target for its first trials is 2025.[5]

Towards this end, a Russian octocopter has been made by NELK Company, equipped with hydrogen-air fuel cells developed by Yuri Dobrovolsky and a team at the laboratory of solid object Ionics of IPCP RAS (Russian Academy of Sciences), RAS Institute of Problems of Chemical Physics (IPCP). In January 2017 the Russian UAV set a world record for the duration of flight in open spaces during tests in Chernogolovka near Moscow. The 12 kg (26 lb.) octocopter remained airborne in poor weather conditions for 3 hours 10 minutes. With its 1.3 kW power plant, the fuel-cell drone can carry up to 0.5 kg (1 lb.). The record duration of the octocopter flight was achieved through the special design of membrane-electrode assemblies, which generate electricity through electrochemical reaction of hydrogen and oxygen and operate at extreme temperatures from -60 to +40 °C. The hydrogen-air fuel cells are formed on the basis of these assemblies. The development of innovative fuel systems that allowed the unit to stay in the air for such a long time had begun in 2015 in partnership with the Central Institute of Aviation Motors n. a. P.I. Baranov (TsIAM) and the United Aircraft Corporation (UAC). After the flight, Sergei Korotkov, chief designer for UAC, noted that this technology might be used in newer aircraft—medium-range MS-21 airliners and wide-body aircraft, developed in cooperation with foreign partners. Six months later, at MAKS-2017, the 13th International Aviation and Space Salon, held at Zhukovsky near Moscow, Korotkov, announced that a range of agreements on batteries to provide planes with new energy had indeed been signed.

In China, an aircraft similar to the stalled Yuneec GW430, the two-seater RX1E Ruixiang, with high-wing cantilever construction and long slender wings as developed by Shenyang Aerospace University in Shenyang City in Liaoning Province, northeast China, was ready for production as certified by the Civil Aviation Administration of China (CAAC). Liaoning Ruixiang General Aviation Co. Ltd. had already built four RX1Es which had logged up over 240 hours total flight time. Similar to Pipistrel's Alpha Electro, the RX1E has six 10 kW.h Kokam battery packs that can be removed for recharging. This feature allows flight schools to have pre-charged packs ready to swap once a student has made a final landing following a flight of 40 minutes at speeds of 150 kph (93 mph), and the ability to climb to 3,000 meters (10,800 feet) at a maximum takeoff weight of 480 kilograms (1,056 pounds). A battery pack can be recharged in an hour and

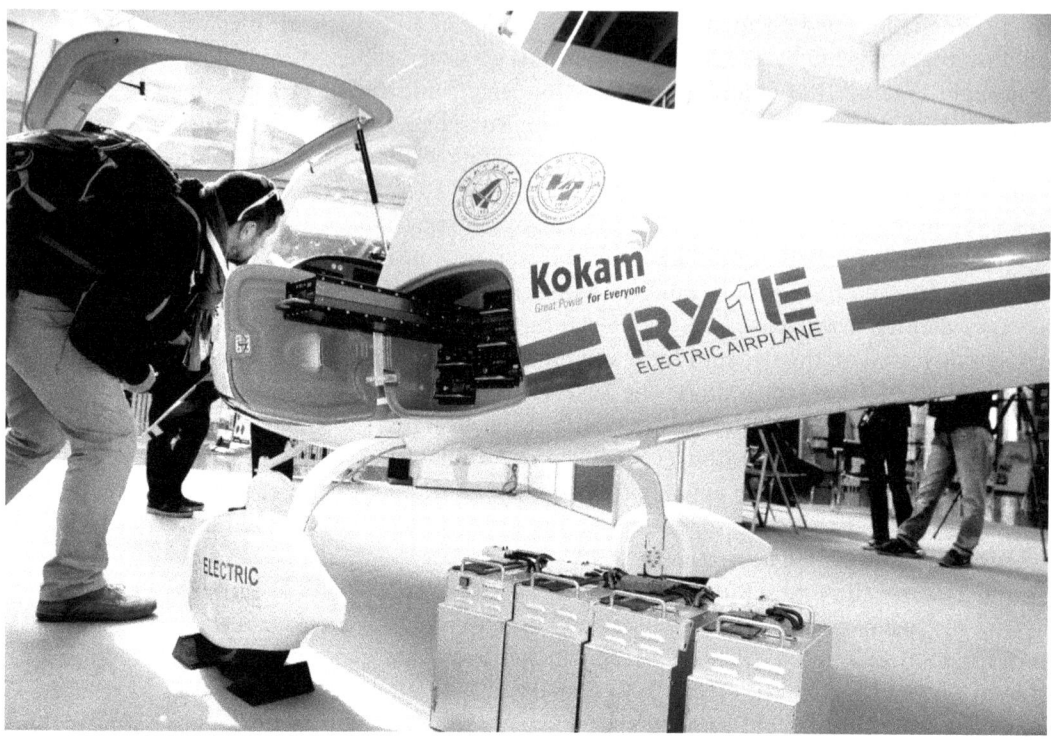

The Chinese RX1E, developed by Shenyang Aerospace University in Shenyang City in Liaoning Province, northeast China, created a lot of interest at Aero Friedrichshafen (photograph: Jean-Marie Urlacher).

a half. LRGA announced that they were ready the manufacture the airplane, starting with 20 more, then potentially increasing to 100 per year within three years.[6] On November 1, 2017, with improved batteries, Shenyang Aerospace University's two-seater RX1E-A made a two-hour flight from Caihu airport in northeast China's Liaoning Province.

However, large corporations have also become essential in developing the research needed to accompany the development of electric planes.

As recounted in Chapter One, in 1887 Siemens provided the electric motor for the Tissandier brothers' Parisian aerostat. Now, 130 years later, they re-entered the challenge—to achieve a power-to-weight ratio of five kW/kg in a large electric motor. The R&D was headed by Dr. Frank Anton and Claus M. Zeumer in the eAircraft department of Siemens Corporate Technology. Support also came from the German Aviation Research Program LuFo in a project of Grob Aircraft of Tussenhausen-Mattsies in Germany. In 2013, Siemens, Airbus and Diamond Aircraft were able to successfully flight-test a series hybrid-electric drive in a DA36 E-Star 2 motor glider for the first time. The test aircraft had a power output of 60 kW.

Research continued from every conceivable angle. They began with some of their existing motors, testing all of the components individually, and reducing materials whenever possible. Engineers found that the aluminum endshield, the part of the motor housing that supports the bearing and protects the motor's internals, was quite heavy. To reduce its weight, Siemens developed a sophisticated computer model of the endshield. The software represents the endshield as 100,000 separate parts and then simulates each

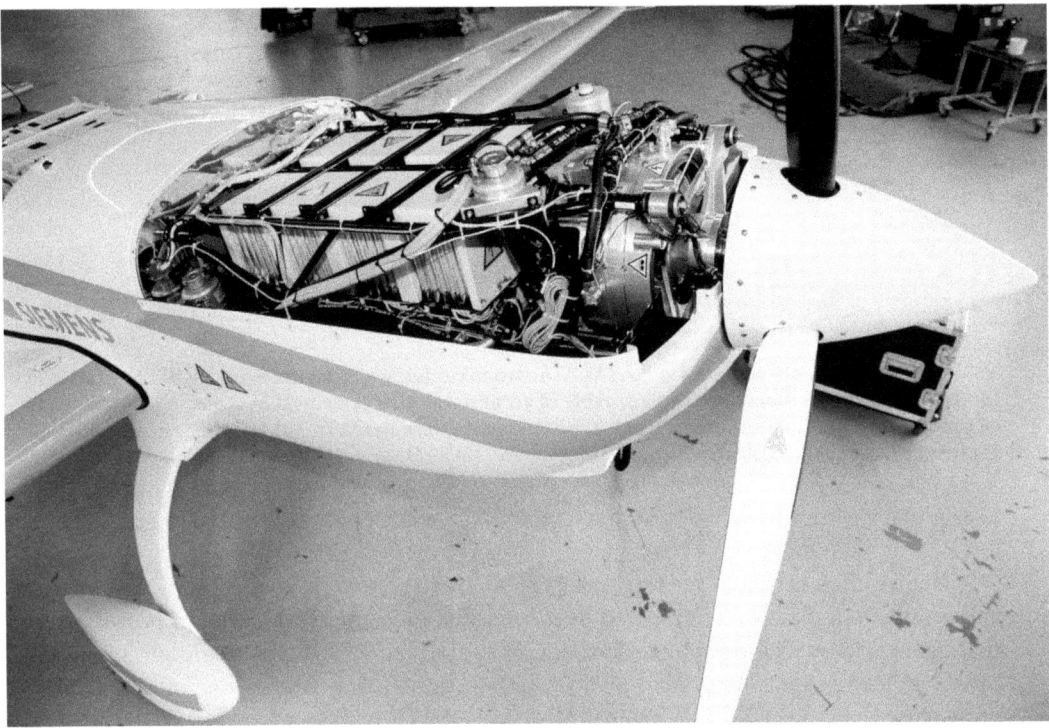

The revolutionary Siemens engine weighs just 50 kg (110 lb.), and delivers a continuous output of about 260 kW (photograph: Jean-Marie Urlacher).

element's performance under various force conditions. At that point the algorithm conducts millions of trial-and-error simulations, eventually finding components that can be eliminated or reduced. The process helped engineers redesign the endshield, turning it into a filigree (lattice-like) structure with the same performance at less than half the weight. They also developed a prototype made from a carbon-fiber composite, to reduce the weight by another factor of two. Inside the motor, the rotor's permanent magnets are configured into a Halbach array, which produces a stronger magnetic field with less material. The stator is made of an easy-to-magnetize cobalt-iron alloy. The motor's windings are surrounded by a special cooling liquid that conducts heat but not electricity. Lead engineer Dr. Frank Anton said, "We use direct-cooled conductors and directly discharge the loss of copper to an electrically non-conductive cooling liquid—which in this case can be, for example, silicone oil or Galden."

By 2015, Siemens had arrived at a unit weighing just 50 kg (110 lb.), and delivering a continuous output of about 260 kW—five times more than comparable drive systems. The electric motors of comparable strength that are used in industrial applications deliver less than one kW per kg. The performance of the drive systems used in electric vehicles is about two kW per kg. Since the new motor delivers its record-setting performance at rotational speeds of just 2,500 revolutions per minute, it can drive propellers directly, without the use of a transmission. The motor began flight-testing at the end of 2015. In the next step, the Siemens researchers planned to boost output further. Frank Anton stated, "We're convinced that the use of hybrid-electric drives in regional airliners with 50 to 100 passengers is a real medium-term possibility. The technology will apply first

X57's DEP system has been tested at NASA's Armstrong center in California using the HEIST (Hybrid Electric Integrated Systems Testbed), capable of accommodating systems that use up to 100 kilowatts of power, mounted to a modified truck. The first tests were made in 2014 on the energy-efficient Pipistrel *Electro Taurus* electric propulsion system (NASA).

to regional airliners flying 50 to 60 passengers over stage lengths of up to 600 miles (1,000 km)."

Before long, as part of the Hypstair project, managed by Pipistrel, a Siemens e-motor and a Rotax combustion engine had been installed on a Pipistrel Panthera prototype at that firm's factory in Ajdovščina, Slovenia. On February 9, 2016, the world's first power-up of a hybrid propulsion system on a four-seater took place. Additional partners in the project are the University of Maribor (Slovenia), the University of Pisa (Italy), and M.B. Vision (Italy). The motor was run in electric-only mode, using battery power; generator-only mode; or hybrid mode combining the two. For the initial testing, a five-blade, low-rpm propeller was attached to the motor.

Another airplane to benefit from the Siemens engine was the *eFusion*, a two-seater, side-by-side low-wing monoplane with nonretractable tricycle landing gear. Built by Magnus Aircraft led by ImreKatona at Kecskemét, Hungary, *eFusion* made its maiden flight at Matkópuszta airfield on April 11, 2016. The empty weight of 410 kg includes the batteries and the ballistic recovery system. The aircraft has a maximum takeoff weight of 600 kg. Siemens designed a safe and robust battery system for aviation use and optimized the electric propulsion system for application in the cost-sensitive segments of Very Light, Light Sport and Ultra Light Aircraft. Frank Anton of Siemens saw the *eFusion* as a flying test bed for our further battery system optimization, developed by their Budapest-based subsidiary in close cooperation with the German colleagues at Siemens's headquarters. *eFusion*'s aerobatic capability will contribute to the training of the much needed new generation of airliner pilots.

In September 2016, it was announced that Tianshan Industrial Group, based in Shijiazhuang City, Hebei province of China, had formed a joint venture with Kecskemét's Magnus Aircraft to manufacture the eFusion 212. Under the €30m deal, the two companies plan to create a joint enterprise at the central Hungarian city of Kecskemét, where Magnus Aircraft is based, to build light planes under license for sale in China. Initially, the new partnership is expected to construct and to sell a total of 1,500 of the eFusion 212 aircraft by 2020. It also includes a greenfield investment scheme with the construction of a medium-sized airport in Kecskemét, an assembly and maintenance plant, and the Magnus Pilot Academy training center network. Magnus Aircraft anticipates employing up to 600 people at its factory in Hungary and an additional 250 in China. In May 2017 a center was also built at Gillespie County Airport, Fredericksburg, Texas, for both pilot-training and distribution of the Magnus eFusion 212. It is anticipated to export the eFusion airplane to China from 2018. Within six years, it predicts it will attract sales revenues totaling more than €162m through the latest deal as the plant gears up for capacity expansion.[7]

In June 2017, Solar Ship linked up with Chris Heintz, CEO of Zenith Aircraft Co. in Midland, Ontario, to convert the existing Zenair STOL (Short Take-Off and Landing) CH750 aircraft into an electric bush plane. This new aircraft would provide extreme short takeoff and landing (XSTOL) capability enabling pilots to take off in areas without runways. The aircraft is recharged by either a battery swap or electric vehicle rechargers. It will not use any fossil fuel and will be available as a bush plane, float or amphibious. The electric bush plane project is part of Zenair and Solar Ship's ongoing partnership since 2011, when Zenair developed the fuselage for Solar Ship's Zenship 11. Established in 1974, Zenair has designed more than 15 aircraft with sales in more than 50 countries.

In August 2017, Boeing confirmed a small experimental "X-plane" hybrid-electric demonstrator planned for the early 2020s could signal an unprecedented push into the commuter market. The X-plane plan is being evaluated as part of the company's EcoDemonstrator technology testbed series and, if successful, could open the door to a new generation of small Boeing airliners seating 12 to 50+. It would use a Brazilian-produced biofuel blend made up of 10 percent biokerosene and 90 percent fossil kerosene, the maximum mixture according to international standards. Studies have shown that sustainably produced aviation biofuel emits 50 to 80 percent lower carbon emissions through its life cycle than fossil jet fuel emissions.

But the research is not only useful for smaller aircraft: Airbus now announced that it had signed an agreement with German industrial conglomerate Siemens to develop hybrid planes that can carry up to 100 passengers. The aircraft would use a combination of electric power and conventional fuel. The hybrid Airbus planes would consume 25 percent less fuel and would be almost silent during takeoff and landing, when running on electric power, according to Siemens. At cruising altitudes, the planes would be powered by jet fuel. The planes are expected to have a range of about 620 miles (1000 km), enough to fly from New York City to Detroit. As a start, the Euro duo set about building E-Aircraft System House, a large development and test facility near Munich, where the new systems would be developed by 200 engineers. Ground was officially broken on the facility in spring 2016, enabling the start of construction in early 2017 and a planned opening by late 2018.

As part of the Goodwood Festival of Speed, held between June 29 and July 2, the *eFusion*, piloted by Frank Anton with Siemens UK chief executive Juergen Maier on board, took to the skies in the first-ever extended all-electric propulsion 30-minute flight over a United Kingdom airfield. Several flights were made.

Frank Anton and the Siemens team had also been working with Extra Flugzeugbau (Extra Aircraft Construction), a manufacturer of aerobatic aircraft, directed by Walter Extra. Extra, born in 1954, trained as a mechanical engineer, then began his flight training in gliders, transitioning to powered aircraft to perform aerobatics. He built and flew a Pitts Special aircraft and later built his own Extra EA. Extra began designing aircraft after competing in the 1982 World Aerobatic Championships. His aircraft constructions revolutionized the aerobatics flying scene and still dominate world competitions. In April 2016, Extra unveiled the aerobatic 260 kW Siemens-engined 330LE (D-EPWR) at Aero

Friedrichshafen. Two months later, the 330LE, weighing 1,000kg (2,200 lb.), its battery management specially prepared by a Pipistrel team, made its first 10-minute flight, then on July 4, its first public flight, at Schwarze Heide Airport near Dinslaken, Germany. It was then used as an ongoing flying test bed to test its limits for the Siemens system.

Others would soon install a Siemens in their airplanes. The *Hamilton aEro* aerobatic plane is sponsored by the Hamilton Watch Company, part of the Swatch Group, a watch company based in Bienne, Switzerland. On October 2, 2016, the *Hamilton aEro*, piloted by Red Bull air racer Nicolas Ivanoff, Hamilton's brand ambassador, took off from Raron airfield 75 kilometers south-southeast of Berne in Switzerland, for its first public flight. The project's founders are Air Zermatt pilot Thomas Pfammatter and aerobatic paragliding champion Dominique Steffen of Hangar 55, along with former *Solar Impulse* members Sebastien Demont and Gregory Blatt. They took a Silence Twister aircraft and fitted it with a Siemens electric motor, its energy coming from 160 kg of Renata high-capacity 200 Wh/kg batteries. In the sponsor's orange-and-white livery, the tailplane and the wings of the *Hamilton aEro1* carry the logo of a black watch face with white hands. Maximum flight time is said to be 60 minutes, with 30 minutes of aerobatic flight plus a reserve. Soon after, Hangar 55 announced its goal to use the Twister in the 2017 Red Bull Air Race as an exhibition. They speculated that this might be a prelude to an all-electric junior class with all pilots using identical Twisters. In April 2017, after *aEero 1* had completed 50 flying hours, it was announced that *Solar Impulse* co-pilot André Borschberg had joined the team. H55 is also supported by the Canton of Valais, in particular the Ark

The *Hamilton aEro* aerobatic electric aircraft has a maximum flight time of 60 minutes, with 30 minutes of aerobatic flight plus a reserve (photograph: Jean-Marie Urlacher).

Foundation, and the Federal Office of Civil Aviation (FOCA). On September 21, 2017, the Hamilton. H55 Silence Twister made its maiden flight at Raron in the Valais, 130 km (80 mi) east of Geneva.

Max Vogelsang of MSW Aviation in Wohlen, in the canton of Aargau in Switzerland, has teamed up with the Bern University of Applied Science in Burgdorf to produce the detachable-wing MSW *Votec Evolaris*, Switzerland's first electric aerobatic aircraft. With 221 kW of power, the *Votec 221* has planned autonomy of 20 minutes but a roll-rate of 460° per second. Takeoff roll distance is 50 m and rate of climb when fully charged is 2.5 m/s. *Evolaris* can run at full throttle for 11 minutes on one charge. First prototype tests took place in spring 2017, with aerobatic tests in the autumn.

While Marc B. Corpataux of Alpin Air Planes GmbH, who is also the Swiss distributor for Pipistrel, presented his concept for an airfield network for training and travel in Switzerland, Lojze Peterle, a member of the European Parliament who is also a pilot, confirmed that Brussels supports the move towards cleaner energy and will work towards changes that will enable electric flight and make it simpler. Corpataux received the E-flight award from Willi Tacke, the man behind the AERO e-flight-expo and the editor of *Flying Pages*. Following this, the first-ever application for a certification of an electric aircraft was filed, for the Pipistrel Alpha Electro in the category CS LSA. Events such as the first E²Flight Symposium held at the Stuttgart Airport further encouraged the exchange of ideas.

Meanwhile, the Siemens-engined Extra 330LE had been tuning up its act. On November 25, 2016, Walter Extra took off from the Dinslaken Schwarze Heide airfield and reached a height of 3,000 meters (1,000 feet) in just four minutes and 22 seconds—equivalent to a climbing speed of 11.5 meters (37.7 feet) per second. In this he beat the previous world record set by Chip Yates in 2013 by one minute and 10 seconds.

Siemens is not alone. In 2014, Joachim Geiger and his team in Bamburg installed one of their HPD 25D engines in an 11-m wingspan *Elektra One* aircraft from PC-Aero. This engine, weighing just 10 kg (22 lb.), provides a maximum 30 kW and 26 kW continuous power. Tandem rotors on either power plant can run independently, enabling economical flight on one rotor, or a "get-home" mode at half power that will at least stretch the glide in the event of one rotor or controller failing. Based on this unit, in 2016, Acentiss of Ottobrunn created two aero engines developing 32 kW and 40 kW. By November, they had installed one in a solar-electric ultralight called Elias, complete with retractable tricycle landing gear. Elias is optionally piloted carrying test gear so that it can also be monitored by an Acentiss Ground Control Station (GCS). A dSPACE MicroAutoBox is used as the onboard flight guidance computer, because it provides a direct interface to MATLAB Simulink so modifications to the control algorithms can be implemented quickly. Depending on whether a pilot is on board, the system can monitor and control, by visual or data transmissions, the mission based on pre-set plans or manual overrides. It can check the multi-spectral, high-definition sensors on board Elias and record their output.[8]

A Belgian start-up, Green Tech Aircraft (GTA) of Leuven, announced the development of their *Ypselon GT*, a build-it-yourself electric airplane for flying enthusiasts. With its rear-mounted pusher propeller, the *Ypselon GT* will have a maximum capacity of two people and an extra carrying capacity of 220 kg (485 lb.). It may achieve speeds of up to 320 kph (200 mph).

In France, a team working at ONERA (the French National Aerospace Research

Laboratory), led by Claude Le Tallec, began to develop a Distributed Electric Propulsion airplane, called the *Ampère*, after André-Marie Ampère (1775–1836), physicist and mathematician, who gave his name to the amp. The 6-passenger French version will be equipped with 32 electric engines powered by a fuel cell. A one-fifth-scale model was built by Aviation Design of Milly-la-Forêt, with a full-scale roll-out due in the mid–2020s.

But France was not alone. Dr. Mark Moore of NASA wrote a paper, "The Coming Era of Distributed Electric Propulsion—and What It Means":

> NASA Langley is pioneering the integration of a new propulsion technology that has the potential to transform aircraft capabilities, the missions they fly, and the way we interact with aviation. The largest aerospace technology shift since the invention of the turbine engine is taking place, and will quickly sweep through the shorter range aviation markets. The result will be a step change in the degrees of freedom available to aircraft designers to achieve digital aircraft systems that achieve capabilities long dreamt of—but previously out of reach. Advanced concepts utilizing this technology will be showcased as vision vehicles that incubate technologies along two vibrant convergent frontiers, electric propulsion and autonomy. Study results will focus on the application of electric propulsion, not only as a new propulsion technology, but as a mechanism for achieving highly coupled digital aircraft systems that ride the wave of self-driving vehicles, sensor fusion, smart materials, and multi-functional systems. The opportunities are unfolding for a new aviation renaissance where multi-disciplinary synergistic coupling opens up entirely new design space to explore.[9]

The realization of this approach has been taking the form of the NASA SCEPTOR X-Plane X-57, named Maxwell after 19th-century Scottish physicist James Clerk Maxwell. It will be the first manned Distributed Electric Propulsion aircraft, capable of achieving a 5x reduction in energy consumption of a general aviation aircraft at the high-speed cruise condition. To realize this, in 2014, NASA teamed with two small think-tank-style technology companies in California: Empirical Systems Aerospace (ESAero) in Pismo Beach, and Joby Aviation in Santa Cruz. An earlier NASA collaboration with Joby had produced the *Lotus*, an all-electric UAS with a set of two-blade rotors at each wingtip and one hinged propeller on the leading edge of the vertical tail. Project Sceptor (Scalable Convergent Electric Propulsion Technology and Operations Research) would be a test aircraft, evaluating the use of distributed electric propulsion (DEP). It involves replacing the wings on a twin-engined Tecnam P2006T (a conventional four-seater light aircraft) built in Capua, Italy, and replacing them with a Leading Edge Asynchronous Propeller T.

In November 2014, a facsimile of the concept vehicle's wing, mounted on a truck bed for dynamic testing of its low-speed lift systems, arrived at NASA's Armstrong center in California. The acronym department came up with HEIST, for Hybrid Electric Integrated Systems Testbed, aka Airvolt. The Peterbilt truck provides a platform that can test the towering contraption at the speeds typical of landing and takeoff technology incorporating a (Leaptech) aerofoil equipped with 14 electrically driven five-bladed propellers. Made of steel and aluminum, it is 13.5 feet (4 m) tall. The first tests were made on the energy-efficient Pipistrel Electro Taurus electric propulsion system, which is typically used for motor-gliders. The Pipistrel motor is powered by lithium-polymer batteries and produces 40 kilowatts of power, which are monitored by the Airvolt that is capable of accommodating systems that use up to 100 kilowatts of power. The test stand can also withstand 500 pounds of thrust. Next up for the Airvolt are tests during late summer on the Joby Aviation JM-1 motor that will provide information for modeling simulations of the electric propulsion elements.

EIGHT. Towards Commercialization

During takeoff and landing, Maxwell will make use of all 14 motors to create sufficient thrust, but once it's up in the air it will only use the two larger cruise motors located on the tips of the wings. The contoured lift propellers will fold inward and fit snugly against the prop shaft recesses in their extended hubs. In that position, the cruise propellers would be immersed in the wingtip vortex, which will increase their efficiency. R&D for a 500 kW nine-passenger aircraft was scheduled from 2017 through 2019. NASA Administrator Charles Bolden stated, "This will be NASA's moonshot for aviation." In the U.S. President's Financial Year 2017 budget, NASA received $790 million to fund New Aviation Horizons, among other similar green-aviation initiatives. That summer of 2016, NASA continued testing the wild new wing technologies.

Maxwell has already won converts. Cape Air, a Barnstable, Massachusetts–based independent regional airline, is working with NASA and the Italian manufacturer to incorporate practical considerations in the design. Cape Air operates a fleet of mainly nine-seat Cessna planes flying short routes, such as from Boston to Nantucket, Massachusetts.

"They have almost perfected traditional commercial jet engines as far as they can go," said Cheryl Bowman, co-technical lead of the aircraft gas-electric propulsion program at NASA's Glenn Research Center in Cleveland, Ohio. In 2015 NASA wrapped up a six-year initiative called the Environmentally Responsible Aviation project, in which researchers worked to document several possibilities for improving fuel efficiency of aircraft. Among the proposals were switching to a lighter composite material for building the body of the planes, and shifting turbines to the back of the plane as part of a wing-streamlining shape. That's not taking into account the possibility of hybrid propulsion systems. Some of the work in that area goes so far as to propose demonstrating all-electric propulsion systems in smaller, aviation-class planes. "The power can either come from something like a battery or a fuel cell or something like that, or it could come from a generator run by a turbine," said Ralph Jansen, the other co-technical lead on the research project at Glenn Research. NASA is working with a long list of corporate partners such as Boeing, General Electric and Rolls-Royce, as well as Ohio State University, the University of Illinois, and Georgia Institute of Technology, to develop different concepts for improving aircrafts' fuel economy, according to Bowman. One concept is a commercial plane about the size of a Boeing 757 that adds extra electricity-powered propulsors along the length of the aircraft. Researchers have put the fuel savings of the design at 7 to 12 percent.

In July 2016 the Tecnam P2006T, wearing a NASA livery, was uncrated and slowly rolled out. By October, in order to hit its ambitious ten-year goal, NASA announced a new research wing at the NASA Glenn Research Center's Plum Brook Station, a 6400-acre (2,600-hectare) remote test installation site near Sandusky, Ohio. The unique NEAT (NASA's Electric Aircraft Testbed) began with a 600-volt power source to test an electrical system that could realistically power a small one- or two-person aircraft. The short-term goal was to turn NEAT into a flexible testbed that could build and test power systems for even larger passenger aircraft without having to crash anything in the process. The long-term goal, however, would be to create a 20-megawatt power system that will be light, yet powerful enough to actually get off the ground. Dr. Roger Dyson and the team for hybrid gas-electric propulsion began work to make the testbed more efficient and lightweight.

As part of the X-57 program, a six-percent scale model of a Boeing BWB (Blended

Wing Body) was tested for six weeks in the 14-by-22-foot subsonic tunnel at NASA's Langley Research Center in Hampton, Virginia. The BWB is triangular in shape: the wings are merged into the body, and there is no tail. Its traditional name is the Flying Wing, theoretically the most aerodynamically efficient (lowest drag) design configuration for a fixed-wing aircraft. The first flying wings were developed over one hundred years before. The top of this NASA BWB was painted in nonreflective matte black to accommodate laser lights that swept across the model in sheets. NASA and Boeing researchers used those laser sheets combined with smoke in the technique known as particle imagery velocimetry, or PIV. This would map the airflow over the model. From December 2016, NASA test pilots and engineers began to "fly" an interactive simulator designed to the innovative specifications of the X-57 Maxwell. Meanwhile, the Tecnam P2006T was undergoing conversion at Scaled Composites in Mojave, California. The aircraft's two inboard engines were being modified to feature an electric system, and could undergo taxi tests in early 2018.

But with the inauguration of Donald J. Trump as 45th President of the USA in January 2017, it was announced that the incoming administration planned to strip NASA's earth science programs of funding as part of a crackdown on "politicized science," which may well include the X-57 program. On February 20, the U.S. Senate passed legislation cutting funding for NASA's global warming research. The House was expected to pass the bill.

On June 8, 2017, Sean Clarke, X-57 Principal Investigator, NASA Armstrong Flight Research Center, speaking at the American Institute of Aeronautics and Astronautics' Aviation 2017 conference in Denver, stated that the battery packs, providing 47 kWh of energy and weighing close to 900 lb. (400 kg), an increase of more than 10 percent over the original goal, were posing the biggest challenge so far. "They have used up our mass margin," Clarke reported.

Mod IV, which is currently being reviewed for funding, is where the X-57 completes its transformation to a full-fledged efficient electric aircraft. After Mod III flight tests, electrical engineers will work to pull off the dummy pylons and replace them with the 12 high-lift electric motors, similar to what were on the semi back in Mod I. With 14 motors total, 12 just to provide lift on takeoff and landing, the X-57 Maxwell will be ready for test flights over the dry lakebed at Edwards as soon as 2020 or 2021. If these prove successful, the X-57 could result in a five-time reduction in the energy required for a private plane to cruise at 175 mph (280 kph). While waiting for batteries with sufficient energy density, the X-57 may use diesel-sourced hydrogen fuel cells, taking it from 40 minutes of flight to about three and a half hours.

Meanwhile, the Defense Advanced Research Projects Agency (DARPA) is sponsoring the Vertical Take-Off and Landing Experimental Aircraft (VTOL X-Plane) program to demonstrate an electric VTOL aircraft design that can take off vertically and efficiently hover, while flying faster than conventional rotorcraft. In April 2016, Aurora Flight Sciences of Manassas, Virginia, was selected, having demonstrated their scale model *LightningStrike*. Led by founder John S. Langford III, Aurora, working with Rolls-Royce and Honeywell, is currently aiming at completion of the full-scale aircraft. The plane has a bullet-shaped fuselage and two boxy wings holding a bank of 24 ducted fans—18 distributed within the main wings and six in the canard surfaces, with the wings and canards tilting upwards for vertical flight and rotating to a horizontal position for wing-borne flight at more than 400 mph (640 kph). Following proof of concept, work on the full-scale XV-24A went ahead, with ground and flight tests taking place during 2017 and flight testing in September 2018.

And it is not just propulsion that is attracting research. Others are looking to make a "more electric" aircraft. SPEC (Safran Power Electronics Center) is looking an aircraft that is more economical, more reliable, and less polluting. The idea behind a more electric aircraft is to gradually introduce electrical systems to replace onboard hydraulic and pneumatic systems used to power the landing gear, brakes, flight controls and thrust reversers, as well as for cabin pressurization and to start the engines. A key area of research for more electric involves the switch from alternating-current (AC) to direct-current (DC) power distribution, to allow exchanges of energy between equipment. For this a modular rig called Copper Bird ("Characterization & Optimization of Power Plant & Equipment Rig") has been set up in Paris to test the stability of onboard electrical networks, and simulate the integration of different electrical systems and equipment. It also measures power quality and network stability. In more general terms, it can be used to demonstrate the maturity of systems and technologies developed for "more electric" aircraft.

In May 2017, Centennial College in Toronto, Canada, received $2.3 million in funding from the Natural Sciences and Engineering Research Council of Canada (NSERC) to collaborate with Safran towards the next generation of electric-actuated landing gear for energy-efficient aircraft. Safran has also developed an electric taxiing system, which will enable aircraft to taxi at airports without using their jet engines or requiring special tractors. By September, Safran Landing Systems was meeting with airlines to present its electric green taxiing system; for Airbus, the system would be used for the A320 family. Safran Center of Expertise, Safran Power Units is also developing a fuel cell for the PIPAA project (for fuel cells for aeronautical applications) for the power supply of aircraft systems, including electric ground taxiing solutions. It should be completed by 2019–2020.

From 2010 the Japanese aero engine manufacturer IHI, formerly known as Ishikawajima-Harima Heavy Industries Co., Ltd., teamed up with Boeing to carry out research into regenerative fuel cell technology to provide electrical power for airplanes. The technology, part of the More Electric Architecture for Aircraft and Propulsion (MEAAP) project, would reduce the load of the aircraft's onboard electrical supply and allowing for smaller, lighter power generation systems. This in turn could potentially reduce weight, fuel burn and CO2 emissions. There was a further environmental benefit, as the only by-product of regenerative fuel cells is water. When climbing or cruising, more electrical power than is required is generated. Regenerative fuel cells use that surplus energy to break water down into oxygen and hydrogen, which is then stored and used to produce electricity when supply falls short. IHI anticipated that applications would include power for galleys, pumps and lighting, but that a prototype regenerative fuel cell would be ready for ground testing within two years. It planned to carry out in-flight tests using regenerative fuel cells to provide auxiliary power by the end of 2013, and an all-electric system for the engine and aircraft of the future within the next decade or two.

The realization of this research began to emerge in the modifications to Boeing's 787 Dreamliner. Virtually everything that had traditionally been powered by bleed-air from the twin engines had been transitioned to an electric architecture, with electrically powered compressors and pumps, while completely eliminating pneumatics and hydraulics from some subsystems, e.g., engine starters or brakes. Electric brakes significantly reduce the mechanical complexity of the braking system and eliminate the potential for delays associated with leaking brake hydraulic fluid, leaking valves, and other hydraulic failures. The total available on-board electrical power is 1.45 megawatts, which is five times the power available on conventional pneumatic airliners. It is also enough electricity to power 31 average American households with all lights, ovens, furnaces and water heaters turned on to full, or 1,160 homes at average loads. The most notable electrically powered systems include: engine start, cabin pressurization, horizontal stabilizer trim, and wheel brakes. Like all aircraft, the Dreamliner has a triple-redundant system to move its control surfaces. Unlike other aircraft, two of them are electric, using motors in the wings to move the control surfaces. The third is a hydraulic system, compartmentalized with valves, and pressure is maintained using electrical compressors. The air conditioners use electrical heaters and electrical compressors. Electricity from the airport's ground power supply, or the aircraft's new massive lithium-ion battery banks, is used to electrically spin up the engines before they can be started. Wing ice protection uses electro-thermal heater mats on the wing slats instead of traditional hot bleed air. The research is ongoing for the next stage: electric self-taxiing by 2020; electric landing gear actuation system; electro-hydraulic flight control actuator; electro fuel pump system; electro oil pump; and an embedded starter-generator.

That lithium batteries can be inflammable was shown in July 2013, when an Ethiopian Airlines–operated Boeing 787 Dreamliner caught fire while on a remote parking stand at London's Heathrow Airport. According to the report by Air Accidents Investigation Branch, the fire was probably caused by wires for an emergency beacon's lithium-metal battery being crossed and trapped under the battery cover, which probably created a short-circuit. Three months later, a lithium-battery-electric Tesla Model S automobile caught fire after hitting metal debris on a highway in Kent, Washington. Such rare accidents can give a remarkable technology a poor reputation in the public mind.

In the pursuit of making all aspects of aviation electric, some are looking not only to innovation in planes, but in airships. In June 2016, France announced another plan for a commercial electric airship. Descended from France's Tissandier brothers' electric aerostat of 1883, one hundred and almost thirty years later, in 2012, a French firm called Flying Whales, founded by engineer Sébastien Bougon of Levallois Perret, has teamed up with Europe's leading ultracapacitor manufacturer Skeleton Technologies, led by TaaviMadiberk, in a program to build a 60-ton Large Capacity Airship, or LCA60T, for the global transport market. Skeleton Technologies is the only ultracapacitor manufacturer to use a patented nanoporous carbide-derived carbon, or "curved graphene," delivering twice the energy density and five times the power density offered by other manufacturers. The main advantage of the LCA60T airship will be its ability to transport heavy and oversized cargo of up to 60 tons in its 250-ft (75-m)-long underbelly, at speeds of 60 mph (100 kph), with a range of several thousand kilometers per day. The helium-filled, rigid-structure airship will be capable of winching to pick up and unload cargo while hovering, at a fraction of the cost of a heavy-lift helicopter, and for much heavier loads. Without the need to make conventional takeoffs and landings, energy consumption via its hybrid electric propulsion system will be low. Skeleton Technologies will join the program to help design and build hybrid propulsion for the LCA60T's electric power systems. Average operational power is expected to be approximately 1.5 MW with the company's graphene-based ultracapacitors assisting to cover the additional 2 MW peaks for hovering, lifting and stabilization in reasonable and turbulent environments. LCA60T will not require an airport or any kind of runway to operate, opening up new markets across the world for industries that require heavy-lift or oversize cargo options, across terrain lacking in infrastructure. It will be able to transport logging timber from remote locations, but that also means being able to deliver large items like wind turbines or electricity pylons in one piece to the side of a mountain, for example. It could also move prefabricated houses or building modules across undeveloped terrain or transport large aircraft components from one supply chain location to the next.

The program is part of the French government's "Nouvelle France Industrielle" plans for future transport, with the country's forestry agency highlighting the need for LCA60T to extract timber cargo. Other plans by the NFI include building at their Future Aeronautical Factory and selling 80 two-seater electric trainer aircraft by 2020.

In 2015, Chinese Prime Minister Li Keqiang and French Prime Minister Manuel Valls oversaw the signing of a cooperation and investment framework agreement between Flying Whales and the China Aviation Industry General Aircraft (AVIC General) company, which is to become a Flying Whales significant shareholder. The project will involve a consortium of about 30 companies and labs to cover the research and development, engineering, industrialization and manufacturing phases of the program. The first phase of engineering was completed by 2016. Industrial production is expected to start in 2021.[10]

As of 2018, French Air Base 125 at Istres–Le Tubé, northwest of Marseilles, and its gigantic hangar Mercure (160 m long, 38 wide and 25 high) will be ready to receive the first prototypes of the LCA60T. Mercure hangar will also see the construction of the Stratobus, an autonomous stratospheric airship, 100 meters long by 33 meters maximum diameter. The airship will be positioned at an altitude of about 20 kilometers (12 miles) over its theater of operations, in the lower layer of the stratosphere, which offers sufficient density to provide lift for the balloon. Winds at this altitude are moderate and stable throughout the entire zone between the tropics, at not more than 90 kph (55 mph), allowing the airship to remain stationary by using its electric propulsion system. Stratobus will carry payloads to perform missions such as the surveillance of borders or high-value sites, on land or at sea (video surveillance of offshore platforms, etc.), security (the fight against terrorism, drug trafficking, etc.), environmental monitoring (forest fires, soil erosion, pollution, etc.), and telecommunications (Internet, 5G). With €17 million in French government funding, Thales Alenia Space, a joint venture between the French aerospace group Thales and the Italian group Leonardo, is the lead company, and also in charge of systems integration, avionics, solar arrays and certification. CNIM (Construction Navale Industrielle de la Méditerranée), located in Seyne-sur-Mer, will build the structure and associated equipment, the ring and the nacelle, while Solutions F, based in Venelles, will provide the electric propulsion system, Airstar Aerospace the fully-dressed envelope, and Tronico-Alcen the energy conditioning system. In addition to these French partners, Cmr-Prototec of Norway will supply the energy storage system and MMIST of Canada the parachutes. Stratobus has been endorsed by the Pégase competitiveness cluster, in charge of launching a dirigible industry in France. At the June 2017 Paris Air Show, Thales Alenia Space announced that it had entered into a minority shareholding agreement in Airstar Aerospace, the European leader in the design and manufacture of stratospheric balloons, among others. Stratobus would gather the expertise of several small or medium groups, especially from Provence Alpes Côte d'Azur. Thus Solution F would be in charge of electric propulsion, while CNIM would be in charge of the equipped structure, the ring and the nacelle. As a reminder, the R&D program was launched in April 2016. Models will now be constructed, tests carried out for definitions, the objective being to be ready for a first demonstrator by 2018. The demonstrator will be small scale, 40 meters (130 ft) long by 12 meters (40 ft) in diameter, against 100 meters (328 ft) long and 33 m (100 ft) in diameter in its definitive version.

The Stratobus program kicked off officially on April 26, 2016, with three major program milestones. The first is the SRR, or System Readiness Review, which aims to consolidate the concept's specifications. The SRR was passed in December 2016. The next phase, the PDR, took place in mid–2017, resulting in a concept that meets the defined specifications. At the June 2017 Paris Air Show, Thales Alenia Space announced that it had entered into a minority shareholding agreement in Airstar Aerospace. The definitive version should take to the air by 2020 for a five-year flight.

Across the Pond, in 2013, DARPA and the USAF selected Lockheed's Skunk Works division over a rival bid from Northrop Grumman Martin to build Isis, a one-third scale unmanned airship demonstrator, powered by solar cells and fuel cells, that would be capable of operating at 21 km (13 mi) altitude for up to 10 years at a time, with radar technology so powerful it could spot a car hidden under a canopy of trees more than 300 km (180 mi) away. The energy cells must be capable of generating enough power to operate the radars, navigation system, communications gear, and the electric motors that will turn the airship's giant propellers. Both demonstrator sensors are significantly smaller than the envisioned operational system, which is expected to occupy an area 6,000 square meters (2.32 mi^2) across. Such an airship could revolutionize the way oil and mining companies haul equipment to the Arctic and other remote areas without roads. The airship should be certified by the Federal Aviation Administration by late 2017, paving the way for initial deliveries in 2018.

In Ontario, Jay Godsall, a biotechnology entrepreneur, has set up Solar Ship to build solar-powered triangular blimp-like flying machines that could serve as lifelines to the world's most remote places. Solar Ships range from an 11-meter-wide wingspan-envelope to a projected 100-meter (328 feet)-wide monster that could carry up to 30,000 kilograms (66,000 pounds) for a minimum of 2,000 kilometers (1,240 miles). While Solar Ship Inc.'s head office is in Toronto, its operations are in Brantford, Ontario, Cape Town, Lusaka, Kampala and Shenzhen. In September 2016, Manaf Freighters purchased four Solar Ship aircraft. On June 29, 2017, Solar Ship completed the first in a series of flights to develop a fossil fuel–free transport and logistics system for Canada's North. Solar Ship is working with Defence Research and Development Canada (DRDC) as part of project to support the Department of National Defence.

On July 26, 2016, Dr. Peter Harrop, Chairman of IDTechEx, gave a Webinar titled "Electric Aircraft Reach a Tipping Point." He observed: "About 20 companies make or will soon make electric aircraft. Nearly all are pure electric and fixed wing, the motorized hang glider and the self-launching sailplane being typical with one hour endurance. A bigger value market being addressed is training planes and bigger still will be hybrid fixed wing and vertical take-off aircraft hybrid and pure electric with the pure electric ones only managing 30 minutes. In this webinar we discuss possible uses, improvements and other types too."

It is now clear that the manned electric aircraft (MEA) business will be around $24 billion as soon as 2020, but the new analysis by IDTechEx sees truly hybrid and pure electric aircraft being a $24 billion business in 2031. Half of that will be relatively low-priced craft such as leisure and small work aircraft, and the high-priced half will be a mix of such things as helicopters, military aircraft and feeder aircraft, according to

IDTechEx projections, with large airliners not quite there. "Manned Electric Aircraft 2016–2031" reveals how much of this will no longer be a reworking of land-based technology but will be based on such things as superconducting power distribution and traction motors with at least four times the kW/kg and Distributed Electric Propulsion (DEP) along the full length of the wing. However, new concepts being developed first on land, such as supercapacitor body work and some other structural electronics, may have a place in these new ultra-lightweight aircraft.

Spreading the word on the new craft and new technologies has seen the growth of more air shows around the world, enabling the public to learn of the advances being made for the industry to share ideas.

EAA's annual AirVenture took place at Oshkosh in July 2016, but visitors hoping to see electric aircraft flying were disappointed. While manufacturers or would-be manufacturers of such light aircraft—including Pipistrel, Aero Electric Aircraft Corp. and Airbus Group—exhibited on the ground, the lack of suitable certification rules in the U.S. and Europe seems to be impeding progress. Airbus exhibited their E-Fan "Plus," painted in the colors of the American flag! But instead of the second rear seat, in its place was a small two-stroke avgas engine (Solo Aircraft 2625) with 41-liter capacity, a series produced in Germany for sailplanes, to drive a range-extending generator, doubling the airplane's flight endurance to around two hours plus a 30-minute reserve. E-Fan Plus would use its batteries and electric motors during takeoff and landing and the gasoline power plant during cruise. Following Air Venture at Oshkosh, the prototype returned to Saint-Sulpice-de-Royan, France, for further testing during the autumn of 2016.[11]

In November, Didier Esteyne traveled to Oslo, where he joined the Norwegian Airbus engineer Nils-Harald Hansen during the Zero Conference at Youngstorget to present the E-Fan 1:1 to KetilSolvik-Olsen, Minister of Transport and Communications; Dag Falk-Petersen, CEO of Avinor; John Eirik Laupsa, Secretary General of Norwegian Air Sports Federation; and Marius Holm, the head of Zero Conference. This fitted into Norway's dynamic of adopting electric transportation in all its forms: land, air and water. The northernmost county of Finnmark may become a giant testing ground for Norwegian and international green flights.

In March 2017 Airbus abruptly dropped its involvement in the all-electric E-Fan, while announcing that it would continue to pursue the development of a 90-seater short-haul hybrid called the E-Fan X. This was part of an austerity plan which would make over one thousand employees redundant, including closure of their R&D site in Suresnes and Pau. The new factory in the south of France would be used to explore how to introduce future electric-hybrid production concepts, eventually including commercial airliners. Several partners, including Daher-Socata and electric motor maker Siemens, are involved in the project.

On Thursday, March 23, 2017, taking off from the Dinslaken Schwarze Heide airfield, Walter Extra piloted the 330LE to a top speed of around 337.50 kph (209.7 mph) over a distance of three kilometers (1.86mi)—13.48 kph (8.38 mph) faster than the previous record, set by Bill Yates in 2013. The World Air Sports Federation (FAI) officially recognized the record flight in the category "Electric airplanes with a take-off weight less than 1,000 kilograms." In a slightly modified configuration with an overall weight exceeding one metric ton, test pilot Walter Kampsmann then flew 330LE to a speed of 342.86 kph (213 mph). On March 24, 2017, the Extra 330LE became the world's first electric aircraft to tow a glider into the sky. Walter Extra took a type LS8-neo glider up to a height of

600 meters (2,000 ft.) in only 76 seconds. Frank Anton, head of eAircraft at the Siemens venture capital unit next47, commented, "Just six such propulsion units would be sufficient to power a typical 19-seat hybrid-electric airplane. By 2030, we expect to see the first planes carrying up to 100 passengers and having a range of about 1,000 kilometers (600 miles)."[12]

In early April 2017, with more than 600 exhibitors from 35 countries, 33,000 visitors and 600 journalists from around the world, the AERO Friedrichshafen in Germany saw further developments in the e-flight-expo, organized in cooperation with Flying Pages GmbH. The "e" in e-flight stood for ecological, electrical, and evolutionary to promote ecological sustainability and progress in aviation. Companies such as Extra, Siemens, Pipistrel, Yuneec, Alisport, Axter Aerospace, Geiger Engineering and many more, as well as institutions such as GAMA or the University of Ulm, presented the latest e- and he-airplane developments. One day prior to the start, on Tuesday, and on closing day on Saturday, several aircraft with electric propulsion systems flew their rounds silently in the sky over the fairgrounds—Fabian Gabor in the *Magnus eFusion*, Walter Extra in *Extra 300 Elektro*, Edouard Maitre in the *Volta Elektro* Helicopter, Len Schumann in the *e-Genius*, and Jochen Polsz in the *Antares 23E*. At an oil spill exercise in the afternoon on Lake Constance, firefighters demonstrated an operation with drones. In addition, there were e-flight lectures about the history and development of electric flight.

Electric air shows still have a way to go compared to aviation meetings of a century ago. For example, in August 1909, a mere 6 years after the Wright Brothers took off, La Grande Semaine d'Aviation de la Champagne (The Great Champagne Aviation Week)

2017: Visitors to AERO at Friedrichshafen saw the Siemens-engined *Extra* and *E-Fusion* fly past in silence (Messe Friedrichshafen/AERO Friedrichshafen).

From left, Tine Tomažič, Frank Anton of Siemens and Ivo Boscarol, key players in Pipistrel's development (Pipistrel).

was held near Reims in France. Almost all of the prominent aviators of the time took part. No fewer than 38 airplanes were entered for the event, though in the end only 23 actually flew, representing nine different types. Eighty-seven flights of more than 5 kilometers (3 mi) were made.

From June 5 to 9, 2017, the AIAA Transformational Electric Flight Workshop & Expo was held at the Sheraton Denver Downtown Hotel. It brought together speakers and attendees from industry, government and academia worldwide. One of its features was the 4th Joint Transformative Vertical Flight Workshop. Among the speakers was Paul Eremenko, CTO of Airbus, who described the CityAirbus that the French original equipment manufacturer (OEM) is currently researching. It will be a four-seat, all-electric vertical takeoff and landing (VTOL) aircraft, and the "flagship" of the Airbus urban mobility division. The first flight of CityAirbus is expected to occur in 2018. It is currently being developed at the E-Aircraft Systems House, which is capable of testing power systems in excess of 20 megawatts. Eremenko said the ultimate goal, though, is the development of a completely new, single-aisle aircraft, powered by hybrid electric propulsion technology. In pursuit of that goal, Airbus continues to work on the E-Fan X. "For the first time since the jet age of passenger aviation ... we're really thinking about opening up the design trade space, beyond the tube and wings configurations," Eremenko said. "Hybrid propulsion enables us to think about distributed thrusters, creating a blown wing effect that could allow shrinking the wing area, or using differential thrust to control the yaw of the aircraft, reducing the vertical tail surface, or boundary level ingestion to re-ingest the wake at the tail of the airplane, cutting the overall drag by up to 10%."[13]

In September 2017, René Meier and his colleagues organized the world's first electric aircraft fly-in at Grenchen airport in Switzerland (Markus Jegerlehner).

One way ahead for Airbus's E-Fan X single-aisle airliner project was to take the BAe 146/Avro RJ regional jet airliner, nicknamed the Whisperjet for its quiet operation, and modify its 2-megawatt-class turbogenerator and batteries powering electric propulsors replacing one or two of the aircraft's four turbofans. But as the system matures and is demonstrated to be safe and, presumably, as battery costs come down, provisions will be made toward replacing a second turbine with another 2MW motor. If sanctioned, the aircraft would fly by 2020.

On the weekend of September 9–10, 2017, a team led by René Meier, ex-colonel of the Swiss Air Force, organized the SmartFlyer Challenge, Europe's 1st Fly-In (as opposed to Air Show) for electric-powered aircraft at Grenchen Regional Airport (LSZG), Switzerland, with both static and air displays of electric and hybrid-electric powered aircraft. The required electricity was produced by the airport itself, because every hangar roof is equipped with solar cells. The airport itself can also afford the three charging stations, each with 200 amps required for the Smartflyer Challenge. Grenchen Airport produces around 340,000 kWh of solar power per month. No fewer than seventeen electric airplanes and component manufacturers exhibited their products. In order to show the full range of electric mobility, electric bicycles and motorcycles were exhibited, Tesla was represented, and there was an 18-ton full-body truck on the ground. In addition to stands where developers and companies from Switzerland, Germany, France, Slovenia, the Czech Republic, Lithuania and Norway presented their products, a number of lectures and presentations were given in the existing training rooms of Grenchen Flying School: speakers included Dr. Frank Anton; Jean-Luc Charron, who reported on the state of the French project, where an electric airplane was integrated into the normal club operation; and Tine Tomažič, together with his colleague Paolo Romagnolli, presented the Pipstrel *Alpha Electro* aircraft. Lange Aviation presented the *Antares E2* fuel cell aircraft. While the old established aircraft manufacturers were noticeably absent, this was made up for by the forward-looking start-ups.

On Friday, September 8, the first electric aircraft to land at Grenchen airfield was the Magnus *e-Genius* piloted by Frank Anton coming over from Biel. But because of rain, Saturday became a static show, and spectators could walk around the passenger octocopter, *Whisper*, developed by Yves Pearcy of the French company Electric Aircraft Concept. Also present was the *SmartFlyer* being developed by two Swiss Airline pilots, Rolf Stuber and Daniel Wenger. The four-seater hybrid-powered aircraft with a 12-m (40-ft) wingspan has been developed in Grenchen; its potential range is 4 hours/800 km, and its 46 kg Siemens e-engine is linked to a two-cylinder piston engine. The project with a budget of 1.2 million Swiss francs over a period of five years is supported by 72 percent of the sustainability fund of the Federal Office of Civil Aviation.

Rain might have completely stopped play, but on Sunday afternoon, between the black clouds above the Jura and the blue sky above the Seeland, a perfect weather window opened. Over twelve e-airplanes, from ultralights to sailplanes to aerobatics, were able to take off and give more than three dozen demonstration flights, a world first.[14] Aviatrix Cornelia Ruppert from Wald, Switzerland, flew the family company's new *Archaeopteryx*, called the Elec'teryx, a self-launch, single-seat high-wing pod-and-boom microlift glider.[15]

2017: Roger Ruppert of Switzerland, engineer of the e-assisted Archaeopteryx sailplane; his aviatrix wife Cornelia has made 126 hours flying the Electric Archaeopteryx (courtesy Ruppert).

In 2017, a first: three electric airplanes in formation. Frank Anton pilots the *Magnus e-Fusion* (foreground), followed by the *D-14 Phoenix* motorglider, and the *e-Genius*. To the left, see the other e-aircraft parked below (photograph courtesy Jean-Marie Urlacher).

There was a never-before formation flight with three different e-aircraft photographed from a 1943 Piper L-4 chase-plane, itself towing a hang-glider, while from left to right flew the *eFusion*, the Storm Composite's *D-14 Phoenix*, and the *eGenius*. There was also a formation display with the *eFusion* in the air, then the Tesla Model S and Tesla Model X chasing it below.

In the international world of aviation, the event was recognized as an historic and pioneering success. The second edition of the Smartflyer Challenge will take place on the weekend of September 1 and 2, 2018, again at Grenchen Airport (LSZG).[16]

While the aircraft are being prepared, the airports themselves have been preparing the infrastructure. In 2009, Minneapolis-St. Paul International Airport installed ten 1 kW AeroVironment wind turbines on top of the airport fire station to harness the power of prevailing northwest winds. The turbines generated 10 kilowatts of electricity an hour and have since been used to recharge their fleet of Cushman utility vehicles.

In 2016, Air New Zealand announced that it was to have the country's largest electric vehicle fleet on the ground. The airline bought 36 fully electric cars from BMW and 12 plug-in hybrid Mitsubishi SUVs for its sales staff. It also directly imported 28 fully electric vans from Renault for use at airports. Air New Zealand said in a statement that it believed its fleet of 76 electric vehicles would be the largest in New Zealand and would save about 65,000 liters (17,000 gallons) of fuel per year. Alongside the New Zealand initiative, Australia's Sydney Airport has begun its initiative towards an environmentally friendly ground transport technology, with a fleet of six Electric Blu buses, built by Gemiland Coachworks and BYD and operated by Carbridge. Geneva Airport in Switzerland has installed fast recharging stations in Parking P1 for the users of electric vehicles. John Holland-Kaye, CEO of London-Heathrow Airport, drives to work in a Tesla, says his customers will soon ride on electric, driverless buses—and swears his plans for a new runway are carbon-neutral. Dag Falk-Petersen, the chairman of Norwegian airport operator Avinor, also thinks it's "highly realistic" that electric planes will fly commercial flights by 2025, although in the medium term, Falk-Petersen believes biofuels are key in reducing the industry's emissions.

The Chinese government is establishing 1,500 charging facilities at both Hongqiao and Pudong international airports. At Frankfurt, the E-Port "Green Gate" initiative has brought together Fraport AG, Lufthansa Group, the State of Hesse and the Rhine-Main model electro-mobility region to work towards aircraft ground services which in the future can be largely carried out with hybrid aircraft tugs and electric pallet trucks, as well as information panels and animations, including solar-powered passenger airstairs and electric conveyor belt vehicles.

Airbus has been testing out drones to make safety inspections of commercial airliners instead of humans making their way around the aircraft and using small cranes to make sure everything is in order. The drone is equipped with an Intel RealSense camera, which independently circles an A330 aircraft while rapidly snapping photos. These images are then applied to a 3D model of the aircraft that allows inspectors to get a close, detailed look and assess if the wing is primed, or prime to fall off. Airbus is currently experimenting with this process on its A350 family of aircraft. Inspection time is reduced to only 10 to 15 minutes.

Inspection drones are being used elsewhere. In 2012, University of Michigan engineering graduates and UAV enthusiasts Danny Ellis and Tom Brady founded Michigan Autonomous Aerial Vehicles (MAAV), then SkySpecs at Ann Arbor, Michigan, to use robotic drones to make 12-minute inspections of large structures. These included onshore and offshore wind turbines, surveying all sides of three blades, and collecting high-resolution images that identify cracks, erosion, lightning strikes and other anomalies. Data is sent to a web portal for viewing, annotating, and reporting. SkySpecs is a member of AWEA, WindEurope and the Commercial Drone Alliance. During the 30 days of May 2017, SkySpecs drones inspected a total 483 turbines or 1,449 individual blades, with an average of two operational systems at any given time.

One device that could make enormous savings is the WheelTug®, an electric motor roller attached outside of the nose wheel that would allow airliners to taxi from the runway to the gate without having to use a plane's main engine, thereby cutting operating costs and greenhouse gas emissions. WheelTug (U.S. Patent 20090261197) is the brainchild of a team led by Isaiah Watas Cox of Gibraltar-registered Borealis Technical Ltd. It is engineered around their multiphase Chorus Meshcon AC induction electric motor drive, which can produce five times the startup torque of a same-sized three-phase conventional motor. It is ideal for low-speed overload such as those with start-stop and traction requirements, providing high torque at low speeds without compromising high-speed operation and performance. In June 2005, Boeing Phantom Works and Chorus Motors ground-tested the WheelTug concept on an Air Canada 767 at the Evergreen Air Center at Pinal Air Park in Marana, Arizona. Further tests were conducted at Prague Airport in November 2010 in snowy and icy conditions, with the first fully "in-wheel" demonstration unit tested at the same airport in June 2012. By January 2015, WheelTug received an airframe from Canadian-based carrier Air Transat for development and testing at Montreal's

Trudeau International Airport. Once convinced, Air Transat decided to deploy the new system full-time from 2018. Cameras may also be installed on the fuselage to help pilots backing up without the need for ground crews and plane tugs. WheelTug estimates the system could save more than $1 million U.S. annually per plane and has received letters of intent for almost 1,000 aircraft from 22 airlines. WheelTug has not been the only approach. In 2016, Honeywell and Safran ended their partnership to develop its Electric Green Taxiing System for Airbus A320s. The system used a motor heavier than Wheel-Tug's. L-3 Communications also ceased similar partnerships with Lufthansa Technik and Crane Aerospace.[17]

On March 17, 2016, Tony Seba of Stanford University presented a paper at the Swedbank Nordic Energy Summit in Oslo titled "Clean Disruption—Why Energy & Transportation will be Obsolete by 2030." Seba boldly asserts that among the outcomes of the Clean Disruption, by 2030, all new vehicles will be electric; all new vehicles will be autonomous (self-driving); and all new energy will be provided by solar (and wind).

If long-haul e-airliners are limited in battery range, another approach would be to revitalize short-haul regional airports. The name Zunum comes from the Mayan word for hummingbird. The plan is to develop 10- to 50-seat hybrid-electric aircraft that can pave the way for what founder Ashish Kumar calls a "golden age of regional travel." Kumar has recruited Matt Knapp and Petek Saracoglu to realize his dream. With 35 U.S. patents, several more global, Haran is in charge of Advanced Drives. Zunum Aero has been working with the Center for Power Optimization of Electro-Thermal Systems at the University of Illinois on a technology approach that blends battery storage with engine-generated power. An onboard software system would calculate when the engine needs to be on, and for how long, depending on the route. Zunum's development plan calls for aircraft with a range of 700 miles (1,000 km) to be ready by the early 2020s, with that range extended to 1,000 miles (1,600 km) by 2030. "You need to do the range very carefully to get to the disruptive economics, which is what we are focused on," Kumar said. Having patented its concept, Zunum has won backing from the Boeing Co. as well as from JetBlue Technology Ventures, a subsidiary of JetBlue Airways. Bonny Simi, president of JetBlue Technology Ventures, was similarly bullish. She said Zunum was likely to "light up a vast network of underutilized airports and re-invent regional travel." Kumar has cited figures suggesting that only 2 percent of the nation's more than 5,000 airports account for 96 percent of today's air traffic. That means the bigger airports are jammed, while scheduled service from the smaller airports is costly, if it exists at all.[18] In June 2017, Zunum won an $800,000 grant from the state of Washington's Clean Energy Fund. By September, Zunum had announced an initial 12-passenger hybrid aircraft, using either Tesla or Panasonic batteries to power twin bypass fan e-motors. Flight services would begin in 2022. By 2030, they expect to have a plane that can carry 50 passengers up to 1,609 kilometers (1,000 miles).

A microcosmic version of this is already taking place in California, where Joseph Oldham, an experienced 64-year-old pilot and director of the San Joaquin Valley Clean Transportation Center for CALSTART, has launched the Sustainable Aviation Project: eFlightEverywhere. This is a joint effort between the City of Reedley, the City of Mendota, Reedley College, Mazzei Flying Service, the Fresno Business Council, and Calstart, with the goal of getting the public familiar and comfortable with electric propulsion in aircraft as a step toward the advanced electric-powered aerial sky taxi vehicles of the future and larger commercial electrically powered aircraft. The strategy, like Zunum, is to create a network of electric airplane charging stations, none more than 45 nautical miles from any three others, so allowing a fleet of four Pipistrel Alpha Electro trainers to depart the traffic pattern, and thus remove the principal limitation that has so far prevented flight schools from investing in larger numbers. Those plans are being paid for with a $1 million grant administered by the Fresno Council of Governments. Waiting for the arrival of their first G2, new hangars and electrical circuits have been installed at the Reedley Municipal Airport and Fresno Chandler Executive Airport.

World electric airspeed record-holder Chip Yates's latest venture is to transform his Long-ESA into a 150-seat electric airliner for short-haul flights. With Jeff Engler, a venture capitalist and formerly founder of Podimetrics, an early warning system to predict and prevent diabetic foot ulcers, Yates called their startup Wright Electric—a play on words "right electric" and as a tribute to pioneer aviators the Wright Brothers. Provided lithium batteries reach the required energy density, the idea for the Wright Flyer would be to use modular battery packs for quick swap installed in the same cargo container as in a regular airplane, which would give it a 300-mile range. The Wright Electric is not only backed by Y Combinator, Silicon Valley's most highly regarded start-up incubator program, but has entered a partnership with budget British airline EasyJet as a client for its "Wright One."[19] The average flight time of EasyJet's 200-strong fleet is under two hours.

EViation is an Israeli project to produce a three-engine distributed-propulsion aircraft that would feature one electric motor at the tail and one on each wingtip for unheard-of aerodynamic efficiency. Based in Kadima, Zoran, the start-up was founded by Omer Bar-Yohai and two other Israeli entrepreneurs. It will be using a 400Wh/kg, 330kWh hybrid lithium-ion and aluminum-air/zinc-air battery developed by Aviv Tzidon at Phinergy Ltd. in Lod, based on technology originating at Bar Ilan University. In conventional aluminum-air batteries, aluminum reacts to oxygen and produces electricity. Phinergy's innovation uses water and recycles the hydrated aluminum oxide to create an anode, a process that enables a closed and sustainable life cycle or recharge. The company has patented a "nano-porous silver-based catalyst," which allows oxygen into the electrode and the cell while at the same time blocking carbon dioxide. Phinergy claims that this invention enables its battery cathode to sustain 25,000 working hours, while significantly reducing weight. An EV fitted with these batteries motored for 1,000 mi. (1,600 km) before needing a recharge. For security, these batteries are spread out in more than a dozen places. EViation is planning two models: in the luxury Alice ER, six to nine passengers will be flown over distances of up to 800 mi. (1,300 km), at a cruising speed of over 220 kt. (253 mph); in the Orca, the unmanned variant, designed to a maximum takeoff weight of 250 kg. (550 lb.), a range of over 497 mi. (800 km) is envisaged at a cruising speed of 144 kt. (166 mph), remaining airborne for more than eight hours, and carrying a payload of 50 kg (110 lb.). It can be summoned using a smartphone application. Designed for VSTOL (very short takeoff and landing), it only requires a 99-ft.-long (30 m) runway.

EViation is using Stratasys 3D printing to accelerate its R&D process for everything from prototyping to tooling and evaluation of production parts. The company has saved several hundreds of thousands of dollars and months of workforce hours with Stratasys 3D. For example, EViation 3D-printed its wingtip motors in a matter of hours, enabling swifter design and functional evaluation, while waiting for the final motors to be shipped. Another key aspect of EViation's design is its ability to reduce interference drag on the exterior of the aircraft by employing smooth, curved surfaces. EViation was able to create the required strong, geometrically complex, lightweight parts to support these surfaces by 3D printing a composite lay-up tool in ULTEM 1010 polyether imide (PEI) material, which was then covered with carbon fiber.

While the prototype was unveiled in Static Display A8 of the 52nd International Paris Air Show in June 2017, EViation announced its teaming with Magnaghi Aeronautica SpA, the manufacturer of the SkyAero aircraft, and FBM, an Israeli producer of carbon-based composites, which manufactured the all-composite prototypes. The first Eviation production model is slated to arrive in 2020, using electric motors supplied either by Yasa of England or Siemens of Germany, as well as 6,000 lb. of batteries (2,700 kg).

Another more popular commercial approach is with an e-airplane produced in kit form which enthusiasts can buy and assemble like a model aircraft, using guidance from an ongoing website. In 2013, Dan and Carol Johnson of Adventure Aircraft Inc. in Corning, California, teamed together with a number of companies including Quicksilver Aircraft to create the low-cost EMG-6 electric ultralight motor-glider trainer to be used as a legal part 103, which can be built from plans or built from a number of different kits. As of July 2017, nineteen different airplane engineers across the U.S. were building the 15 kW Plattenburg 37-engined EMG-6. The design incorporates match-drilled aluminum skins and CNC manufactured components.

On September 21, 2017, when California politician and former action star Arnold Schwarzenegger presented the world's first electric Hummer by Kreisel Electric in Rainbach, Austria, he remarked, "If Kreisel keeps it up at this pace, I will soon be able to fly here from LA in an electric airplane."[20] Given that many Hollywood stars own and drive Tesla S and X autos, Schwarzenegger may soon be followed into the sky.

In October 2017, *Industry Today* published a report titled "Global Electric Aircraft Market accounted for $18.35 billion in 2015 and is expected to reach $24.9 billion by 2022 growing at a CAGR of 4.5% during the forecast period." In 2015, North America accounted for the highest market share as compared to other regions. The Asia Pacific region is expected to grow at the highest CAGR.

On February 7, 2018, the World Flying Electric Vehicle Summit will be held in Geneva. The event, covering all aspects of flying electric vehicles, will show how, in just one decade, advances have been made in aircraft of every size from kit planes though gliders and single-seaters to commercial airliners, taking in airships for both passengers and freight. Electric flight is also affecting not only the aircraft and how it works, but the very nature of airports as well.

Taking the whole range of electric aircraft, while this is largely a book about prototypes or small editions, where they *have* scored commercially in a very major way is with those smaller units known as UAVs or drones, which began full-scale and then diminished to small scale down to micro and/or back to full scale, as the next chapters will recount.

NINE

UAVs (Generation Gas)

In his book *The Unknown Tomorrow*, published in 1910, Anglo-French journalist and writer William Tufnell Le Queux described a civilization where "nearly everybody possesses his own aeroplane." It was a dream delayed by two world wars, but that vision may soon be coming true, if in rather a roundabout manner.

To understand the unmanned vertical takeoff electric airplane, we must follow two historical threads, both of which begin again in France. The origins of the controlled unmanned aircraft begin in Paris, with Octave Détable, general clerk at the Paris Law Courts and vice-president of the French Kite-Flying League. In 1894, Détable had built and flown a kite with divergent cones to give it stability, an innovation he improved in 1897 with his 19 m² (200 ft²) kite. By 1914 Détable and his son Pierre had progressed to a low-powered flying machine with Lieutenant Max Boucher, trained by Louis Blériot in Etampes, as his test pilot. These tests were interrupted by the declaration of war, but three years later, Captain Boucher, following the advice of General Ferrie, a wireless pioneer, and in collaboration with a team of lieutenant engineers, installed a remote-control transmitter system into a 150 hp Voisin aeroplane. On July 2, 1917, the pilotless plane, carrying a sack of sand weighing 95 kg (210 lb.) to replace the pilot, made a test flight. With only 2 liters (½ gal.) of gas on board, after 1 kilometer, the plane landed, having run out of fuel.[1]

Given that UAVs are remote-controlled, mention should also be made that in the 1890s, hydrogen-filled model airships, controlled by a spark-emitted radio signal, were flown as a music-hall act around theater auditoriums.[2] Then in 1898, watched by thousands of spectators, Croatian inventor Nikola Tesla had demonstrated the application of Herzian waves when he radio-controlled his 4-ft. (1.2 m) "teleautomaton" model electric boat up and down a pond at New York's recently completed Madison Square Garden during an electrical exhibition. Tesla was able to control the boat's motor, sending it zipping around the pond seemingly under its own control. He even installed lights on it that he could blink at a distance from his control box. Many observers, unfamiliar with radio waves, thought that the device must have a brain of its own or that somehow Tesla was controlling it with his mind. When it was first shown, "it created a sensation such as no other invention of mine has ever produced," Tesla would later write. He had taken out U.S. and UK patents for it.[3]

Twelve years later, in 1910, an experiment was carried out with a full-scale boat on the Dutzendteich Lake near Nuremberg, Germany. Christoph Wirth, an electrophysicist, having patented a wireless-controlled current distributor, decided to test it out in a boat.

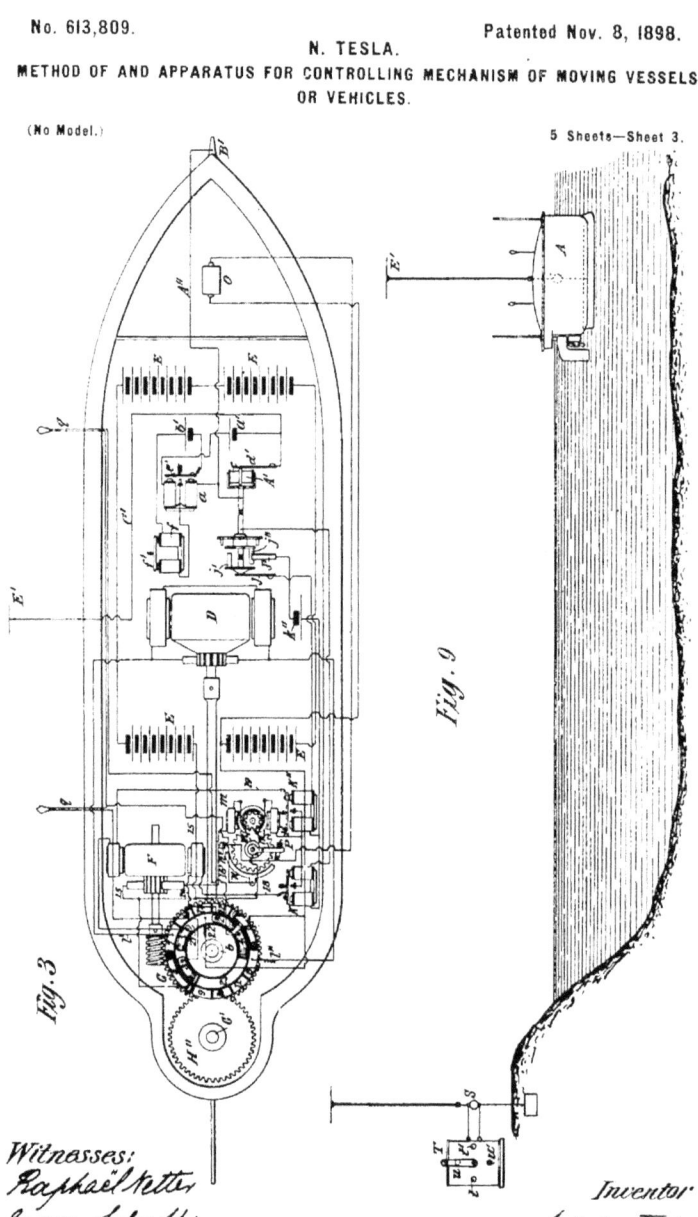

Nikola Tesla's patent for the remote control of his 4-foot "teleautomaton" on a pond at the recently completed Madison Square Garden, New York, in 1898.

Together with the manufacturer Beck and a merchant named Knauss, Wirth set up the demonstration of a 33-ft (10-m) boat called the *Prinz Ludwig* with a 3 kW motor and an accumulator battery of 80 volts and 300 ampere-hours. The experiment began with the pilotless boat positioned out in the center of the lake in front of the clubhouse of the Kaiserliche Marine (Imperial German Admiralty). At the given signal, a gunshot, the *Prinz Ludwig* silently accelerated to a speed of 10 mph (16 kmh). It was then made to turn right or left or to stop completely and start again by the controlling operator in obedience to the requests from members of the club. Each order was executed within one to five seconds, and signal lights flashed back the receipt of the impulses. The maneuvers were continued for several hours. In the weeks that followed, "das Geisterschiff" ("The Phantom Ship") attracted many spectators—military, scientific, and the simply curious—who paid a small sum for admission to the immediate neighborhood of the operator on the bank. In 1912 a boat 50 feet (15 m) in length was exhibited in Berlin, again at the invitation of the Kaiserliche Marine Club. An antenna of four wires was stretched between the cupola of the Kaiser Pavilion and the restaurant on the shore of the Wannsee. The transmitting apparatus, which was installed at the restaurant, was of the induction coil type, and was of about 100 watts capacity. The various operations performed on the boat were accomplished by sending impulses by means of a Morse key. The boat was equipped with an antenna of four wires about 15 feet (4.5 m) high, a radio receiver capable of adjustment to different wavelengths from the transmitter, a distributor or selector, electric steering apparatus, signal guns, lights, and fireworks apparatus. The tuning of the apparatus could be altered by sending a long signal; this was for the purpose of evading interference. The implications of such a device for controlling the direction of a torpedo were not to be ignored.[4]

Over in Britain, Archibald Montgomery Low, a consulting engineer, was preparing his Aerial Target (AT), an unmanned flying machine remotely controlled by wireless telegraph. Its main task was to train apprentice pilots to shoot without risking shooting down the plane towing the target. Assisted by Captain Poole and Lieutenant Bowen, Low demonstrated a working prototype. Impressed, General Sir David Henderson (Director-General of Military Aeronautics) ordered that the Royal Flying Corps Experimental Works should be created to build the first proper Aerial Target, complete with explosive warhead. As head of the Experimental Works, Low was given about 30 picked men, including jewelers, carpenters and aircraftsmen in order to get the pilotless plane built as quickly as possible. The plane, named the Ruston Proctor AT from its manufacturer, was designed by H.P. Folland.

It had its first trial on March 21, 1917, at Upavon Central Flying School near Salisbury Plain, attended by 30 to 40 allied generals. The AT was launched from the back of a truck using compressed air (another first). Low and his team successfully demonstrated their ability to control the craft before engine failure led to its crash landing. A subsequent full trial on July 6, 1917, was cut short, as an aerial had been lost at takeoff. At a later date, an electrically driven gyro (yet another first) was added to the plane, but ultimately the Aerial Target project was not followed up after the war, due to the shortsightedness of military planning. Low later came to be known as the "father of radio guidance systems" due to his pioneering work on guided rockets, planes and torpedoes.

In the United States, Lawrence Sperry, a Curtiss Flying School graduate and son of Elmer Sperry, had developed a lightweight gyroscopic stabilizer apparatus that could be coupled to control surfaces to maintain the flight axes of aircraft. Having fitted it onto a

Curtiss C-2 biplane with a hydroplane fuselage, in June 1914 Sperry Jr. entered and won the Concours de la Sécurité en Aéroplane (Airplane Safety Competition) held on the banks of the Seine River by the Aero Club of France. During World War I, Sperry Instruments Corporation introduced advanced aircraft instruments, including compass, air speed indicator, altimeter, bank indicator, angle of attack and stall warning. In 1917, with Peter Cooper Hewitt, Sperry worked on a radio-controlled airplane project, the Hewitt-Sperry Automatic Airplane. An adapted Curtiss C9 Seaplane, it included the Sperry automatic gyroscope to control and maintain an airplane in any desired position, and also to free a pilot for aiming and dropping bombs.

Back in France, Captain Boucher continued his experiments with automatic stabilization, controlled flight, programmed flight, and remote-controlled flight. On September 14, 1918, a Voisin BN3 plane was flown for 51 minutes on a journey of 100 km (60 mi). It flew from Chicheny, to Méréville, Pussay, Chicheny, Villesauvage, Chalou, and landed back at Chicheny. The following day Captain Boucher received the order to cease operations and to lay off his team. War was soon over. Despite this, George Clemenceau, then president of the Senate Commission of the Army, launched an unmanned airplane contest. So Boucher persisted. Working with engineer Maurice Percheron, on April 17, 1923, he succeeded in flying a new unmanned aircraft from a field in Etampes, but this was not enough to obtain continued support from the military. Boucher died, virtually forgotten, in 1929.[5]

Over in the USA, another unmanned aerial vehicle was developed in 1918 as a secret project supervised by Orville Wright and Charles F. Kettering. Kettering was an electrical engineer and founder of the Dayton Engineering Laboratories Company, known as Delco, which pioneered electric ignition systems for automobiles and was soon bought out by General Motors. At GM, Kettering continued to invent and develop improvements to the automobile, as well as portable lighting systems and refrigeration coolants, and he even experimented with harnessing solar energy. When the U.S. entered World War I, his engineering prowess was applied to the war effort. Under Kettering's direction, the government developed the world's first "self-flying aerial torpedo," which eventually came to be known as the "Kettering Bug." The bug was a simple, cheaply made 12-foot (3.66 m)-long wooden biplane with a wingspan of nearly 15 feet (4.58 m) that, according to the National Museum of the U.S. Air Force, weighed just 530 pounds (240 kg), including a 180-lb. (80 kg) bomb. It was powered by a four-cylinder, 40-horsepower engine manufactured by Ford. Kettering believed that his Bugs could be calibrated for precision attacks against fortified enemy defenses up to 75 miles (120 km) away—a much greater distance than could be reached by any field artillery. During the 1920s, what had become the U.S. Army Air Service continued to experiment with the aircraft until funding was withdrawn.

The Royal Aircraft Establishment Larynx (from "Long Range Gun with Lynx engine") was an early British pilotless aircraft, to be used as a guided anti-ship weapon. Started in September 1925, it was an early cruise missile guided by an autopilot.

Geoffrey de Havilland was not only an aircraft designer and builder. He was a passionate lepidopterist or lover of butterflies and moths. His airplanes, built at Stage Lane Aerodrome, Edgeware, North London, were therefore named Hermes Moth, Genet Moth, Gipsy Moth, Cirrus Moth and Giant Moth. In 1935, de Havilland produced a pilotless version of their Tiger Moth DH.82 which was called the Queen Bee. But their noisy, slow and heavy flight rather made them look like male *drones* rather than queen bees. These aircraft retained a normal front cockpit for test-flying or ferry flights, but a four-bladed wooden windmill in the propeller slipstream on the fuselage port side drove an air pump to provide compressed air for the gyro unit and servos. The Queen Bee was first flown, manned, at Hatfield in 1935, then remotely controlled at Farnborough later that year. A total of 412 were built between 1933 and 1943.

In December 1935, the Chief of U.S. Naval Operations, Admiral William H. Standley, attended the London Disarmament Conference, after which he was given a demonstration of the Queen Bee. On his return, Standley assigned an officer, Lieutenant Cmdr. Delmer S. Fahrney at the Radio Division of the Naval Research Laboratory, to develop a similar system for U.S. Navy gunnery training. Fahrney adopted the name "drone" to refer to these aircraft in homage to the Queen Bee. Drone became the official U.S. Navy designation for target drones for many decades.[6]

But this output pales into insignificance when compared to the U.S. Navy's autonomous aircraft target, the U.S. Radiophone OQ-2, otherwise known as Target Drone Denny 1 (TDD-1), of which several thousand units were built. A follow-on version, the OQ-3, became the most widely used target aircraft in U.S. service, with over 9,400 being built during World War II. The U.S. Navy also launched a new program, called Operation Anvil, to target deep German bunkers using refitted B-24 bombers filled to double capacity with explosives and guided by remote control devices to crash at selected targets in Germany and Nazi-controlled France.

The Third Reich's ground/air radio-controlled missiles V-1 and V-2 had a flight schedule preset before takeoff. But the "Wasserfall" V-2 could be remotely controlled by a microwave radar system called "Rheinland." Other radio-controlled missiles with names such as Enzian and Rheintochter were also developed, as well as no fewer than 5 radio control systems (Burgund, Franken, Elsass, Brabant, Ganza). Air/ground radio-controlled missiles were also developed such as the Hs 293 and Fritz-X, which were piloted by a Strasburg-Kehl radio control system using a 48–50 MHz waveband.

Drones were also developed during the Korean War and the Vietnam War for observation and surveillance missions in enemy territory. In 1993 the Japanese cult Aum Shinriko looked at using a radio-controlled helicopter to spray nerve gas.

Our second thread, the vertical takeoff multi-rotor airplane, or copter, can be traced much further back. It was in 1493 that the Italian polymath Leonardo da Vinci, also cited in Chapter One of this history, made a sketch in his notebook of a screw-like machine, commenting: "If this instrument made with a screw be well made—that is to say, made of linen of which the pores are stopped up with starch and be turned swiftly, the said screw will make its spiral in the air and it will rise high."[7]

Four hundred years later, in 1863, the Viscount Gustave du Ponton d'Amécourt, president of the Society for the Encouragement of Aerial Locomotion by Means of Heavier than Air Machines, published a forty-page monograph in Paris, titled "The Conquest of the air by propeller. Account of a new system of aviation." In this he put together the Greek words *helico* and *pteron*, meaning "spiral" and "wing," to make the word hélicoptère (helicopter). For d'Amécourt and his friends were among those very few who were passionately convinced that the future of flying was with *heavier-than-air* machines. To prove their point, having built a fragile flying machine model driven by clockwork springs, they watched it ascend vertically for a few seconds to a height of less than 3 meters (10 feet), only to crash back to the ground. Stimulated by the brief success of his clockwork models, in 1868, the Viscount built a hélicoptère powered by a small steam engine with an aluminum boiler. This model proved a failure inasmuch as it only lifted a third of its own weight.[8]

Still in France, forty years later, Louis Charles Bréguet and his younger brother Jacques were working on AC electric motors for submarines at the family business in Douai, France, when they came into contact with the respected physiologist Professor Charles Richet. He challenged them to design a gyroplane with flexible wings. The Bréguet-Richet had an uncovered open steel framework with a seat for the pilot and a 34 kW (46 hp) Antoinette water-cooled piston engine at the center. Radiating from the central structure were four wire-braced tubular steel arms, each bearing a superimposed pair of four-bladed rotors. In essence, a quadcopter. To eliminate the torque effect, two rotor sets were driven clockwise and two counterclockwise. On September 29, 1907, *Gyroplane No. I* was given a static test by engineer Maurice Volumard in the family factory and then flown in a field at La Brayelle, albeit to an elevation of only 0.6 meters (2.0 ft) for a couple of minutes! It was not a free flight, as four men were used to steady the structure. It was neither controllable nor steerable, but it was the first time a rotary-wing device had lifted itself and a pilot into the air. It later flew up to 1.52 m (4.99 ft) above the ground. The design was improved and *Gyroplane No. II* appeared the following year. *No. II* had two two-blade rotors of 7.85 m (25.75 ft) diameter and also had fixed wings. Powered by a 41 kW (55 hp) Renault engine, it was reported to have flown successfully more than once in 1908. *No. II* was damaged in a heavy landing and was rebuilt as the *No. II bis*. It flew at least once in April 1909 before being destroyed when the company's works were badly damaged in a severe storm.

Papin and Rouilly's "Gyroptère" of 1914, resembling a giant boomerang, was meant to work in reverse of a falling maple seed (©D.R. / Coll. musée de l'Air et de l'Espace, Le Bourget).

Another French flight design was inspired by the motion of the maple seed spinning as it falls to the round. In 1913, Alphonse Papin and Didier Rouilly of Paris obtained French and world patents for their "Gyroptère," characterized in the contemporary French journal *La Nature* in 1914 as "un boomerang géant" (a giant boomerang).

The machine weighed 500 kg (1,100 lb.) including the float on which it was mounted. It had a single hollow blade with an area of 12 square meters (130 sq ft.), counterweighted by a fan driven by an 80 hp Le Rhone rotary engine spinning at 1,200 rpm, which produced an output of just over 7 cubic meters (250 cu ft.) of air per second. The fan also propelled air through the hollow blade, from which it escaped through an L-shaped tube at a speed of 100 m/s (330 ft/s). Directional control was to be achieved by means of a small auxiliary tube through which some of the air was driven and which could be directed in whatever direction the pilot wished. The pilot's position was located at the center of gravity between the blade and the fan. Testing was delayed due to the outbreak of World War I and did not take place until March 31, 1915, on Lake Cercey on the Côte-d'Or. Due to the difficulty of balancing the craft, a rotor speed of only 47 rpm was achieved instead of the 60 rpm which had been calculated as necessary for takeoff. In addition, the rotary engine used was not powerful enough; it had originally been planned to use a 100 hp car engine, which proved unobtainable. Unfortunately, the aircraft became unstable and the pilot had to abandon it, after which it sank. As we will see, this design has also been resurrected by drone engineers.

Still in France, then the cradle of innovation, Etienne Edmond Oehmichen was a keen naturalist studying the flight of birds and insects. He earned his bread and butter as an electrical engineer, first at the Electrical Department of the Mechanical Construction Company of Alsace in Belfort, then with Peugeot at Beaulieu-Valentigney, where he innovated electric lighting and a starting system for automobiles. He also patented an electric stroboscope for inspecting auto parts. In 1920, Oehmichen published a book, *Nos maîtres les oiseaux, étude sur le vol animal et la récupération de l'énergie dans les fluides*, published by Dunod (*Our masters, the birds, a study on animal flight and recovering energy in fluids*).

In 1921, financed by Peugeot, Oehmichen assembled his first "hélicostat," fitted with two large propellers and a hydrogen balloon to ensure stability, and on February 18, made a flight of 1 minute at a height of 10 meters. On November 11, 1922, he first flew Hélicoptère Oehmichen No. 2, fitted with twelve propellers (four for lifting and eight for direction). He then progressed to four vertical axis rotors distributed on either side of the fuselage, while five small propellers ensured horizontal stability, completed by a steering prop and two forward props propelled by an 88 kW Rhône motor. On May 4, 1924, Oehmichen made a flight of 1 km around a closed triangular circuit on the Arbouans flying field, next to Peugeot's auto factory. The VTOL was born. The feat enabled him to obtain a grant of 90,000 francs from the Service Technique de l'Aéronautique (STAé) with which he was able to pay back Peugeot. The same year, he made a stationary flight of three minutes, then another with two passengers on board. Oehmichen continued to develop his hélicostats, integrating the technologies of rotary wing and airship, but was unable to win over those responsible for aeronautics. In his conference of May 20, 1937, at the French Colonial Institute, Oehmichen stated his firm belief that the hélicostat was the only solution to ensure aerial safety.

Across the Atlantic, at the same time, George de Bothezat and his assistant Ivan Jerome, both refugees from the Russian Revolution, were also working on an experimental quad-rotor helicopter. Having written and lectured extensively on rotorcraft theory, de

In 1922, Etienne Edmond Oehmichen designed and personally test flew his experimental "hélicostats" (©D.R. / Coll. musée de l'Air et de l'Espace, Le Bourget).

Bothezat received a contract from the United States Army in 1921 for the construction of an experimental quadcopter they called the "Jerome-de Bothezat Flying Octopus." Built by the U.S. Air Service, it made its first flight in October 1922. Although its four massive six-bladed rotors allowed the craft to successfully fly, it suffered from complexity, control difficulties, and high pilot workload, and was reportedly only capable of forward flight in a favorable wind. About 100 flights were made by the end of 1923. Designed to take a payload of three people in addition to the pilot, it was supposed to reach an altitude of 100 meters (330 feet), but the highest the Octopus ever reached was about 5 m (16 ft)! The Army canceled the program in 1924, and the aircraft was scrapped.

Both the Oemichen and Bothezat designs were propelled by propellers perpendicular to the main rotors and were therefore not true quad-rotors. Since these early designs suffered from poor engine performance and could only reach a couple of meters in height, not much was done to the quad-rotor design during the following three decades.

One rare example was the disc-copter, designed by a polymath sculptor, carpenter and blacksmith called Alexander G. Weygers of Oakland, California. Fascinated with the concept of flight from an early age, as Weygers watched dolphins skimming the waves at the bow of ships, he wondered how they could swim so fast, for such a long time. When he realized they were actually surfing the cushion of water pushed in front of the ship, he was inspired to apply this principle to the designs for his flying machine. The disc-copter was designed to float on a stream of air from internal rotors, housed within the body of the machine. He was working on his idea of what we now call a "flying saucer" from 1927 and sent detailed plans to all the branches of the U.S. military. He was eventually told that they were intrigued by the concept and the design of the craft, but were not prepared at that time because the war effort superseded its development. He was only granted U.S. Patent US2377835 in June 1945.

In January 1928, the U.S. Patent Office granted Nikola Tesla what would be his final patent, for a "novel method of transporting bodies through the air" (1,655,114). Calling it a helicopter-plane, he described and sketched an open box-type craft with a tilting propeller and wing that theoretically would enable the vehicle to rise vertically and fly horizontally, though he also suggested a design in which two propellers "coaxially or otherwise disposed" would "revolve in opposite directions," powered by his turbine engine. By 1928, Tesla was without a laboratory or funding and headed toward impoverishment, passing much of his time tending to the pigeons in New York City's Bryant Park. He died alone and penniless in 1943 at 86.

In 1938 during a Philadelphia Conference on VTOL, test pilot James G. Ray proposed that "A vehicle that can take you from your home to your office, to your country club, or your bank or to your friend's house, by air or by road, whichever is most convenient, will have a vast usefulness." In 1931, Ray, who later envisioned a convertiplane in every garage, landed his Pitcairn PCA-2 autogiro on the White House lawn, showed it to President Hoover, and took off in it again.

In May 1951, *Modern Mechanix* published a picture story, titled "Flying Saucer," about another German, Walter Otto Galonska of Frankfurt-am Main, who, for advertising purposes only, had spent a year building a small oval flying machine powered by two contra-rotating props. Since the postwar building of free-flying machines was forbidden to Germans by the Western Powers, Galonska had tethered it with an electrical supply steel cable. Further research has revealed that some six years before, on April 2, 1945, Galonska, who had developed plans for a device for controlling unmanned airplanes on his estate in Breslau, was offered a contract by the Soviet Union to build a free-flying working model, but the Americans heard about it. Despite U.S. Military Government Law No. 23, Article III, Paragraph b: "Applied scientific research is prohibited in the fields of aerodynamics," in July 1948 Golanska was allowed to build Germany's first postwar model aircraft in Frankfurt, as seen in the American magazine. In 1958, Galonska received U.S. Patent 3059428 for an internal combustion engine supercharging turbine for liquid fuels and coal dust, although nothing more was heard of his unmanned helicopter design, or drone.

In 1956, David H. Kaplan designed and test flew the first true quad-rotor, the 1-ton heavy Convertawings Model "A," which was able to hover and maneuver over Long Island using its two 90 horsepower engines. The four rotors were positioned in an "H" configuration, and the design incorporated simplified hubs with strap-mounted blades, a form of "hinge-less" rotor. Convertawings proposed a Model E that would have a maximum weight of 42,000 lb. (19 metric t) and to be able to fly with a payload of 10,900 lb. (4.9 metric t) to over 300 miles (480 km) and at up to 173 mph (278 kph). Despite successful testing and development, military support for the quad-rotor ceased following cutbacks in defense spending.

Also working on this configuration was Richard Vogt, former Nazi engineer and aircraft designer, recruited in post–Nazi Germany and taken to the U.S. for government employment, at the end of World War II. While working for Hamburger Flugszeugbau, Vogt had designed Seedrache (English: Sea Dragon), a long-range maritime reconnaissance flying boat. The aircraft was powered by three engines. The center engine was mounted above the "hanged" fuselage. By the late 1950s Vogt, in his sixties, was chief designer at Curtiss-Wright's Santa Barbara Division (formerly the Aerophysics Development Corporation) when he was asked to design a flying jeep-type light VTOL utility vehicle, designated the VZ-7. It was of exceedingly simple design, essentially consisting of a 16-ft. (5-m) rectangular central airframe to which four vertically-mounted propellers were attached in a square pattern. The central fuselage carried the pilot's seat, flight controls, fuel and lubricant tanks, and the craft's single 425shp Turbomeca Artouste IIB shaft turbine engine. In essence, it was a quadcopter. The aircraft performed well during tests, but was not able to meet the Army's standards; therefore it was retired and returned to the manufacturer in 1960. Vogt's Patent U.S. D189462 S, granted in December 1960, would become the ancestor for today's quadcopter innovations.

United States Patent Office

Des. 189,462
Patented Dec. 20, 1960

189,462
AIRCRAFT
Richard Vogt, Santa Barbara, Calif., assignor to Curtiss-Wright Corporation, a corporation of Delaware
Filed Feb. 19, 1959, Ser. No. 54,681
Term of patent 14 years
(Cl. D71—1)

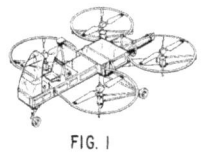

FIG. 1

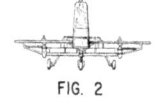   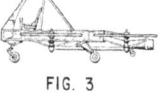

FIG. 2   FIG. 3

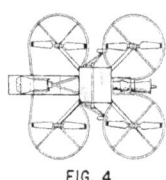   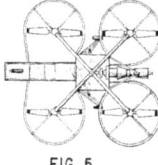

FIG. 4   FIG. 5

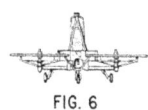

FIG. 6

Figure 1 is a perspective view of an aircraft showing my new design;
Figure 2 is a front view thereof;
Figure 3 is a side elevational view thereof;
Figure 4 is a plan view thereof;
Figure 5 is a bottom view thereof; and
Figure 6 is a rear view thereof.

I claim:
The ornamental design for an aircraft, substantially as shown.

References Cited in the file of this patent
Product Engineering, Sept. 9, 1957, page 29, top aerial jeep, by Aerophysics Development Corp.

Richard Vogt's 1960 patent was the ancestor of today's quadcopter variations.

In 1959 Edward G. Vanderlip, working for the Piasecki Aircraft Corporation in Essington, Pennsylvania, applied for a patent for an "Omni-directional, vertical-lift, helicopter *drone*." From 1942 to 1946, Vanderlip had already worked for the U.S. government on the first stabilization system for guided missiles, and as vice-president at Piasecki he now created the automatic pilot for this drone—a 6-cylinder Lycoming gas-engined quadcopter called "Sea-Bat" for VTOL from U.S. Navy ships with a top speed of 92 mph. In order to reduce the controller's need to remotely fly the drone, Vanderlip's ingenious system and rotor configuration were designed to allow the drone's center body to keep a constant azimuth heading regardless of direction of travel. While Piesecki went onto the develop the "Mud-Bat" and "Ice-Bat," Vanderlip and John J. Schneider (formerly of Curtiss-Wright) left Piasecki to create the Vanguard Air & Marine Corporation in Radnor, Pennsylvania, to further develop what they called the *Omniplane* in cooperation with the USAF Wright Air Development Center and NASA's Ames Center.

It was in the late 1950s that the U.S. Air Force, concerned about losing pilots over hostile territory, began planning for unmanned flights. Following a Soviet Union shootdown of the secret U-2 aircraft in 1960, the highly classified UAV or drone program was launched under the code name "Red Wagon." Modern-era UAVs got their first use during the August 2 and 4, 1964, clash in the Tonkin Gulf between the U.S. and North Vietnamese navies. After Chinese photographs surfaced of downed U.S. unmanned aircraft during and after the Vietnam War, the official U.S. Air Force response was "no comment." However, by 1973, the U.S. military officially confirmed that they had been utilizing UAV technology in Vietnam, stating that during the war, more than 3,435 UAV missions were flown, of which about 554 were lost in combat.

During the 1973 Yom Kippur War, aided by the U.S. government, Israel developed the first UAV with real-time surveillance, after Soviet Union surface-to-air missiles used by Egypt and Syria dealt heavy damage to Israel's fighter jets. The images and radar decoying provided by these UAVs helped Israel to neutralize Syria's air defenses at the start of the 1982 Lebanon War, resulting in no pilots lost. By 1987, Israel had developed proof-of-concept capabilities in tailless, stealth-based, three-dimensional thrust-vectoring flight control, jet steering UAVs for the first time. Interest in UAV technology grew during the 1980s and 1990s—being used during the Persian Gulf War in 1991—and became cheaper and more capable fighting machines. While most drones of the earlier years were primarily surveillance aircraft, some carried munitions. The General Atomics MQ-1, which utilized an AGM-114 Hellfire air-to-ground missile, was known as an unmanned combat aerial vehicle (UCAV).

At least 45 combustion-engine VTOL, short-take-off-and-vertical-landing (STOVL) and vertical-or-short-takeoff-and-landing (V/STOL) aircraft able to fly faster than helicopters were built and tested in the last half of the 20th century. But only four ever went into production. The ones scrapped had four or six—or as many as 10—engines to distribute propulsion. But the capability was never compelling enough to justify the cost, complexity and poor reliability of all those engines, transmissions, driveshafts and gearboxes.[9]

The next step was electric propulsion.

TEN

UAVs (aka Drones) Go Electric

Nearly all the UAVs described in the previous chapter had been avgas fueled. As we have already shown in Chapter Four, the radio-controlled electric model airplanes designed and flown by Fred Militky and the Boucher brothers in the early 1960s could be considered as hobby drones, whose targets were only altitude, speed and range. Then, in the early 1970s, development of electric and solar drones began in earnest

In the spring of 1973, Gerry Sayer, a founding partner in Developmental Sciences Corporation, developed the *SkyEye*, one of the first battlefield UAVs. He flew the first prototype R-4E-10 with a 2 hp pusher prop with a tricycle gear at El Mirage Lake, a dry lake bed in the northwestern Victor Valley of the central Mojave Desert, in California. They then tested the 10 hp electrically-propelled R-E-20.

Bob Boucher of Astro Flight has insisted:

> In 1974, our patented model 7212 built for the military demonstrated the ability to carry a six pound payload (camera) for 1 hour and 20 minutes at speeds of between 50 and 75 mph (over 100kmhr). Model 7404 (Sunrise) proposal to the military was for a solar UAV to carry a 50 lb (23 kg) payload up to 90,000 feet and remain aloft indefinitely. These could not be considered model airplanes in that they were too large and too heavy to qualify (over 10kg gross weight). Flight tests took place at Camp Irwin, Bicycle Lake, California, in October 1974, and four with the 7404–02 at Nellis AFB in September 1975. As far as I know we were first in the world to build such vehicles. When we started Astro Flight we dreamed of building thousands of UAV for the military and competing with the giants of the industry like Boeing, Lockheed, Northrup [sic], etc. Alas this was not to be. But the motors I designed and built were incorporated in many surveillance UAV and in underwater vehicles.[1]

Between 1980 and 1982, the Flight Dynamics Laboratory (FDL) at Wright-Patterson AFB tested out electric propulsion for RPVs (remotely piloted vehicles) by adapting an XBQM-106, a tail-stabilized, pusher-propeller teleplane or mini-drone, normally powered by a Herbrandson DH220, 18 hp, 2-cycle gas engine. This was replaced by a samarium cobalt brushless DC unit developed by Timothy F. Glennon of Sundstrand Corporation in Rockford, Illinois. Two batteries were to be evaluated: a Honeywell lithium-thionyl-chloride battery (Li/SOCL2) and a Yardney silver-zinc alkaline battery, both of which differed considerably from the more familiar lead-acid battery and to a certain extent from other alkaline batteries such as nickel-cadmium and nickel-iron. Tests were conducted both at the Clinton County Airport, Wilmington, Ohio, and at Eglin AFB, southwest of Valparaiso, Florida.

On September 14–15, 1982, a workshop on aircraft electric secondary power was held at NASA's Lewis Research Center in Cleveland, Ohio, to discuss the technologies related to aircraft power systems with a view toward aircraft in which all secondary power would be supplied electrically. This would later become known as an "all electric aircraft." Among the speakers were Timothy F. Glennon of Sundstrand Corporation, who spoke about 400-Hz aircraft power generation systems, and Robert C. Webb of General Electric, who described the application of a cycloconverter to a permanent-magnet generator which could either be integrated with the engine or mounted on the accessory gearbox.

It was inevitable that when Paul MacCready built solar-powered man-carrying airplanes (see Chapter Five), he too would focus on UAVs or drones. A solar-powered UAV could in principle stay aloft indefinitely, as long as it had a power-storage system to keep it flying at night. The aerodynamics of such an aircraft were challenging, since to reach high altitudes it had to be much lighter per unit area of wing surface than the *Solar Challenger*, and finding an energy storage system with the necessary high capacity and light weight was troublesome as well.

Ray Morgan recalls: "After Challenger's flights in Phoenix, a US Government Agency, whose name I cannot reveal, got interested in what we were doing and asked us at AeroVironment for what they called an atmospheric satellite to stay in orbit 12 miles above a certain spot. We did a study on paper and calculated that if we had a certain kind of airplane it could do it. It would mean a lot of wing area, low drag. They called it High Altitude Solar (HALSOL)." On paper, and based on projected technology, the mission could be achieved with an aircraft that had a span of 300 feet, and a wing-loading (weight divided by wing area) of approximately 1 pound per square foot. An approach for the design was laid out by Ray Morgan, Peter Lissaman, Bob Radkey and Bart Hibbs, which used a very flexible, aerodynamically pitch stable wing with distributed mass along the wing span as well as distributed overhung moments (the engine pylons) that matched aerodynamic moments along the span, thus minimizing structural weight. Analytically, it showed itself to be feasible with solar cells over 30 percent efficient and an energy storage system that weighed less than 1 kg for 600 watt-hours of energy stored. However, to prove out such a radical conceptual design, a series of progressively larger models were planned to be built and test flown.

A series of very small models were built and flight tested, some simply gliders and some powered. A total of six were built and tested before building the first that incorporated a sub-segment of the full-sized airplane. This first partial full-scale aircraft was a single 100-ft (30 m) wing (slightly longer than a Boeing 737) intended to be covered with 8 kW maximum of solar panels and a power source of six 16 1.50.6 kW electric motors, paired in parallel into 8 pylons and propellers, composing propeller speed pitch variations for directional thrust control, allowing differential thrust and drag to provide directional control. Its propellers and wing chord were the same size as the ultimate full-span aircraft would be. It was planned to fly larger span models, adding motor pylons as the span was increased.

It flew using silver-zinc batteries in tests, and a nominal solar array for evaluation. However, the intended lithium-based, regenerative fuel cell (for storing energy to fly overnight) and the so-called "ribbon" solar cells failed to meet the claimed performance, and the HALSOL program was retired after nine test flights—up to 3,000 feet (914 m) above ground and accumulating about 25 hours flight time in 1983 at a remote test facility in Nevada. The aircraft showed good aerodynamic performance and very good stability and control, but the power source could not meet the requirements. Consequently, the program was canceled after a corroborative study conducted by the Jet

Propulsion Laboratory. The aircraft was mothballed and put in storage in 1984 in the Nevada desert.

Interest by Col. Dale Tietz in the Ballistic Missile Defense Office of the Pentagon and Doctors Lowell Wood and Nicholas Collela of the Lawrence Livermore National Laboratories (LLNL) in developing a Boost Phase Intercept (BPI) platform reawakened the concept about a decade later, about the same time as the HALSOL program was declassified, and the aircraft was offered to AeroVironment. Project Pathfinder was resurrected in 1992. This led to the de-mothballing, refurbishment and upgrading of the HALSOL airframe with digitally controlled, electronically commutated motors (derived from Sun-Raycer and Impact experience), improved propellers (no moving parts—with electronic motor commutation, props were changed to fixed pitch, and no gear reduction was required), and eliminated gear drives. To this were added improved and more complete solar arrays (although still terrestrial grade to save costs, initially), as well as a more modern, digital flight control (with autonomous navigation capability), stability augmentation, and datalink system with redundancy. LLNL invested in regenerative, hydrogen/oxygen fuel cell technology as an improved energy storage system as well as developing potential payloads for the mission. Pathfinder was successfully test flown at low altitude (below 500 feet) with terrestrial grade solar cells at Rogers Dry Lake Bed under NASA DFRC's oversight in 1994. Unfortunately, funding priorities for BMDO changed in 1994, ending their support for Pathfinder. However, through the efforts of Dr. Nicholas Collela of LLNL, AeroVironment was able to move the airframe into a cooperative development program between NASA Aeronautics and industry, some twelve years later, for NASA's Environmental Research Aircraft and Sensor Technology (ERAST) program.

This program was initiated by the acting associate administrator for aeronautics, Rich Christiansen, and the program manager was Jenny Baer-Riedhart at the Dryden Flight Research Center, co-located with Edwards AFB in California. The purpose of this program was to tap into existing efforts to develop unmanned, high-altitude aircraft which could provide atmospheric science measurements up into the stratosphere, with a target of reaching 100,000 feet, at which point the density is very similar to that of the atmosphere near the ground on Mars. A secondary goal was to demonstrate multi-day flight in the stratosphere. Three other UAV developers joined AeroVironment as "Alliance B Members," but only AeroVironment was pursuing solar-powered flight to meet the mission. (It was intended that data gathered during this program would be shared with all participating members. This worked to varying degrees, depending on the individual company cultures.) On paper, based on projected available technology for solar cells and energy storage, a larger version (200 feet/60 m in span) with little or no energy storage, but 22 percent efficient solar cells, could reach the 100,000 feet altitude. This model was named Centurion. A 250- to 300-foot (70 to 90 m) version with high specific energy storage should meet the multi-day operation above 65,000 feet (20,000 m) goal. This model would be called Helios (Greek for the sun).

On September 11, 1995, Pathfinder set an unofficial altitude record for solar-powered aircraft of 50,000 feet (15,000 m) during a 12-hour flight from NASA Dryden. Due to its success, the Pathfinder was put on display at the Edwards AFB open house and air show the following month. It shared the hangar with two (then) classified aircraft, the B-2 Bomber and the F-117. Unfortunately, after the air show, when the Air Force decided to move out these two aircraft at night (for security reasons), there happened to be a windstorm sufficiently strong to blow over portable restrooms. When three of the hangar sides were opened for the entry of the aircraft tugs and towing of the two classified planes, the Pathfinder was caught in a wind gust of about 30 knots and blown into the F-117. Damage was severe. Much of the array and about half of the wing structure were destroyed, as well as several pylons. Fortunately, Dryden was able to come up with sufficient funds to repair the damage, and AeroVironment was able to upgrade the arrays replaced with more efficient SunPower terrestrial grade solar cells producing about 20 percent efficiency.

After further modifications, the aircraft was moved to the U.S. Navy's Pacific Missile Range Facility (PMRF) at Barking Sands on the Hawaiian island of Kauai, to take advantage of the lower latitude, a longer flight season, and a much larger range, that would allow the team to maximize the altitude envelope by optimizing the sun angles on the array during the climb.

On June 9, 1997, Pathfinder made its first flight from Kauai and reached an altitude of over 67,000 feet (20,400 m). Then, on its second flight, July 7, 1997, Pathfinder raised the altitude record for solar-powered aircraft to 71,530 feet (21,800 m), which was also the record for all propeller-driven aircraft. It also demonstrated the ability to re-fly to the stratosphere after a two-day turnaround, and did two more flights over the Hawaiian Islands flying multi-spectral imaging and other environmental payloads over the littoral waters, inland forests, and agricultural fields of sugar and coffee.

Pathfinder was followed by Pathfinder Plus, with the wingspan increased another 20 feet by replacing the middle section of the wing with a longer span that had an airfoil designed for 100,000-foot altitudes (30,000 m), new, higher efficiency solar arrays, new propellers, and various avionics improvements, as well as more solar cells. The ERAST goal for Pathfinder Plus was to reach 80,000 feet and validate the new wing airfoil, new

As part of NASA's Environmental Research Aircraft and Sensor Technology (ERAST) program, on August 6, 1998, AeroVironment's *Pathfinder Plus* raised the national altitude record to 80,201 feet (24,445 m) for solar-powered and propeller-driven aircraft (NASA).

motors, and new propellers intended for the 100,000-ft goal. On 6 August 1998, Pathfinder Plus raised the national altitude record to 80,201 feet (24,445 m) for solar-powered and propeller-driven aircraft.

This was followed by Helios, which, on August 14, 2001, remotely piloted by structural engineer Greg Kendall, set an altitude record of 96,863 feet (29,524 m)—the record for FAI class U (Experimental / New Technologies), and FAI class U-1.d (remotely controlled UAV: mass 500 kg to less than 2,500 kg). In fact, this altitude is higher than any non–rocket powered aircraft has flown in level flight to date. Ray Morgan retired from his position as vice-president of AeroVironment and director of the Design Development Center in 2000, but subsequently was hired as a consultant to NASA Dryden to serve as their technical representative for ongoing flight tests and demonstrations of the solar-powered aircraft developed for ERAST. At this point, to save money due to budget constraints, it was realized that both the high-altitude flight and the long-endurance flight could be achieved, and Centurion was changed to Helios (now 250' in span), after its initial, battery-powered flight tests at Dryden.

It had been initially planned that this long-endurance version of Helios would be modified with two to three regenerative fuel cells, comprising a hydrogen-oxygen fuel cell and an electrolyzer with gas storage tanks integrated into the tubular composite spars—resulting in a mostly span-loaded distribution of mass. However, funding limitations for the regen fuel cell system prevented reaching the goals before the program was slated to end in 2003. Consequently, AeroVironment, with NASA concurrence, compromised the design, and instead integrated an available automotive hydrogen-air fuel cell and two external hydrogen gas storage units, estimated to be sufficient to demonstrate nighttime flight, with a two-day total flight endurance in the stratosphere. This required placing a single, 500-lb (220-kg) fuel cell system in the center of the wing, with composite hydrogen gas tanks mounted near each tip to counterbalance this point load, which was beyond all experience to date with the span-loaded flying wing design. Further, it increased the gross takeoff weight about 40 percent. It was hoped that modifications to the flight control system could compensate for this change in load distribution and relative stiffness.

On the first flight of the Helios in this configuration, the aircraft seemed to be reasonably stable and controllable in benign conditions, but a leak in the gas handling system forced an early landing. On its second flight (June 26, 2003), the winds were much stronger than prior experience. Because of its slow climb rate at this higher gross weight, it encountered a much stronger shear line as it exited the wind "shadow" of the mountains of Kauai about 2,000 feet (600 m) lower than previous flights, which, coupled with the highly center-loaded wing, caused the tips to bend up much higher than the center wing, resulting in an oscillation in airspeed and pitch. A relatively inexperienced pilot erroneously turned off the airspeed hold system (thinking it had failed), rather than increasing the gain of that system (as planned in the emergency procedures), resulting in an unmitigated dive, over-speeding the aircraft and ripping off the leading edge, causing the Helios to crash into the waters between Kauai and Niihau, ending the program.

A follow-on flight research program funded by NASA, outside the ERAST goals, evaluated the effects of turbulence on the Pathfinder Plus, still in flight status, during low altitude flights at Rogers Dry Lakebed (part of the Edwards/Dryden complex). The Pathfinder Plus was outfitted with very sensitive turbulence-measuring devices and also load-measuring devices for the airframe for these tests, which were conducted in 2005.

Morgan was able to invite Captain Steve Ptacek (previously one of *Solar Challenger*'s pilots, now flying 777s) back to ride in the chase car during these low-altitude flight tests. Subsequent to these last flights, the Pathfinder Plus was hung in the Udvar-Hazy Air and Space Museum at Dulles Airport outside Washington, D.C.

In 2005–6, Ray Morgan participated as a subject matter expert in a NASA Langley-led study considering options available with near-term technology for flying above hurricanes to take measurements, track the eye, and predict the track more accurately. The conclusion was that current technology readiness levels for solar arrays and energy storage systems were inadequate to allow an airplane to maintain position above hurricanes through the required months, even to just 30 degrees of latitude, although a solar-powered blimp may be able to do so (provided the ground speed was sufficiently low).

In early 2008, Morgan teamed with Dr. Vince Castelli to convince Defense Advanced Research Projects Agency to create a program to study potentially available (3- to 5-year time frame) technology advances that would permit an essentially geostationary stratospheric satellite to perform militarily useful missions in the future. The program was accepted by the head of DARPA, Dr. Tony Tether, and Morgan was tasked to lead an internal feasibility study before finalizing requirements for a Broad Area Announcement (a solicitation for the aerospace community to propose concepts). The result of this feasibility study showed that it may be possible to hold station against probable stratospheric winds year round approximately 95 percent of the time up to about 35 degrees of latitude. In order to make it "DARPA-hard," however, Dr. Tether required that the announcement ask for a target of 99 percent station keeping up to 45 degrees of latitude year-round for a total flight time of 5 years on a single aircraft. Multiple companies bid, with three participating in the phase I conceptual design study—Aurora Flight Sciences, Lockheed-Martin, and Boeing. This pushed the design flight speed requirement to around 80 knots, the solar cell efficiency requirements to around 40 percent, and the specific energy and round-trip efficiency of the batteries (energy storage system) beyond state of the art. In Phase 2 only one performer was still funded, Boeing, which had the highest probability of meeting the energy storage requirement, but funding was reduced dramatically, turning the project from a system design to a technology advancement study. In the fall of 2017, Boeing purchased Aurora and its autonomous LightningStrike XV-24A technology in the race to develop air taxis.

Alongside this, AeroVironment had been working on the development of a small hand-launched remote-controlled unmanned aerial vehicle (or SUAV). Named the RQ-11 or Raven, this was a half-scale derivative of the Pointer, which had been developed under Ray Morgan's direction in the 1980s:

> I'd first proposed its development internally to Paul McCready in 1984, based on then available technology and a design derivative of one of Martyn Cowley's world-class, competitive free-flight glider designs that was both very efficient, inherently stable, and capable of "deep stall" landings. In 1986, we got approval from Peter Lissaman (who was technically my boss then, before I was promoted to VP) to develop a prototype with a black and white video camera on board, Ni-Cd batteries driving one of Boucher's electric model motors, and about 9 feet in span. The airplane came apart in six, plug-together pieces, that, with its ground control and video display fit into a single plastic backpack that I then took to Washington, D.C., and flew demos for various government organizations and military folks, including walking through the halls of the Pentagon carrying the fuselage in my arms while a lt. col. watched the video at his desk. After about a year of doing this, we got an order for four aircraft and a ground station from an organization that wished to stay unknown, as well as funding to build 6 airplanes and two ground stations for a demo system to be evaluated by the Marines over about a

year. In 1988, we let a couple of enlisted Marines demo the system to the Marine Commandant, including letting him fly and land it with one minute's instruction, which resulted in a million-dollar contract to build forty aircraft and 8 ground stations (with militarized backpacks) for several years of test and evaluations by Marines, Army, DEA, and other governmental organizations. In the mid '90s, we knew the state of the art for batteries, motors, but, especially the payloads (IR, EO, night-vision cameras) and navigation systems would allow reducing the size of the Pointer by about a factor of 2, but lacked internal funds to develop it on our own, and couldn't get any more from outside. Some of these were taken to the middle east for the first Gulf War.[2]

The Raven made its first flight in October 2001. It can fly up to 6.2 miles (10.0 km) at altitudes of approximately 500 feet (150 m) above ground level (AGL), and over 15,000 feet (4,600 m) above mean sea level (MSL), at flying speeds of 28 to 60 mph (45 to 97 kph). The Raven was the winner of the U.S. Army's SUAV program in 2005, and went into Full-Rate Production (FRP) in 2006. Shortly afterwards, it was also adopted by the U.S. Marines, and the U.S. Air Force for their ongoing FPASS Program. After 9/11, they started buying Pointers, and then AV got the internal funding for developing the Raven, with improved performance and capability and reduced size and weight, in extreme quantities. It has since been adopted by the military forces of many other countries. More than 25,000 Raven airframes have been delivered to customers worldwide to date. Alongside the Raven, AeroVironment of Monravia, led by Kirk J. Flittie, developed a family of military drones: RQ-12 Wasp, RQ-20A Puma, and Shrike VTOL.

Perhaps one of the most remarkable UAV programs began in 2001 when the British Ministry of Defense (MOD) split its Defense Evaluation and Research Agency (DERA) in two. DERA was privatized and renamed QinetiQ (as in Kinetic). One of QinetiQ's first projects was to find a way of filming an attempt to break the world altitude record in a balloon. Chris Kelleher was chosen as designer, technical director and flight operations manager to lead the Farnborough-based team.

Born in 1958, Christopher Charles Kelleher was as keen on aeronautic research as on flying aircraft. The son of a wartime RAF pilot, he studied at Queen Mary University of London, and also attended at the RAF College in Cranwell. A member of the RAE Aero Club, he flew light aircraft as a hobby and in occasional air displays, being a three-time winner of the British Aerobatics Association Advanced Level. He was also an Approved Person for airworthiness flight tests (including prototype). At the same time he had also worked on the UK's 4 Skynet military communications satellites, and also on two spacecraft for NATO, developing himself an international reputation for orbital dynamics and satellite operations used by both the UK and U.S. governments. So he understood the potential for a loitering stratospheric vehicle that could be a low-cost complement or alternative or "an eternal solar platform."

Bringing together a 20-strong team, Kelleher now began to work on a series of UAVs named Zephyr. For Zephyr 2, a proof-of-concept aircraft weighing less than 7 kg flew in both free and tethered modes off the Clifton suspension bridge in Bristol. By 2002, a 12 m (40-ft.)-long, 15 kg (33 lb.) UAV called Zephyr 3 had been created. It was supposed to fly tethered to a manned balloon attempting a world record altitude of 132,000 ft. (4,023 m), but unfortunately the balloon had a technical problem, and thus neither the balloon nor the aircraft ever flew. Following on from this, the Zephyr 4 project was created in order to carry out critical technology risk reduction and concept development work. Zephyr 4 also had a 12 m wingspan, but weighed approximately 17 kg, and was designed to be launched by the use of a helium balloon. In February 2005, Zephyr 4 underwent a test flight in Woomera, South Australia, where it flew for one hour after being launched at 30,000 ft. (9,000 m) by the balloon. In December 2005, Zephyr 5 flew in New Mexico for four hours, then for six hours, demonstrating successful ground launch, ascent, cruise and descent. In July 2006 the two aircraft were flown again in the USA with Zephyr 5-1 reaching 36,000 ft. (11,000 m) on an 18-hour flight, including 7 hours overnight. Collectively the two aircraft had a flying time of 35 hours. Next came Zephyr 6 with an 18-meter wingspan but weighing less than 30 kg (66 lb.) as it was constructed out of ultra-light carbon fiber. In July 2007 Zephyr 6 flew for 54 hours in New Mexico at a peak altitude of over 58,000 ft. (17,000 m), thus flying in the targeted 50,000–60,000-ft. range and validating the thermal and performance models. In August 2008, Zephyr 6 flew for over 87 hours above Yuma, Arizona, reaching an altitude of 61,000 ft. (18,600 m) and surpassing the previous world record of 30 hours for the longest unmanned flight, set by Global Hawk in 2001. This was homologated by the Fédération Aéronautique Internationale.

On July 9, 2010, monitored as usual by Chris Kelleher and his team, the Zephyr 7 took off from the Yuma Proving Ground and stayed airborne for over 2 weeks (336 hours), at an altitude of 70,742 feet (21,562 m), setting new world records, and was dubbed the first "eternal plane." Of this flight, Kelleher observed, "*Zephyr* will transform the delivery of current services such as communications and lead to many new applications that are not possible or affordable by other means."

Since 2008, Airbus had been working with its subsidiary Astrium on High Altitude Pseudo Satellites (HAPS). In 2013, Astrium acquired the Zephyr assets from QinetiQ, integrating it into the Airbus High Altitude Pseudo-Satellite program. As further proof, in 2014, an 11-day flight was launched from British-controlled Ascension Island in the South Atlantic during equal day–equal night conditions just south of the equator. This explored the ability of the Licerian battery packs provided by Sion Power to store enough solar energy to power the Zephyr through the hours of darkness, without significant loss of altitude. This was also the first flight controlled by a satcom datalink. There was also a flight in 2014 from Dubai that explored the Zephyr's integration into civilian airspace. The UAV was equipped with Mode S ADS-B so that it could report ascending through a "bubble" of airspace that other traffic could avoid. The Zephyr 7 takes up to 12 hours to reach its operating altitude above the jet stream. It has almost no ground speed when flying into wind. When on station at 65,000 feet (20,000 m) or above, it loiters by GPS control. It is hand-launched by a five-man ground crew, and recovered by belly-landing. Zephyr 7 has flown for over two weeks continuously—eight times longer than any other UAS.

Tragically Chris Kelleher, the man behind the Zephyr program, died on August 22, 2015, of natural causes while cycling near his home in Church Crookham. Three months later, in November 2015, in the House of Commons, British Prime Minister David Cameron laid out plans during the 2015 Strategic Defence Review (SDR) to further enhance Great Britain's intelligence, surveillance and reconnaissance (ISTAR) capacity. In the speech, he stated that the UK was to field a "British-designed unmanned aircraft that will fly at the edge of the earth's atmosphere and allow us to monitor our adversaries for weeks on end, providing critical intelligence for our armed forces." The British Ministry of Defence later confirmed the purchase of three Zephyr 8 platforms. The Zephyr 8 aka Zephyr S pseudo-satellite would have roughly 24 kg (53 lb.) of lithium batteries and a 5 kg (11 lb.) payload, and be 30 percent lighter. At an altitude of 20 km the aircraft has a visual range of over 400 km, offering a data collection and high bandwidth communications relay capacity to areas in excess of 1,000 km^2. The UK MOD gave Airbus D&S a $14 million (£10.6 million) contract to provide two larger UAVs for an operational concept demonstration in 2015. Further still, the Airbus Zephyr T, with its twin tails to support the additional weight, would weigh up to 140 kg (309 pounds) and carry a payload of 20 kg (44 pounds) on a 33-meter (108-foot) wingspan. Full-scale flight testing is scheduled for the Zephyr T in 2018 aimed at providing a maritime surveillance and communications capability. The lithium sulphur (Li-S) batteries being developed by OXIS of Abingdon, Oxfordshire, for the Zephyr program have a theoretical energy density five times greater than lithium-ion, with lighter weight and longer life-cycle characteristics. If OXIS arrives at the specific energy target of 425Wh/kg, the Zephyr will be able to fly above any weather in the troposphere and remain aloft for three months without needing to land. All the various Zephyr demonstrations to date have logged more than 1,000 hours.

In 2015, Bill Fredericks and a team at NASA's Langley Research Center in Hampton, Virginia, developed the GL-10 *Greased Lightning*, an unmanned hybrid-electric aircraft with a wingspan of 10 feet (3 meters), that can swivel its wings and engines—into the vertical position for vertical takeoff and landing (VTOL), and then horizontal for conventional forward "wing-borne" flight. GL-10 made its first tethered flight on August 19, National Aviation Day, performing transitions between vertical and horizontal flight. Its VTOL could be used for small package delivery, or long-endurance surveillance for agriculture, mapping and other applications. The plan is to demonstrate that the GL-10 is four times more aerodynamically efficient in cruise than a helicopter.[3] A full-scale version would involve 26 hybrid-electric propulsion fans distributed on the aircraft. In July 2017 the *Greased Lightning* concept was licensed to Fredericks's Virginia-based startup, Advanced Aircraft Co. (AAC); its first product, the hybrid-electric 65-inch (165-cm) octocopter Hercules is currently in its flight-testing phase, but with the ability to fly for 19 hours with a 5-pound payload. A two-stroke gasoline engine balanced by battery power provides fail-safe power supply to ensure safe flight operations. Deliveries of the Hercules were planned for December 2017.

A team led by Gregory W. Walker at Silent Falcon UAS Technologies in Albuquerque, New Mexico, has developed a thin film photovoltaic (TFPV), carbon-fiber fixed-wing, long-range unmanned aerial vehicle (UAV) with the ability to stay in the air for extended periods of time—up to five hours depending on flying conditions. Once the Silent Falcon reaches 100 meters, it's effectively undetectable.

On September 14, 2017, the Tailwind hybrid UAV developed by Troy Mestler, Robert Karol and Ivan Qiu of Skyfront, Menlo Park, made a flight of 4 hours and 34 minutes in winds that were between 8 and 10 mph, with gusts up to 15 mph. Menlo Park, of course, was once the headquarters of Thomas A. Edison, who had envisaged electric flight some 130 years before.

Drones for Deliveries

Alongside their primary role in high-altitude observation, UAVs have recently entered the realm of delivering goods.

In December 2013, German postal and logistics group Deutsche Post DHL used their Parcelcopter, a Microdrone md4-1000 drone, to deliver medical products from a pharmacy across the Rhine River. It was the first civilian package delivery via drone. A more challenging location was next selected for a 2014 trial: the North Sea island of Juist, again to be supplied with time-sensitive goods and urgent medicines. During early 2016, DHL directly integrated a parcelcopter logistically into its delivery chain for the Bavarian community of Reitim Winkl.

From October 2016, organized by Arthur Draber of Zipline, a Californian-based start-up, drones began regularly delivering batches of different blood groups to an increasing number of village hospitals in southern and western Rwanda, up to a range of 150 km (90 mi). Following a command by SMS to Zipline's Gitarama base, each trip was followed on electronic charts with deliveries arriving within a predictable radius of 5 meters. In early 2018, Zipline's Nest 2 would go into service for 20 hospitals located in eastern Rwanda, while negotiations were in hand for deliveries in Costa Rica and Tanzania.

As for fast-food deliveries: on May 11, 2014, Francesco's Pizzeria used a drone to deliver a pizza to a friend of the restaurant CEO on an apartment rooftop in Mumbai, India. A Russian pizza chain called Dodo Pizza conducted six drone pizza deliveries on June 21, 2014, but was later fined by Russian authorities for the illegal flights. Matt Sweeney of Flirtey in Nevada has organized Smartphone-related drone delivery service. The drone carries items weighing up to 5.5 lbs. (2.5 kg) on round-trip journeys of up to 10 miles (16 km). To avoid hitting trees or power lines, they hover about 40 to 50 feet (12 to 15 m) off the ground and lower their deliveries by cable to the waiting customer below. The cable-lowering method also keeps customers at a safe distance from the drone's whirling rotors. In November 2016, Domino's Whangaparaoa store in New Zealand used a drone to deliver a piri-piri chicken pizza and a cranberry chicken pizza to a local customer. And in the United States, 7-Eleven said in July 2016 that it had delivered Slurpees, doughnuts and other food to a customer in Reno, Nevada. Google has tested drone delivery of chipotle burritos on the Virginia Tech campus. Among those bigger giants seeing the advantages of drone delivery are companies like Amazon, Google, FedEx, UPS, and DHL. In September 2016, United Parcel Service (UPS) used a CyPhy drone to deliver a medical inhaler to an island near Boston.

From November 2016, one of China's biggest e-commerce companies, JD.com, founded by Liu Qiangdong as Jingdong Mail and headquartered in Beijing, had a fleet of drones flying autonomously on round trips of a maximum of 15 miles to reach rural communities (though a person still takes the package on the last leg of its journey to the recipient). Developed via its in-house drone team called JDX, the drone delivery program has so far licensed more than 30 flight routes in four different bases: Beijing, Jiangsu Province near Shanghai, Shaanxi Province in northern China, and Sichuan Province. Since China's Air Force controls the air space, JD.com is leasing the drone flight routes from the government on an annual basis and has to notify officials for each flight. When a rural customer orders goods online, the order is flagged when it is suitable for drone delivery. Once the parcel reaches the delivery station, a team fixes it to a drone that will follow a fixed route avoiding populated areas, highways and signal towers. Most can travel 10 km in one trip. During a recent flight test with the media, the model used, called M-TC2, was able to travel at a maximum speed of 100 kph. The drone hovers above a designated area in the destination village, drops the parcel down, and leaves immediately after. Fragile items are clearly not suitable for drone delivery. JD has invested CNY 20.5 billion to implement an in-depth cooperation with Xi'an National Civil Aerospace Industrial Base at Shaanxi over the next five years, to develop and manufacture seven different types of delivery drones. These range from drones that can fly up to 60 mph (100 kph), delivering packages weighing from 5 kg to 30 kg (10 lb. to 65 lb.) to drones which can carry as much as 1,000 kg (2,200 lb.), or one metric ton, with a maximum distance of approximately 100 km (60 mi) before recharging. JD, with its own nationwide network of thousands of delivery stations manned by 65,000 employees and 235 million regular customers, will establish three headquarters, three platforms, and four major industries. The three headquarters include JD's global logistics headquarters, global unmanned system industry headquarters, and JD Cloud's Shaanxi big data operating headquarters. The three platforms include a fusion smart logistics platform, a flight transport platform, and a big data operating platform. The four major industries cover smart manufacturing, smart logistics, cloud computing, and "characteristic town."

On December 7, 2016, Amazon made its first commercial drone delivery of an Amazon Fire streaming device and popcorn to a customer identified only as Richard B., in Cambridgeshire, England. The flight took off from a nearby Amazon warehouse and lasted 13 minutes, covering about two miles.

At the end of September 2017, Matternet, based in Menlo Park, California, began to deliver toothbrushes, deodorant and smartphones 8 to 16 kilometers (5 to 10 miles) above Zurich, Switzerland (population 391,000), to awaiting Mercedes-Benz electric delivery vans in populated areas such as congested urban streets and beyond natural barriers such as Lake Zurich. Customers ordered products using an on-line commerce startup called Siroop. The initiative has been approved by Switzerland's aviation authority, as has a separate Matternet project to carry medical supplies between Swiss hospitals. At PACK EXPO 2017 in Las Vegas, PMMI (Association for Packaging and Processing Technologies) collaborated with electric truck and *HorseFly* octocopter maker Workhorse to give hourly demonstrations of how, across a virtual landscape of 15,000 square feet, they would be delivering packages up to 10 lb. from stationary electric vans to the mailboxes and doorsteps of mock houses.

A team led by Przemyslaw Kornatowski at École Polytechnique Fédérale de Lausanne (EPFL) in Switzerland has developed a delivery-focused drone with an origami-inspired foldable carbon-fiber cage that protects the aircraft and its cargo in case of a collision or fall. The origami design means that the frame can be folded and unfolded in a single movement, reducing the volume by 92 percent, and can conveniently fit in a rucksack. Furthermore, the quadcopter can take off and land vertically and carry a package weighing up to 500 grams (1.1 pounds) over a distance of 2 kilometers (1.24 miles). The rounded protective cage is not only safe for regular handling of the cargo drone, but is also useful in emergency situations, where there is no landing spot, where instead the drone can be safely grabbed as it approaches the recipient. The EPFL team completed more than 150 test flights around the institute's campus and is already planning improvements such as an emergency parachute for breakdowns.

Indeed, in July 2016, Amazon.com filed a patent on "docking stations" for its delivery drones that will be built on tall structures such as lampposts or churches as drone "perches" which will allow the unmanned machines to recharge and pick up packages.[4] Another Amazon patent is for delivery whereby the drone would deploy packages at altitude and monitor and adjust (radio-control) package trajectory during descent using either a parachute, a compressed air canister, or a landing flap. The package could carry multicolored markings so that a camera on the drone could tell exactly how it's oriented on the way down.

In the fall of 2017, Amazon opened a new 60,000 sq. ft. (5,500 m²) site at One The Square in the CB1 business district of Cambridge to research artificial intelligence (A.I.) and drone delivery. Among the 400 mathematical experts, scientists and engineers, teams working on products such as Kindle, Echo and Amazon's voice assistant Alexa also moved into the new Cambridge location. Amazon partnered with the UK government to test drone deliveries and recently designed a UAV with robotic wingtips and legs that act as landing gear to help them touch down on uneven surfaces.

In October 2017 Amazon patented a system whereby a drone could deliver battery energy to an uncrewed electric automobile running short of energy (U.S. 9,778, 653 B1). Servers relay energy requests to drones, which would then be dispatched to meet vehicles on the road. The drones would authenticate the connection, attach to a vehicle connector on the roof or door, and start fueling—even if the car is in motion. The passenger could even make the help call to Alexa, Amazon's personal assistant. The battery would be independent of the drone's own battery supply to fly back to base.

Skysense, originally from Trento, Italy, has developed a drone-charging pad made up of modular 50 cm² pads, a tessellation of conductive tiles that can be dynamically activated. Whenever a drone lands, two spring-loaded contacts power the on-board Li-Po balance charger to fast recharge the batteries. The pad can be plugged into an electrical socket or powered by solar panels, and works even in rain and extreme weather conditions.

In an approach to resolve the issue of recovering the delivery module, Saul Griffith and an ex–MIT team at Otherlab engineering laboratory, San Francisco, have developed an industrial-grade paper airplane as a low-cost disposable aerial delivery vehicle based on DARPA's ICARUS program (for Inbound, Controlled, Air-Releasable, Unrecoverable Systems). They call this APSARA (for Aerial Platform Supporting Autonomous Resupply Actions). Once the goods have arrived, the drones biodegrade in a matter of days. And because it is a glider without motors and rotors, it means that all of the onboard electronics, courtesy of DARPA's VAPR program, go with it. Whilst it has its commercial application, several hundred APSARA gliders loaded with critical medical supplies could be used in emergency zones.

But beware! As with any other form of transport, according to a report by Intel's McAfee Labs, cybercriminals are threatening to target regular drones used for deliveries, law-enforcement or camera crews, in a crime known as "dronejacking." Once a package delivery drone is overhead, the drone could be sent to the ground, allowing the criminal to steal the package. The culprit could also steal expensive photographic equipment carried by drones, in order to knock out surveillance cameras used by law enforcement. Dronejacking toolkits are already being traded on "dark web" marketplaces.[5]

Counter-drone strategy involves detection at up to 200 yards (183 m), interception and neutralization. This can be done by training birds of prey or using mobile jamming systems, stationary jamming systems, drone interception systems, drone capture nets, or anti-drone lasers. Detection methods could use both radar and acoustics to detect and identify the flying drone, displaying the flight speed and direction of the drone. The challenge is to detect the right object, because efficient radar should identify the drone as not a bird. As most drones use the same 2.4 GHz frequency that is used in telecommunications like Wi-Fi, a system of selected jamming has been innovated.

Parrot and the Arrival of the Toy Drone

Ever since Capitaine Max Boucher's experiments with radio-controlled aircraft during World War I, the French have sustained a fine interest in model aircraft. Since childhood, Pascal Zunino and his friend Fabien Paganucci of Les Pennes-Mirabeaux, in the Bouches-du-Rhône Department of southeast France, enjoyed building electro-mechanical gadgets. Both attending the local la Renardière primary school in the 1990s, they read an article in an aeromodeling magazine about a quadcopter, and decided to build one themselves in the Zunino garage. Moving on to the Jacques Monod College, they continued to work together on projects, with the mechanics developed by Fabien and electricity, then electronics, by Pascal. At the Pierre Mendès France secondary school in Vitrolles, Zunino and Paganucci assembled a team to compete in the E=M6 robotics competition. They called their robot "Glouton" (Glutton). The circuit boards built were used to liaise between the robot and the control unit through a PWM modulation, and so command the two propulsion engines and that of the suction turbine. Winning, they were officially presented with the 2000 E=M6 Trophy at the City of Science and Industry in Paris.

In 2001, while Pascal Zunino studied electronics at the University of Aix in Marseilles, Fabien Paganucci studied microtechnology and computer-aided architectural design at the same city. They assembled the red micro-robot Z6 prototype drone, which made its very first flight in the early hours of January 25, 2001. By 2002, they had progressed to their more spindly 70 g (2.5 oz) carbon fiber X4. Its electronics brought together on one card an 8051 microcontroller, a 3-gyro inertial unit, and 4 speeds. They also built an 80 g (2.8 oz) Mini X4 with a 34-cm (13-in.) wingspan and an even smaller 40 g (1.4 oz) Micro X4 with a 16-cm (6-in.) wingspan. This gave them confidence to assemble a team of students at Grenoble INP and to build the foldable-rotor CPX4 for the international miniature drone competition organized by the ONERA and the DGA that September 2005. At the Mourmelon Military Camp near Châlons-en-Champagne, they demonstrated the CPX4 to civil and military professionals in the industry and won 1st prize ex-aequo. It had particularly attracted the attention of the DGA (Délégation Générale de l'Armement, the state organization responsible for armament programs) as being operational and innovative due to its size and previously unseen ease of use: with its hovering capability, the CPX4 could also be very easily fitted with sensors or cameras (day/night, infrared).

In 2006, financially backed by the Michel Caucik Innovative Business Incubator, they created the start-up Novadem Aerial Robotics in Aix-en-Provence with the aim of democratizing aerial robotics. In this they were joined by Pascal's brother, Eric Zunino, himself specializing in information technology, and who had often helped with previous prototypes. Before long, their enterprise came to the attention of Henri Seydoux of the Parrot Company. In 1994, Seydoux (son of the head of the Pathé-Gaumont Cinema empire, Jérôme Seydoux), together with Christine de Tourvel and Jean-Pierre Talvard, had set up Parrot to create a vocal recognition system for the blind in the burgeoning smartphone and tablet market.

To diversify their business, Seydoux was looking into a Wi-Fi remote-control automobile. The product was ready for production. But it was not fun enough. Seydoux was about to cancel the project when he got the idea to turn the car into a drone, formed a team called Commando, and contacted Zunino and Paganucci at Novadem, who would

2010, left to right: Fabien Paganucci, Eric Zunino and Pascale Zunino. Pascal Zunino founded Novadem Aerial Robotics in Aix-en-Provence with the aim of democratizing aerial robotics (NOVADEM).

help Parrot to develop various key aspects of the drone: propeller design, control system and stabilization algorithms, ultrasound telemeter, benchmarks, noise reduction, and testing procedures, while Parrot focused mainly on video. That was the beginning of a time-saving collaboration that would last for three years. The augmented reality AR. Drone was officially unveiled at the 2010 Consumer Electronics Show (CES) in Las Vegas as the first Wi-Fi augmented reality quadcopter drone, a form of flying video game. It could be controlled by mobile or tablet operating systems such as the supported iOS or Android within their respective apps or the unofficial software available for Windows Phone, Samsung BADA and Symbian devices. The drone was loaded with sensors and video cameras. It had debuted just in time to surf the wave of the famous miniature remote-control helicopter PicooZ. In fact, Parrot was not just catching the wave but was extending it. The AR.Drone was launched onto the market just a few months later, with a global launch beginning in August 2010, starting in Hong Kong and France before spreading to the rest of the world. This initial product was a major success: 120,000 AR.Drones would be sold by the end of the year.[6]

Along with AR.Freeflight, the application designed for free operation of the drone, Parrot also released AR.Race, allowing users to take part in solo games or interact with other drones in combat simulations. Inside the airframe, a range of sensors assist flight, enabling the interface used by pilots to be simpler, and making advanced flight easier.

The onboard computer runs a Linux operating system and communicates with the pilot through a self-generated Wi-Fi hotspot. The onboard sensors include an ultrasonic altimeter, which is used to provide vertical stabilization up to 6 m (20 ft). The rotors are turned by 15-watt brushless motors powered by an 11.1-volt lithium-polymer battery. This provides approximately 12 minutes of flight time at a speed of 5 m/s (11 mph). Coupled with software on the piloting device, the forward-facing camera allows the drone to build a 3D environment, track objects and drones, and validate shots in augmented reality games.

In 2012, Parrot acquired 57 percent of the Swiss aerial imaging drone manufacturer of senseFly4 drones and invested more than 2.4 million Swiss Francs in Pix4D, both in Lausanne. In 2015 Parrot invested in drone startups: Airinov, drones for agriculture; Micasense, sensors for agricultural drones; EOS Innovation, robot for inspection; Iconem, drones for archaeology. The same year, Parrot separated its core business into two distinct subsidiaries: Parrot Drone and Parrot Automotive.

In May 2014 at the annual AUVSI conference in Orlando, Parrot announced the AR.Drone 3.0, code-named Bebop, with the option for a Skycontroller, when purchasing the Bebop. The Skycontroller allows the Bebop Drone to fly up to 2 km (1.2 mi). The Parrot Bebop Drone was released in December 2014. In almost a man's world, Christine Caubel, mechanical design engineer at Parrot Drones near Paris, has led several innovations for her firm since 2008. In 2015 she gave quadcopters protection by providing them with removable bumpers for protecting their propellers. Each bumper is connected to the propulsion units on the same side by connection arms, each comprising a pair

2010: Henri Seydoux of the Parrot Company launches the *AR.Drone*; 120,000 would be sold by the end of the year (PARROT).

of elastically deformable elongated blades with a clamp at their end for mounting on a barrel of the motor.

There were many who preferred building their own drones to buying them off the shelf. In 2006 London-born Chris Anderson, then editor of *Wired* magazine, built a drone for his kids for fun. In 2007 he turned his new hobby into a Web-based community called DIYdrones.com to trade hardware tricks and software tips in forums; during the next decade it would grow to over 70,000 members. In October 2007, Anderson flew a remote-controlled aircraft allegedly equipped with a camera over Lawrence Berkeley National Laboratory, causing security concerns when the aircraft crashed into a tree. In 2009 Anderson met Mexican Jordi Muñoz online and together they formed 3D Robotics in Berkeley, California, to design and build UAVs.

In Italy in 2008, Paolo Marras and his team founded Aermatica SpA in Gironico, Como, with the goal of developing solutions to capture data and generate enhanced information through the use of small aircraft to pilot remote systems (drones, UAV or UAS). They developed the variable-pitch Anteos quadcopter and obtained official permission from the Italian Civil Aviation Authority (ENAC) to fly in the nonsegregated civil airspace. They patented their "Aerial Robotic System" in 2009 (WO 2010128489 A3).

A rival to Parrot is DJI, directed by Frank Wang. In 2004, Wang's dream of landing at an elite American university such as MIT or Stanford was thwarted due to his unimpressive academic performance, so he ended up at the Hong Kong University of Science & Technology, where he studied electronic engineering. In 2006, during his senior year, working in his dorm room, Wang developed a helicopter flight-control system. Even though the hovering function for the on-board computer failed just before the presentation, Li Zexiang, his robotics professor, noticed him and brought him into the school's graduate program.

In 2006, Wang and two classmates moved to a manufacturing hub of Shenzhen, Guangdong, China, and started working out of a three-bedroom apartment. Wang funded the venture Shenzhen Dajiang Innovation Technology, or DJI, with what was left of his university scholarship, and modest sales (to Chinese universities and state-owned power companies) were enough to help him pay for a small staff. In January 2013, DJI produced their white *Phantom 1* drone. It was a multirotor that most people could order online and fly without much trouble after reading the manual. It was commonly equipped with a GoPro camera for amateur filmmaking or photography. Its battery life was around 15 minutes with a GoPro.

Eleven months later, DJI released the *Phantom 2*. Upgrades included auto-return, increased flight speed, increased flight time and controllable range, increased battery capacity, compatibility with smartphones and tablets, and a feature called Smart Guidance Control (or IOC), that made the *Phantom 2* ridiculously simple to fly. By the end of 2014, DJI, with a 2,000-strong workforce, had sold an estimated 400,000 units, making Frank Wang the world's first drone billionaire.

The same year, the American Federal Aviation Administration (FAA) proposed some regulations and guidelines to ensure proper operation of these vehicles in the USA, and the regulations came into effect on December 21, 2015. As CES 2017 was opening, news broke that DJI had bought a majority stake in the German camera manufacturer Hasselblad—the company that made the cameras American astronauts took to the moon. Soon after, DJI departed from its white Phantom range to present its "H" drone, a black hexacopter with ST16 remote control with Android software and 25-minute autonomy.

Lorenz Meier, a Ph.D. student in electronics at the ZTH Swiss Federal Institute of Technology Zurich, was curious about technologies that could allow robots to move around on their own, but in 2008, when he started looking, he was unimpressed—most systems had not yet even adopted the affordable motion sensors found in smartphones. So as part of what he called the "sFly Project: Swarm of Micro Flying Robots," Meier built his own system instead: the Pixhawk flight stack platform for autonomous drone control. In 2009 he launched, the results inspiring several drone companies such as Parrot, DJI, Yuneec, and Skydio. Importantly, Meier's system aimed to use cheap cameras and computer logic to let drones fly themselves around obstacles, determine their optimal paths, and control their overall flight with little or no user input. In 2011, he open-sourced it under the PX4 software brand following through with the Pixhawk open hardware autopilot in 2013.

The following year, the Dronecode Opensource UAV Platform, which is a nonprofit organization governed by the Linux Foundation, was formed by Chris Anderson with the goal of using open-source Linux for the benefit of founding members including Qualcomm and Intel. 3D Robotics moved along with Yuneec International providing cheaper, better, and more reliable UAV software. For example, the Yuneec DataPilot™ software, integrated in the new H520 hexacopter, is tightly coupled to the PX4 flight control architecture. The DataPilot™ is a complete solution for planning survey and waypoint-based UAV flight. Over 500,000 units of Pixhawk have been manufactured. In February 2017, Meier helped Yuneec to establish its R&D (research and development) center in Zurich, and since then has become an advisor to the company in high-tech development. In September 2017 Meier was named as one of top under-35 innovators by *MIT Technology Review*.

The Arrival of Scaled-Up Dronecopters, aka PAVs

Although the drone was born by scaling down technology, during the past few years the technology of the drone has been scaled up in the form of light e-airplanes. Tian Yu, CEO of Yuneec (Unique) International Co., is based in Kunshan, Jiangsu Province, China, where his factory employs 1,800 workers to assemble 1 million drones per year. When Tom Peghiny of South Woodstock, Connecticut, developed his Flightstar e-Spyder ultralight, he then went into collaboration with Yuneec to develop the GreenWing International eSpyder (GW280), a single-seat electric airplane sold as a kit for construction as an amateur-built airplane, as well as the two-seat GW430. It has a Yuneec 24-kW electric power system that can lift a 220-lb (100 kg) payload at an initial climb rate of 375 feet (110 m) per minute. Maximum speed is 56 mph (90 kph); economy cruise is 37 mph (60 kph). Flights as long as one hour are possible, with a 30-minute reserve still available upon landing.

The first flight of the GreenWing E430 took place from the Yuneec factory near Shanghai on June 12, 2009. In kit form, it was then shipped for further testing to Camarillo, California. On July 14, 2009, the prototype aircraft was registered in the USA as N386CX, and on July 18, 2009, it was given a Certificate of Airworthiness by the Federal Aviation Administration. Further test flights were carried out, totaling 22 hours. The prototype E430 was then shipped by truck to Wisconsin and displayed at EAA AirVenture Oshkosh in July 2009. The two-seater high-wind composite is powered by a 42 kW

Yuneec motor, powered by a three-pack Yuneec OEM Li-Po battery weighing 13 kg (28.6 lbs.) and producing 66.6V (30 Ah). With the three-pack, the airplane is expected to have an endurance of 1.3 to two hours cruising at about 53 knots. With a five-pack battery configuration, endurance is said to be 2.25 to 2.5 hours. It takes about three hours to charge the batteries from regular AC power outlets. Batteries are expected to last about 1,500 hours. The innovative battery system allows battery packs to be easily and quickly swapped so that the plane can be flown with one battery while another one is being charged simultaneously. Tian Yu announced that his company was building a 260,000 ft^2 (25,000 m^2) factory to produce the aircraft in Shanghai, that was expected to open in October. Yuneec is working on developing a solar-cell installation for the wings that will recharge the aircraft's batteries.

In 2010, the E430 was named the winner of the Lindberg Prize for electric aircraft at AirVenture. In the same year it was named Brit Insurance Design of the Year in the transport category. But by December 2012, a total of only two examples had been registered in the United States with the Federal Aviation Administration. The first one was the initial prototype shipped to the U.S., registered in the Experimental–Exhibition category on July 14, 2009, although its registration expired on March 31, 2012. The second was registered in the Experimental–Research and Development category on January 26, 2011, to Flying Tian of Monterey Park, California. German DULV certification was awarded in February 2013. GreenWing began taking orders at the 2013 EAA AirVenture show in Oshkosh and kits were scheduled to begin delivery the end of 2013. But as of August 2015, GreenWing International had stopped trading.

Almost in parallel, Huazhi Hu, based in the Guangzhou Province of southern China, had already been making camera and hobby drones, such as the Ghostdrone1.0, when he decided to "scale up" and make a passenger-carrying autonomous drone (PAV), which could transport a person via air the same way as Google's self-driving automobile via road. After two of his friends, his CEO Ji Chen and his helicopter coach, were tragically killed in air crashes, Huazhi Hu determined to make an absolutely safe transporter. The idea is that a passenger hops in, enters his destination on a 12-inch (30-cm) touchscreen, and hits the takeoff button. Weighing 440 lb. (200 kg), the EHang 184 AAV looked like a small helicopter although with four "lifter booms," apparently a quadcopter. It was in reality an "octocopter" because each boom had two independent high-power electric motors and composite propellers. Thus, if any of the motors or props were to fail, there was still adequate lift and control to bring the EHang 184 to a safe landing. The electric-powered drone could be fully charged in two hours, carry up to 220 lb. (100 kg), and fly for 23 minutes at sea level. The cabin fitted one person and a small backpack and even had air conditioning and a reading light. It was designed to fly about 1,000 to 1,650 feet (300 to 500 m) off the ground with a maximum altitude of 11,500 feet (3,500 m) and top speed of 63 mph (100 kph). If any components malfunctioned or disconnected, the aircraft would immediately land in the nearest possible area to ensure safety. Its propellers folded inwards as it landed so it could fit in a single car parking space. The communication was encrypted and each AAV had its independent key.

The EHang 184 was unveiled at CES International in Las Vegas on January 6, 2016. U.S. authorities were just starting to lay out guidelines for drone use, and a human-passenger drone seemed certain to face strict scrutiny. On June 8, 2016, the EHang 184 was given clearance for testing by the Nevada Institute for Autonomous Systems (NIAS) and the Governor's Office of Economic Development (GOED) to put the drone through

From 2014, Huazhi Hu, a drone millionaire based in the Guangzhou province of southern China, developed the *Ehang184* (Ehang).

testing and regulatory approval. The program will take place at Nevada's FAA-approved test site, one of six such drone-testing locations across the U.S. There was no mention of whether this testing would cover the vehicle's capacity to carry humans, but the stated intention is to gain the EHang 184 a Certificate of Airworthiness, an FAA document that gives it authority to fly. "I personally look forward to the day when drone taxis are part of Nevada's transportation system," the institute's business development director, Mark Barker, told the *Las Vegas Review Journal*.

During 2016, EHang continued its progress; a handful of 184s made some 200 test flights, some fully autonomous, to further improve the technology, but only in China. Monitoring of these flights was carried out by the very first flight command center in Guangzhou; the building was converted to remotely track live data—including speeds, altitude, individual propeller power, location, drone camera feed and video feed of the passenger—plus communicate with passengers and schedule air traffic. EHang planned 2- and 4-seater versions and large-scale working drones for agriculture. In February 2017 at the World Government Summit, Mattar al-Tayer, the head of Dubai's Roads and Transportation Agency, announced that, following experiments in the skies above their city, a blue and white livery EHang 184 would go into regular operation. This was to be integrated into the project of Sheikh Mohammed bin Rashid Al Maktoum whereby by 2030 one-quarter of public transport will be autonomous. But in an ecosystem where winds can go up to 40–50 knots (46–58 mph), with both sand and fog, certain critics pointed out that its ballistic parachutes would not have the time to deploy at altitudes below 100 ft. (30 m), and even if the chutes did deploy, the PAV could crash into power lines, trees or water. Unfortunately, with the recent lack of sales of its Ghostdrone, EHang had to lay off about 70 employees and appeared to be dealing with fiscal problems resulting in missed payments to suppliers; one of the victims was the EHang 184. When Dubai launched its drone-powered taxi service, EHang did not appear.

During 2015, students from the National University of Singapore (NUS) built and tested the single-seater eVTOL Snowstorm, under the auspices of FrogWorks, a collaboration between the NUS faculty of engineering's Design-Centric Programme (DCP) and the University Scholars Programme (USP). Snowstorm was fitted with 24 motors, each driving a propeller of 76 cm diameter with 2.2 kW of power. Its hexagonal frame was made up of anodized aluminum beams, carbon-fiber plates, and tubes with Kevlar ropes. The pilot seat was positioned at the center of the machine, its weight supported by six landing gear legs, the bottom of which is an inflated ball that adsorbs shock when landing. Three independent rechargeable lithium battery sets provide a total power of 52.8kW. Snowstorm was never commercialized.[7]

The number of near-misses and collisions between drones and aircraft is on the rise. According to the UK's Civil Aviation Authority (CAA), there were 23 drone-related incidents at UK airports between November and April 2017, including 12 near-misses. A clutch of British companies have developed a counter-drone technology, the Anti-UAV Defence System (AUDS), which sends out drone-jamming signals and will soon be tested out at U.S. airports by the Federal Aviation Administration (FAA). According to a recent report, the FAA had close to 600 drone-sighting reports in the period between August 2015 and January 2016. The agency said it receives more than 100 reports every month regarding unmanned aircraft flying dangerously close to an airport or airplane.

Recognition of the importance of the vital importance of controlling the air traffic goes back to the dawn of commercial aviation. In 1921, Croydon Airport, London, was the first airport in the world to introduce air traffic control, using wireless telegraphy. In the USA the number of planes using Cleveland Airport jumped from a few thousand in its inaugural year to nearly 20,000 by 1929. A new terminal building constructed that year contained the world's first air traffic control tower, a tall, glass-enclosed structure with a 360-degree view of the airfield. Soon after its construction, two-way radio was installed in the tower, the first time this had been used in the aviation field. This proved to be an important addition, as in its early years the airport used the "allway" landing mat process, which allowed multiple planes to land simultaneously on different parts of the airfield, a process designed to prevent pilots from having to wait mid-air for space to land. Following the crucial use of radar in World War II to control the arrival and departure of warplanes during the Battle of Britain and other theaters, the system was adopted to monitor and control the busy airspace around larger airports. With the exponential growth of drones and runway-free eVTOL taxis, the pressing challenge has been to monitor traffic at an altitude and even below, halfway between automobile traffic and conventional air traffic at 33,000 ft. (10,000 m) and so prevent accidents.

From late 2015, extensive work was carried out by engineers at NASA's Ames Research Center in Moffett Field, California, on UTM for Civilian Low-Altitude Airspace and Unmanned Aircraft System Operations drone traffic control. The majority of flight testing occurred at Crows Landing, a remote, closed, private-use airfield, 18 miles (29 km) southwest of Modesto, California. Prior to flight test, the team deployed a 100-ft. (30-m) weather tower, small weather stations, microphone, Automatic Dependent Surveillance-Broadcast (ADS-B) in a ground relay station for air traffic feeds, and a radar station for flight test monitoring and data collection. The Technical Capability Level One system will enable UAS operators to file flight plans reserving airspace for their operations and provide situational awareness about other operations planned in the area. Technology Capability Level 2 (TCL2) has focused on flying small drones well beyond the pilot's line of sight over sparsely populated areas at five of the FAA test sites: Virginia, North Dakota, Texas, Alaska and Nevada. The team worked with over 250 partners throughout the industry that plan to use drones in their businesses in the airspace above buildings and below crewed aircraft operations in suburban and urban areas. So NASA's NTX research center is exploring how flight corridors can work without voice interactions. This includes improved "sense-and-avoid" technology that will allow drones to communicate with other passenger aircraft to avoid one other.

On October 25, 2017, President Donald Trump directed Secretary of Transportation Elaine Chao to launch the unmanned aircraft systems (UAS) Integration Pilot Program, an initiative to safely test and validate advanced operations for drones in partnership with state and local governments in select jurisdictions. This program will evaluate a variety of operational concepts, including night operations, flights over people, flights beyond the pilot's line of sight, package delivery, detect-and-avoid technologies, counter-UAS security operations, and the reliability and security of data links between pilot and aircraft. It is designed to provide regulatory certainty and stability to local governments, communities, UAS owners and operators who are accepted into the program. In less than a decade, the potential economic benefit of integrated unmanned aerial systems into the nation's airspace is estimated to equal up to $82 billion and create up to 100,000 jobs.

Another way to tackle the threat posed to commercial air traffic by wandering drones has been tested by Dubai with a Drone Hunter, developed by the UAE–based tech firm Exponent. Once the pilot confirms the location, the hunter is ready to take off in three minutes at a speed of 110 km per hour. Elsewhere, in the Netherlands, the Dutch National Police have taken a lower-tech approach by collaborating with a private company called Guard from Above, which trains birds of prey, such as juvenile bald eagles imported from North America, to intercept drones.

The AgustaWestland Project Zero is a tiltrotor/fan-in-wing airplane. In December 2010, the management of AgustaWestland in Farnborough, England, approved the formation of a team under James Wang with the intention of producing a technology demonstrator incorporating as many innovations as possible on a single airframe. Various companies in Italy, the UK, the U.S. and Japan worked on the design and/or manufacturing of elements of Project Zero, including four different branches of Finmeccanica. Ansaldo Breda designed a custom-built electric motor inverter and accompanying motor control algorithm, while Selex ES provided the High-Integrity Flight Control Computer and the Actuator Control Unit. Lucchi R. Elettromeccanica custom-built the axial flux permanent magnetic motors; Rotor Systems Research LLC worked in conjunction with AgustaWestland on the aerodynamic design of the rotor blades. Lola Composites produced the composite material from which most of the exterior surfaces are made of carbon-fiber-reinforced polymer (CFRP), while UCHIDA manufactured the composite structure for the rotor blades, shrouds, and spokes. Stile Bertone worked to develop the aesthetic and aerodynamic styling of the aircraft. In June 2013, Project Zero was publicly displayed at the Paris Air Show. Flight testing has also continued, but focused on ⅓-scale models rather than the full-scale demonstrator due to the limited flight endurance it possesses. In February 2016, it was announced that, *until batteries increase their energy density*, a hybrid drive system would be installed on the full-scale aircraft; this is aimed at extending the flight endurance from 10 minutes to 35–45 minutes.

Another example of scaling-up has been made by Hirobo Electric Corporation of Hiroshima, well known and respected for over forty years for its line of high-quality radio-controlled helicopter models such as the Hirobo *Eagle* 3 EP. In 2010, Hirobo decided to use their expertise to present a personal electric all-composite helicopter, the Hirobo *HX-1 BIT*, which was launched at the International Robotics Exhibition in Tokyo in 2013. The Japanese company's compact IMU-05 attitude sensor is one such technology that has been developed to enable autonomous flight, and offers the company a serious competitive advantage in that area. The IMU-05 collects a wide range of information such as attitude angle, acceleration, angular rate and magnetic direction, and enables the helicopter to be remarkably stable in blustery winds—it is a key enabling technology of autonomous flight for helicopters. The company is not expecting to deliver the passenger-carrying variant before 2021, citing expected bureaucratic problems.

Joe Ben Bevirt of Santa Cruz, California, whose motto is "To found, build, invest in, and guide companies which improve the world," had already founded Joby Energy, Inc., to develop airborne wind turbines to harness the immense and consistent power in high-altitude wind to provide reliable and low-cost renewable energy. In 2015 Bevirt founded Joby Aviation with aeronautical engineer Alex M. Stoll. Bevirt and Stoll then applied for a patent for an "Aerodynamically Efficient Lightweight Vertical Take-Off and Landing Aircraft with Pivoting Rotors and Stowing Rotor Blades."[8]

Alex Stoll is well qualified. Following a spell in the Aircraft Aerodynamics and Design Group Stanford University (2008–2012), he worked with Mark D. Moore, William J. Fredericks and Nicholas K. Borer of NASA Langley Research Center to investigate drag reduction through distributed electric propulsion. The two-seater Joby S2, powered by lithium-nickel-cobalt-manganese-oxide batteries, uses 12 tilting electric propellers to provide multirotor-style balanced VTOL capabilities. Once it reaches cruising speeds, these rotors fold away into aerodynamic bullet shapes, and the aircraft can reach speeds of up to 200 mph (322 km/h) and ranges of up to 200 miles using four additional cruise-optimized props on the backs of the wings and tail fins. The two-seater Joby S2 Electric VTOL PAV should soon begin trials.

During 2010, as part of his doctoral degree, Dr. Mark D. Moore, the chief technologist at NASA's Langley Research Center, Hampton, Virginia, working with a team from Massachusetts Institute of Technology, the Georgia Institute of Technology, the National Institute of Aerospace (NIA), and M-DOT Aerospace, came up with the concept of a hover-capable, electric-powered, low-noise, personal VTOL technology-concept, proprotor aircraft. Called *Puffin*, it would be capable of flying a single person at a speed of 150 mph (240 kph). Range is expected to be less than 50 miles (80 km) with initial battery technology. The design has a 14.5-foot (4.4-m) wingspan and stands 12 feet (3.7 m) tall on the ground in its takeoff or landing configuration. In August 2010, the one-third-scale model of the *Puffin* was on display at the NASA Langley campus for filming for the Discovery network series *Dean of Invention*. The concept was also presented to the American Helicopter Society conference on aeromechanics.

In 2010, ONERA (the French National Aerospace Research Laboratory), led by Claude Le Tallec, an expert in remotely piloted aircraft systems, launched the idea of the *P-Plane* (Personal Plane Project), essentially a European program. Like the Paris-based Velib (bicycle) and Autolib (electric car) fleets, the *P-Plane* concerns a similar network of ATC electric air taxis. Its R&D brings together a consortium of five research centers, four universities and a range of businesses from nine European countries and Israel.

The Sikorsky Aircraft Corporation of Stratford, Connecticut, also launched Project Firefly to build and flight-test the world's first large-scale all-electric tilt-rotor. Sikorsky announced the existence of their new aircraft on July 19, 2010, at the Farnborough International Air Show in the United Kingdom and displayed it for the first time at the Experimental Aircraft Association's AirVenture 2010 convention in Oshkosh, Wisconsin, on July 26. The *Firefly* is a modified Sikorsky S-300C helicopter with its engine replaced by an electric motor and digital controller from U.S. Hybrid and two Li-Po battery packs. Integrated sensors provide real-time aircraft health information to the pilot through a panel-integrated, interactive LCD monitor. Eagle Technologies LLC executed the custom airframe modifications and assembly of the demonstrator aircraft. The helicopter can hold only the pilot, no passengers, and operate for 12 to 15 minutes. It has a top speed of 80 knots, or about 92 mph. It was expected that the helicopter would make its first flight in late 2010 or in early 2011. The project was put on hold until more energy-dense batteries could be located.

Without the resources of a big corporation, on August 12, 2011, Pascal Chretien, a French-Australian electrical/aerospace engineer and helicopter pilot, built and flew his e-helicopter *Solution F* for 2 minutes 10 seconds up to a maximum height of 1 meter at Venelles, near Aix-en-Provence, France. As a conventional tail rotor drains somewhere between 8 percent and 10 percent of total hover power, Chretien modeled and built a coaxial design with two counter-rotating rotors on top—a torque-balanced design that can fly without the need for a tail rotor to stop the aircraft from rotating out of control—instead, it just needs a simple, lightweight tail fin. This concept was taken from the conceptual computer-aided design model on September 10, 2010, to the first testing at 30 percent power on March 1, 2011—less than six months. In place of the typical cyclic control, which uses an ingenious variable blade-tilting system to control which way the helicopter tilts and advances, Chretien chose an extremely simple weight-shifting system—a big set of handlebars (incorporating the collective control) that literally tilt the main weight of the aircraft underneath the rotors—as his steering assembly. The rechargeable battery cells are Li-Po pouch cells, with an energy density of 160 watt-hours per kg. Chretien initiated a few tethered test flights to test the action and torque balance of the rotor controllers, the weight-shift directional tilt system and the ground effect behavior of the aircraft.

During the following weeks, *Solution F* logged a total 99.5 minutes of flying time in 29 flights. A typical flight lasted four minutes, with a demonstrated maximum of six minutes. Actual forward flight above translational speed (15 to 20 knots) was never experimented. Although very low speed was tried, the test flights were just hovers, outside ground effect. Chrétien and *Solution F* filed patents relating to a "serial hybrid" helicopter concept. The principle is built around an engine that produces electric power via a generator. The generator feeds batteries, which enable a distributed stack of electric direct drives to turn the blades. In 2012 Pascal Chrétien formed a company in the USA based on his patented technology, Tetraero.

The challenge of flying a conventional electric helicopter has been taken on by Philippe Antoine working with the ENAC HCI engineering school with the removable-wing *Volta*, an MC1 microcopter. The maiden flight, lasting a record 15 minutes with Antoine at the controls, took place on December 2, 2016, at Castelnaudary runway near Toulouse, France. Further funds to increase battery energy are en route.

On November 11, 2013, American singer Lady Gaga (aka Stefani Joanne Angelina Germanotta) launched *ARTPOP*, her third studio album, at the ArtRAVE party inside a building at the Brooklyn Navy Yard by piloting "Volantis," the world's first flying dress, designed by Benjamin Males of the London company Studio XO and built by TechHaus. She was strapped into a white carbon-fiber bodice in the shape of a woman, above which six twin-motor fan units were surrounded by white columns arranged hexagonally and connected to a central node above the bodice, which rested on the ground using a circular stand when not in flight. The batteries and associated control and radio link systems were housed in a central hub at the top of the column, in order to minimize the weight of copper used for the electrical power cabling. The hex 12 drone propelled Lady Gaga half a meter (20 inches) off the ground and hovered forward for several meters. "Hopefully, one day you'll own a Volantis of your own," she commented. *ARTPOP* became the ninth best selling album of 2013 with 2.3 million copies worldwide.[9]

Scaling Down the Drone aka MAVs

The use of drones as MAVS (Micro Air Vehicles) has an interesting history. In December 1992, the RAND Corporation conducted a study for the Defense Advanced Research Projects Agency (DARPA) that considered a wide variety of micro-devices for defense applications. This study projected that it would be possible to have flying vehicles with a 1 cm span and less than 1 gm payload in ten years. In 1993 the RAND Corporation performed a feasibility study on very small controlled or autonomous vehicles. A more detailed study followed and was performed at the Lincoln Laboratory in 1995. This led to a DARPA workshop in the fall of 1995. Developing 15.24-cm (6-in.) flying vehicles was proposed in the fall of 1995 by R.J. Foch of the U.S. Naval Research Laboratory (NRL) and M.S. Francis (DARPA). Vehicles of this type might carry visual, acoustic, chemical or biological sensors. They became of interest because electronic detection and surveillance sensor equipment were miniaturized so that the entire payload weighed 18 grams or less.

It was not long before AeroVironment's Paul MacCready had become involved. In 1996, they were funded by DARPA with a Phase I SBIR contract to study the feasibility of a 15.24-cm MAV. They concluded that a vehicle of this size was feasible and received a Phase II SBIR contract in 1998 that resulted in the *Black Widow*, one of the smallest and most successful MAV systems that could carry a useful payload. This vehicle was electrically powered by one 10-watt DC motor with a four-inch propeller; had an aspect ratio of 1.0, a wing span of 15.24 cm, and a total mass of ~80 g; and could carry a color video camera and transmitter. It also had a 3-gram fully proportional radio control system. A pneumatic launcher and a removable pilot's control unit with a 10.16-cm (4-in.) liquid-crystal display, in a briefcase, were also developed to complete the system. In 1999 the AeroVironment MAV team led by Matt Keenon received awards from DARPA and *Unmanned Vehicles Magazine* for the *Black Widow*. The *Black Widow* set several records for an outdoor flight of a micro air vehicle on August 10, 2000, including an endurance of 30 minutes, a maximum range of 1.8 km (2 mi), and a maximum altitude of 234.39 m (770 ft.). The success of the *Black Widow* led AeroVironment to the development of a somewhat larger "flying wing" MAV, the *Wasp*. The *Wasp* had a root chord of 21.33 cm (8.4 in.) and a wing span of 36.57 cm (14.4 in.), and weighed 181.43 g (6.4 oz). It was powered by one 10 w DC electric motor, was designed to fly between 40.23 kph and 48.27 km/h (25 and 30 mph) at a maximum altitude of 91.44 m (300 ft.), and had a color video camera and transmitter. The *Wasp* was hand-launched, and had an autopilot, an endurance of

60 minutes, and a range of 4 km (2.5 mi) line-of-sight. This vehicle eliminated the need for the pneumatic launcher of the *Black Widow* and was easier to fly. A stripped-down RC *Wasp* set an endurance record of 1 hour and 47 minutes on August 19, 2002.[10]

In January 2010, Tamkang University (TKU) in Taiwan realized autonomous control of the flight altitude of the *Golden Snitch*, an 8-gram (0.3-oz), 20-cm (8-in)-wide, flapping-wing MAV. The MEMS Lab in the TKU has been developing MAVs for several years, and since 2007 the Space and Flight Dynamics (SFD) Lab has joined the research team for the development of autonomous flight of MAVs. Instead of traditional sensors and computational devices, which are too heavy for most MAVs, the SFD combined a stereovision system with a ground station to control the flight altitude, making it the first flapping-wing MAV under 10 grams that realized autonomous flight.

In 2008, the TU Delft University in the Netherlands developed the smallest ornithopter fitted with a camera, the *DelFly Micro*, the third version of the DelFly project that started in 2005. This version measures 10 cm (4 in) and weighs 3 grams (0.1 oz), slightly larger (and noisier) than the dragonfly insect on which it was modeled. The importance of the camera lies in remote control when the *DelFly* is out of sight. However, this version has not yet been successfully tested outside, although it performs well indoors. Researcher David Lentink of Wageningen University, who participated in the development of previous models, *DelFly I* and *DelFly II*, says it will take at least half a century to mimic the capabilities of insects, with their low energy consumption and multitude of sensors—not only eyes, but gyroscopes, wind sensors, and much more. He says fly-size ornithopters should be possible, provided the tail is well designed. Rick Ruijsink of TU Delft cites battery weight as the biggest problem; the lithium-ion battery in the *DelFly* MAV, at one gram, constitutes a third of the weight. Luckily, developments in this area are still going very fast, due to demand in various other commercial fields.

Was there a limit to the miniaturization of camera-equipped quadcopter drones? In 2014, Guangdong Cheerson Hobby Ltd., located in Shantou City, Chenghai District, Fengxin industrial zone, presented their *CX YouCute Tiny*. It measured only 2.3 × 2.3 × 2.0 cm (1 in. × 1 in. × 0.8 in.) and weighed only 7g (0.2 oz), being launched from two fingers of a human hand.

For the autonomous navigation of miniaturized robots (e.g., nano/pico aerial vehicles), MIT engineers have designed a computer chip that uses a fraction of the power of larger drone computers and is tailored for a drone as small as a bottle cap. The new methodology and design has been termed *Navion*. The team, led by Sertac Karaman, the Class of 1948 career development associate professor of aeronautics and astronautics at MIT, and Vivienne Sze, an associate professor in MIT's department of electrical engineering and computer science, developed a low-power algorithm, in tandem with pared-down hardware, to create a specialized computer chip. Karaman says the team's design is the first step toward engineering "the smallest intelligent drone that can fly on its own." He ultimately envisions disaster-response and search-and-rescue missions in which insect-sized drones flit in and out of tight spaces to examine a collapsed structure or look for trapped individuals.[11]

At the General Robotics, Automation, Sensing and Perception (GRASP) Laboratory at the University of Pennsylvania, mathematicians, computer scientists, and engineers have been working to create these complex robots and to operate them. The GRASP team prefers the term "robot" to "drone" altogether because of the latter's violent stigma. The GRASP laboratory received a $5 million grant from the DoD to study swarming groups of networked autonomous robots—the Scalable Swarms of Autonomous Robots and Mobile Sensors, or SWARMS project. The goal is to create a swarm of bug-like quadrotors so intelligent it will re-form. Multiple vehicles can fly as a formation on "search and rescue and disaster recovery" missions, from acting as first responders in dangerous situations to gathering intelligence in a hostage crisis. The Pelican, another model, is equipped with seven-inch-long arms and a laser scanner and camera. It can fly into a building, scope out the digs, and construct real-time 3D maps, identifying features like doorways, people, and furniture, estimating its position with respect to these features one hundred times a second, and navigating. Another micro-drone research project is called the Micro Autonomous System Technologies Collaborative Technology Alliance, also funded by the Army Research Lab, thanks to a $22 million grant—the single largest grant in the history of the university's engineering school. The stated intent is "to help create the fundamental networks and technologies that will put unmanned machines on the front lines of battle."[12]

Another example is the Perdix drone developed by MIT engineering students in 2011. Perdix, named after a character from Greek mythology who was changed into a partridge, has a wingspan of 12 in. (30 cm), operates autonomously, and shares a distributed self-healing brain. It carries out its mission without human piloting, but can talk to other drones to collaborate on getting the job done: surveillance. Because every Perdix communicates and collaborates with every other Perdix, the swarm has no leader and can gracefully adapt to drones entering or exiting the team. It operates in cooperative swarms of 20 or more. The technology was first modified by the U.S. military in 2014. The Perdix program became known from March 2016, when the *Washington Post* revealed footage of an F-16 fighter releasing 20 Perdix over Alaska. At the time, however, the *Post* stated the drones had already been undergoing flight testing for two years. On October 26, 2016, the U.S. military launched a swarm of 103 Perdrix drones from an F/A-18 Super Hornet fighter jet during a test over Naval Weapons Station at China Lake in California. Launching was from small pods on hard points on both fighter-plane wings. The Perdix are capable of withstanding ejection at speeds of up to Mach 0.6 and temperatures as low as minus 10 degrees Celsius. On October 26, the drones formed up at a preselected point and then headed out to perform four different missions. Three of the missions involved hovering over a target, while the fourth mission involved forming a 100-meter-wide circle in the sky. The Perdix may revolutionize aerial warfare.

Further progress was made with AIAUVs during the summer of 2017, when Ashish Kapoor, Andrey Kolobov and a team at the Adaptive Systems and Interaction Group at Microsoft Research, Redmond, Washington, tested two artificially intelligent model sailplanes above the desert valley surrounding Hawthorne, Nevada. The Styrofoam machines with their 16.5-foot (5-m) wingspan navigated the skies on their own, guided by computer algorithms that learned from onboard sensors to monitor air temperature, wind direction, altitude, and other metrics, in addition to speed and location data from GPS, which enabled them to predict air patterns. The AI pilot can detect when the sailplane is suddenly gaining altitude, indicating it has located a rising thermal, and plan a route forward. The Microsoft Aerial Informatics and Robotics admitted that prototypes were still dependent on an electric motor to get off the ground in the first place, while the servos that move the sailplane's flaps and ailerons so that it can steer towards and stay within a thermal are all powered by an onboard battery. But eventually solar cells on a larger gilding aircraft's massive wings could provide all the power it needs, as could wind-powered generators incorporated into its fuselage, and enable it to fly indefinitely, or in other words, an infinite soaring machine.[13]

Since 2011, engineers led by Marcus Fischer at the Bionic Learning Network of FESTO, an industrial control and automation company based in Ostfildern in southern Germany, have developed the Li-Po battery 23-watt 3.70 m (12 ft) *SmartBird*, resembling the herring gull and capable of flapping its wings to take off, to fly and to land without the aid of other devices to provide lift. The natural wing beat of a bird was emulated by using bionics technology to decipher bird flight. The Festo Network includes TU Berlin, Delft University of Technology, TU Ilmenau, Friedrich Schiller University Jena, Christian-Albrechts University in Kiel, University of Arts and Industrial Design Linz, University of Oslo and Akershus, Department of Product Design, University of Applied Sciences Ravensburg-Weingarten, University of Stuttgart, CIN University Tübingen, University of Ulm, Fraunhofer IPA.

The honeybee (*Apis mellifera*), which pollinates nearly one-third of the food we eat, has been dying at unprecedented rates because of a mysterious phenomenon known as colony collapse disorder (CCD). The situation is so dire that in late June 2014 the White House gave a new task force just 180 days to devise a coping strategy to protect bees and other pollinators. The crisis is generally attributed to a mixture of disease, parasites, and pesticides. Inspired by the biology of a bee, researchers, led by engineering professor Robert Wood at the Microrobotics Lab of the Wyss Institute, Harvard University, are developing Autonomous Flying Microbots aka RoboBees, man-made systems that could perform myriad roles in agriculture or disaster relief. A RoboBee measures about half the size of a paper clip, weighs less than one-tenth of a gram, and flies using "artificial muscles" compromised of materials that contract when a voltage is applied. To construct RoboBees, researchers at the Wyss Institute have developed innovative manufacturing methods, so-called pop-up micro-electromechanical (MEMs) technologies that have already greatly expanded the boundaries of current robotics design and engineering. A RoboBee can lift off the ground and hover midair when tethered to a power supply. After two years of R&D, in 2016 the Wyss team announced that their RoboBees can now perch on objects from any angle, using an electrode patch and a foam mount that absorbs shock to perch on surfaces and conserve energy in flight—like bats, birds or butterflies. The new perching components weigh 13.4 mg, bringing the total weight of the robot to about 100mg—similar to the weight of a real bee. The robot takes off and flies normally. When the electrode patch is supplied with a charge, it can stick to almost any surface, from glass to wood to a leaf. To detach, the power supply is simply switched off.

But they still need to be able to fly on their own and communicate with each other to perform tasks like a real honeybee hive is capable of doing. The researchers believe that as soon as 10 years from now these RoboBees could artificially pollinate a field of crops, a critical development if the commercial pollination industry cannot recover from severe yearly losses over the past decade. RoboBees will work best when employed as swarms of thousands of individuals, coordinating their actions without relying on a single leader. The hive must be resilient enough so that the group can complete its objectives even if many bees fail.[14]

Another approach has been taken by Anna Haldewang, a 24-year-old industrial design student at Savannah College of Art and Design (SCAD) in Georgia. Haldewang created 50 designs of a bee drone before landing on the final model, Plan Bee, which does not resemble a bee at all. The drone consists of a foam core, a plastic-shell body and two propellers. There are also six sections of the drone that meet at the bottom, all of which have tiny holes that let the machine gather pollen while it hovers over plants. It can then release the pollen at a later time for cross-pollination. Haldewang noted that Plan Bee is still in its early stages, but she has filed a patent for the technology and design. Its application in backyards as a teaching tool has potential, but the drone could conceivably be used in large-scale farming, even in hydroponic farming.

During 2016, a team led by Mirko Kovac at Imperial College London's Aerial Robotics Lab at the South Kensington campus, again taking inspiration from nature, developed an aquatic micro aerial vehicle (AquaMAV) which dives like a gannet and launches like a flying fish to collect water samples. The drone only weighs 200 grams (7 oz) and can currently achieve speeds of around 30 miles per hour from a starting point beneath the water. It can make the aerial leap even if conditions on the surface are rough. The researchers state that using waterjet propulsion and energy from a 3.5 g (0.1 oz) Li-Po battery, AquaMAV can currently fly around five kilometers to and from an analysis. The team says the aerial range would enable those analyzing the samples to be at a safe distance away from a potentially hazardous situation. The Imperial College team has also developed another nano-quadcopter which acts like a spider, whereby tools can be attached to the drone's moving arm, which could allow it to repair and examine buildings in hard-to-reach or dangerous areas; it can even spin a simple web of fine thread.

In agriculture, farmers are now using infrared camera–carrying drones to pinpoint problem spots with insects and aphids in vast fields and ranchlands. Based on the mapping, another drone then drops a "cocktail" of predatory insects, transported in a sock attached to the underbelly of the drone and containing a mixture of vermiculite and insects, onto grapevines and citrus trees to combat pests. By focalizing pest control, they prevent spread and save money. One example of this is the AgriDrone developed by Saga University in Japan and IT firm OPTiM. Agri Drone uses a suspended bug zapper which delivers a glowing, insect-enticing electric payload to the points at which the pests are congregating in harmful numbers. Primarily used at night, it utilizes infrared and thermal cameras to shoot targeted doses of pesticides where insects are congregating. A similar project has been developed by the Department of Life Sciences and Computing at Imperial College London with agriculture services company Agrii of Grantham.

Researchers at the University of Illinois at Urbana-Champaign and Caltech have developed a self-contained robotic bat—dubbed Bat Bot (B2)—with soft, articulated wings that can mimic the key flight mechanisms of biological bats. The B2 possesses a number of practical advantages over other aerial robots, such as quadrotors. Bats have more than 40 active and passive joints and the B2 robot has only 9 (5 active and 4 passive). The compliant wings of a bat-like robot flapping at lower frequencies (7–10 Hz vs. 100–300 Hz of quadrotors) are inherently safe because, replaced by rapid-spinning propellers, their wings comprise primarily flexible silicone membrane materials and are able to collide with one another, or with obstacles in their environment, with little or no damage.

Instead of designing microdrones to mimic insects, an alternative is to equip live insects with electronic navigation systems. In 2009, DARPA funded a joint project by the University of California, Berkeley, and Nanyang Technological University in Singapore to control wirelessly the flight of the giant flower or rhinoceros beetle using a 6-cm (2.4-in.), 8-gm (0.3-oz) backpack powered by a 3.9 volt lithium battery connected to six microelectrodes implanted into the insect's wing-folding coleopteran muscles. In 2017, a team led by Jesse J. Wheeler at the Charles Stark Draper Laboratory, Inc., in Cambridge,

Massachusetts, announced their DragonflEye project whereby, in conjunction with the Howard Hughes Medical Institute, they had "equipped" live dragonflies with aoptrode navigational systems, even smaller than optical fibers. In addition, real dragonflies are unbelievably nimble and quick, compared to clumsy man-made drones, with the ability to maneuver turns as sharp and fast as 9-Gs. Another advantage is that, as long as they have food, water and sunlight, they have a greater autonomy than battery-powered mechanical drones. Draper is also planning to use the same system on bumblebees for pollination.[15]

ELEVEN

Dronomania!

There seems to be no stopping the proliferation of multicopter drones in every walk of life and they come in all shapes and sizes, be they UAVs, MAVs or PAVs.

In contrast to the less-than-one-hour duration and short distance of a drone race, others have still been working on the eternal UAV, first developed by Paul MacCready. In April 2014, Google acquired Titan Aerospace, which had developed catapult-launch drones with 200-ft. (60-m) wingspans called *Solara 50* and *Solara 60*, capable of flying at a reported altitude of 20 km for impressive periods of over 5 years. By early 2016, Google was developing the *Skybender*, their latest solar-powered UAV, in a large warehouse in New Mexico. The goal was to use millimeter wave radio transmissions to bring Internet speeds of up to 40 times faster than those provided by 4G LTE systems.

The Autonomous Systems Lab of the Swiss Federal Institute of Technology Zurich (ETHZ), in cooperation with industry partners, developed their *AtlantikSolar*. It weighs just 6.8 kg (15 lb.) and has a 5-m (16-ft.) wingspan. On July 14, 2015, the *AtlantikSolar* was hand-launched at the Rafz RC-model club. Despite winds of up to 60 kph (37 mph) and thunderstorm clouds in the last two hours of the flight, the UAV flew continuously for 81 hours, breaking the world endurance record for unmanned aerial vehicles under 50 kg (110 lb.). Given its name, the next flight was planned to be 5,000 km (3000 mi) across the Atlantic Ocean, on a preprogrammed route from Boston to Lisbon over seven days. That flight will follow a preparatory 12-hour flight from Belem to the Caxiuana research station in the middle of the Amazon rain forest, covering about 400 km (250 mi.) on solar power alone.

ETH's team leader Philipp Oettershagen, a Caltech Fulbright scholar and UAS engineer, predicts squadrons of these mini-planes deployed to save human lives. In June 2015, the *AtlantikSolar* participated in the EU-funded search and rescue project ICARUS in Portugal, where it tested out victim detection flights over water for the first time. From October 21 to 31, 2015, *AtlantikSolar* had the chance to assist directly in the first real-life (i.e., outside of research projects) disaster support mission. Requested by SIPAM (Brazilian Amazon Protection System, part of the Brazilian Ministry of Defense), *AtlantikSolar* was tasked to perform aerial sensing and mapping around the site of a disaster—a sunken ship that involved over 4,400 dead cattle and 750 tons of oil spill—that had happened 2 weeks before. In June 2017, glaciologists from ETH Zurich used the 6-kg (13.2-pound) *AtlantikSolar* to monitor glaciers in Greenland; after 13 hours in the air, fog rolled in and that drone had to be retrieved. To date, *AtlantikSolar* has made flights up to 81 hours, and the plan is still to make an autonomous flight across the Atlantic Ocean.

After the flights of Paul MacCready in the *Solar Challenger* and Didier Esteyne in *Airbus E-Fan*, it was inevitable that a drone would cross the 35 km (21.7 mi) of the English Channel. This was achieved on a sunny February 16, 2016, by commercial drone mapping/photography firm Ocuair, based in Westbury, Wiltshire, UK. A custom airframe was designed by UK-based UAV manufacturer Vulcan, efficient T-motors and blades provided the lift, while Optipower provided two huge 22-amp hour batteries, Jeti provided the secure and robust control links, and Nottingham Scientific Ltd. provided the GPS tracking devices. The quadcopter dubbed *Enduro1* was piloted by Richard Gill from an accompanying rigid inflatable boat staying within 500 m (1,640 ft.) of the drone. Morning takeoff was from Wissant beach between Boulogne and Calais. The flight was going well as they passed the point of no return, 17 km, from which the distance to the UK was shorter than returning to the original launch position, so there was no escape plan. At 23 km, *Enduro1* suddenly lurched to the left and the pilot had to disable the GPS guidance and take manual control of the flight for the final 20 minutes. Flying without GPS assistance was extremely challenging. After a flight of 72 minutes, the drone landed on Shakespeare Beach in Dover, southeast England, mission accomplished.

To improve the breed of commercial drones, in April, Airbus Group SE teamed up with Local Motors to launch the Caro Drone Challenge Contest. The starting design of this competition was Airbus's Quadcruiser hybrid concept, combining the VTOL and hovering capabilities of the well-known quadcopter design with the speed and cruise efficiency of a fixed-wing aircraft by using an additional pusher motor. There were 425 entries. At a ceremony held in July 2016 during the Farnborough International Airshow, five winners were announced: Alexey Medvedev from Omsk in Russia; Harvest Zhang from Mountain View in California; Dominik Felix Finger from Aachen in Germany; Finn Yonkers from North Kingstown, Rhode Island; and Frederic Le Sciellour from Pont De L'Arn in France. They shared an overall prize pool of well over U.S. $100,000. As a next step Airbus and Local Motors have been building a demonstrator version of the winning design concept, again through an open collaboration involving the community, including potential customers and end-users.

On July 21, 2016, Facebook announced that it had flown *Aquila*, its solar-powered drone, for more than 90 minutes in a test over Arizona, calling it a "big milestone" for its connectivity plans to build drones, satellites and lasers to deliver the Internet to everyone, whether living in a world city or in a remote area of a developing country. The video was watched by almost 3 million viewers. Ascenta, headed by Andy Cox, is a small drone maker in Bridgewater, Somerset, England. Facebook acquired the team behind Ascenta in March 2014 for just under $20 million. The *Aquila* was dismantled and taken in pieces to Arizona. There, it was reassembled for its first flight. The *Aquila* drone has the wingspan of a 737, but can run on the power of three hair dryers. It is designed to fly for up to three months at a time, only consuming 5,000 watts of energy at cruising speed.[1] Ten months later, *Aquila* completed its second flight, during which it flew for an hour and 46 minutes. This time around, Facebook added "hundreds of sensors" to gather additional data; modified the auto-pilot software; installed a horizontal propeller stopping mechanism to support a successful landing; and added "spoilers" to the wings, which increase drag and reduce lift during the landing approach. It is being checked by the U.S. National Transportation Safety Board. Airbus and Facebook have now agreed to partner on high-altitude pseudo-satellite development.

Also in July 2016, the Farnborough Air Show in England hosted the very first edition of its UK Drone Show. For the first time, drones had their own dedicated space at the show. There were drone-related activities on the public days, including a drone air display and the UK Drone Racing Masters Championship (see Appendix A).

It was inevitable that fuel cells would be incorporated into drone technology. In June 2015, Michel M. Bitton and a team at EnergyOr Technologies in Montreal flew their fuel-cell multicopter H2Quad 1000 inside a hangar at half a meter above the floor for 3 hours, 43 minutes and 48 seconds. Two years later, EnergyOr shipped the first H2Quad 1000 to the French Air Force's Centre d' Expertise Aérienne Militaire (CEAM) in Mont de Marsan, France.

On the afternoon of January 19, 2016, the Scottish Association for Marine Science (SAMS) completed a test flight above Oban airport in Scotland of a Raptor E1 drone, built and designed by Trias Gkikopoulos of Raptor UAS and using Cella's hydrogen-based power system. The complete system, a Cella gas generator along with a fuel cell supplied and integrated by Arcola Energy, is considerably lighter than the lithium-ion battery it replaced. Although the flight lasted only 10 minutes at a cruising altitude of 260 ft. (80 m), the drone had a potential autonomy of 2 hours. Larger versions of this system will have three times the energy of a lithium-ion battery of the same weight. Cella is also working on aerospace systems with its partner Airbus-Safran Launchers, now ArianeGroup.

In late summer 2016, Protonex, a subsidiary of Burnaby-based fuel-cell developer Ballard Power Systems, delivered its proton exchange membrane (PEM) fuel cells to Insitu, a subsidiary of U.S. aerospace giant Boeing that produces military- and industrial-grade, long-endurance, fixed-wing drones such as the ScanEagle. Insitu has stated that it hopes to roll out a commercial, liquefied hydrogen fuel-cell version of its *ScanEagle* drone by 2017. At the same time, Intelligent Energy of Loughborough, England, took a standard DJI Matrice 100 drone and, by reequipping it with a fuel cell, extended its flight range from 20 minutes to over an hour. Demonstrations were at the InterDrone conference, which ran September 7–9, 2016, at the Paris Hotel in Las Vegas. In China, MMC introduced its carbon-fiber HyDrone 1800 with a flight endurance of 4 hours, or 50+ hours when combined with MMC tethered technology.

Proliferation of UAVs for military and security purposes has continued. The CIA has been flying unarmed drones over Afghanistan, Pakistan, Yemen, and Somalia since 2000. It began to fly armed drones after the September 11, 2001, attacks. Some were used during the air war against the Taliban in late 2001. But on February 4, 2002, the CIA's Eagle Program first used an unmanned Predator drone armed with Hellfire missiles in a targeted killing. The strike was in Paktia province in Afghanistan, near the city of Khost. The intended target was Osama bin Laden, or at least someone in the CIA had thought so. As of 2008, the USAF has employed 5,331 UAVs, which is twice the number of its manned planes. Of these, the Predators have been the most effective. The overall success of the Predator missions is apparent because from June 2005 to June 2006 alone, Predators carried out 2,073 successful missions in 242 separate raids. While Predator is remotely operated via satellites from more than 7,500 miles away, the Global Hawk operates virtually autonomously. Once the user pushes a button, alerting the UAV to take off, the only interaction between ground and the UAV is directional instructions via GPS. Global Hawks have the ability to take off from San Francisco, fly across the U.S., and map out the entire state of Maine before having to return. In February 2013, it was reported that UAVs were used by at least 50 countries, several of which have made their own, including Iran, Israel and China. The U.S. military inventory now comprises more than 12,000 ground robots and 7,000 UAVs.

Off-the-shelf drones began to be used by Islamic State or Daish in 2014. At first they used them to film propaganda videos from the air. Then they became scouts. A drone video of a Syrian military base was released shortly before the base was hit by multiple suicide bombings that targeted its weak spots, suggesting that the drone had been sent in on a surveillance mission. In October 2016, the Islamic State carried out a drone strike of their own. A UAV, its explosive device inside disguised as a battery, reportedly exploded after being shot down by Kurdish forces in Iraq, killing two fighters.

Sky Sapience HoverMast-100, a compact, mobile, electric-powered, intelligence-gathering tethered flying machine, has been developed since 2010 by Brigadier-General (Ret.) Gabriel Shachor, with Shy Cohen and Roonen Keidar at Yokne'am Illit, northern Israel. Weighing 10 kg (22 lb.), it can carry almost its own weight in equipment, such as radar systems or sensors, while tethered to a small vehicle like a car or truck. The cable also carries power generated by an electric power supply unit at the base station to the machine at a height of 50 meters, attained in 10 seconds, and the electric power supply unit offers unlimited operation time. Data collected by the payloads is transmitted through a wide-band communication link connected between the HoverMast-100 aerial system and the ground station. The first HoverMast-100 was ordered by the Israel Defense Forces (IDF) for its Ground Forces Command in August 2012, and the system was delivered in September 2013. One is reminded of the tethered *PKZ1* rotorcraft developed by Stephan von Petróczy in 1917 (see Chapter Three).

During the Great Solar Eclipse of August 21, 2017, researchers from Oklahoma State University and the University of Nebraska used low-flying drones to track changes in atmosphere; the flight was part of the broader Collaboration Leading Operational Unmanned Development for Meteorology and Atmospheric Physics (CLOUD MAP). Later that month, Hurricane Harvey was recorded as the wettest tropical cyclone on record in the contiguous United States. The resulting floods inundated hundreds of thousands of homes, particularly in Texas, displaced more than 30,000 people, and prompted more than 17,000 rescues. While it was dissipating, the U.S. National Hurricane Center began monitoring a tropical storm over the west coast of Africa. Further surveillance of the storm led the NHC to classify it as Tropical Storm Irma on August 30. As the storm picked up speed, Irma took a deadlier turn. As of September 5, Irma was upgraded to a Category 5 hurricane with wind speeds gusting at 175 mph. It goes down in record books as the strongest storm in the Atlantic Ocean.

On September 8, an earthquake of 8.1 on the Richter scale hit Mexico in which over 90 people died. Although weather satellites and NASA were able to show the size and the movement of the two hurricanes, in the aftermath, to establish the extent of the damage, the FAA issued permits to commercial drone operators to assist in a number of different functions that expedited the recovery process. These included identifying victims, delivering rescue ropes and life jackets in areas that were too dangerous for ground-based rescuers to venture into, and observing the damage to buildings, roads and bridges, and power lines; this work was done by small quadcopters to the military-grade catapult-launched Insitu *ScanEagle*, as well as 4G LTE tethered drones to provide wireless service to areas without network coverage by acting as temporary cell towers.

In October, following the nation's deadliest mass shooting on the Las Vegas Strip

(59 deaths and 489 injured), it was suggested that a weapon-armed security drone could have taken off in two minutes and fired an incendiary device into the lone gunman's hotel room on the 32nd floor of the Mandalay Bay Hotel.

The first use of a drone for filming the aerial footage in a major film was in 2012 for *Skyfall*, directed by Sam Mendes. *Skyfall* has a spectacular opening sequence, shot by the Flying-Cam 3.0 SARAH Unmanned Aircraft System, in which James Bond 007 uses a motorbike to chase a terrorist across the rooftop of the Grand Bazaar in Istanbul. The high-speed aerial footage captured by the drone in that scene made a buzz in Hollywood, contributing to the movement that saw, a couple of years later, 6 aerial filming companies get the first 333 FAA exemptions for closed-set filming.

In 2015, a British thriller film, *Eye in the Sky*, highlighted the ethical challenges of drone warfare. It is just one of over twenty recent feature films which feature combat drones. In February 2016, the *Hollywood Reporter* wrote, "There has been no shortage of films dealing with drones over the last few years … audiences have recently had the occasion to explore a form of modern warfare whose true repercussions are yet to be fully understood, let alone divulged to the general public."[2]

In October 2016, the American video production company Stratus Productions bolted a 1,000-watt LED light bar to the underside of a Freefly Alta 8 octocopter drone. Although it has an autonomy of only 10 minutes, the drone was able to light up an entire city block or forest.

Drones have played an increasingly large role in wildlife preservation and filmmaking in recent years. In 2011, a team of filmmakers flew a German-built Microdrone md4-1000 quadcopter above the Masai Mara region of the Serengeti to capture video of all sorts of African wildlife. Researchers in Kenya, meanwhile, have begun using unmanned aerial drones to monitor areas susceptible to rhino poaching. In November 2016, Northrop Grumman engineers and San Diego Zoo Global scientists traveled to the Arctic, where they used a UAV to track polar bear movements over thousands of miles while measuring the ice pack that is critical to the species' survival. The UAV was an all-electric, fixed-wing aircraft with a 14-foot wingspan and a custom fuselage for the accommodation of several optical sensors. It was also equipped with multi-terrain landing gear and environmental packaging for the rough Arctic environment, where temperatures regularly drop below zero degrees Fahrenheit.

Rather than drones which are used to look down, and spy, and bomb, and race, others are becoming an art form to be looked up at. A visual display was created in 2012 by Saatchi & Saatchi's award-winning Jonathan Santana and Xander Smith with Marshmallow Laser Feast of London. "Meet Your Creator: Quadrotor Show" was a live theatrical performance / kinetic light sculpture with quadrotor drones, LEDs, motorized mirrors and moving head spotlights dancing to music by OneOhtrix. Elsewhere, Horst Hörtner and his 15-strong team at ArsElectronica Futurlab GmbH in Linz, Austria, created a similar choreography of "spaxels" (space pixels), luminous colored drones which are remote-controlled to perform a ballet in the night sky. The first ballet took place at home in 2012 when a formation of 50 drones took to the sky, thrilling festival goers in Linz and creating a media sensation worldwide. Spaxels® shows in London, Brisbane and Dubai not only delighted crowds on site; they created a sensation in social networks too. With Intel's interest expressed in a display of coordinated aerial artistry in conjunction with a new Intel campaign, the company supported the technical R&D that aimed to make the flight more secure. The challenge was for four pilots to launch 100 drones and deploy them aloft to paint 3D images and messages to the accompaniment of a live orchestra for maximum impact. "Drone 100," performed in the sky in November 2015 above Flugplatz Ahrenlohe in Tornesch, near Hamburg, Germany, earned a new world record title for the most Unmanned Aerial Vehicles (UAVs) airborne simultaneously. The Spaxels moved in sync to the beginning of the Beethoven's Fifth Symphony, even spelling out the word "INTEL."

Raffaello D'Andrea, a Canadian/Italian/Swiss engineer, artist, and entrepreneur working with Weixuan Zhang and Mark W. Mueller at the Department for Dynamic Systems and Control at the Federal Institute of Technology in Zurich, Switzerland, developed the *Moonspinner*, a drone that can fly with only one propeller. In May 2016 it featured in a Cirque du Soleil show in Broadway, where a fleet of drones disguised as lampshades danced around the human performers as part of their Paramour Show. The *Moonspinner* features no additional actuators or aerodynamic surfaces. It cannot hover like a standard multicopter, but can be launched like a Frisbee.

The Vivid Festival in Sydney provided the ideal setting for the public debut of the Drone 100 project. On five evenings, June 8–12, 2016, spectators in Australia witnessed a performance custom-tailored to the occasion and with musical accompaniment by the Sydney Youth Orchestra. The brilliantly illuminated silhouette of Sydney's world-famous Opera House, the city's architectural landmark on Bennelong Point, lit up the sky above the harbor.

Meanwhile, James Alexander Stark, R&D Imagineer Principal at Walt Disney Imagineering, with Clifford Wong and Robert Scott Trowbridge, had obtained U.S. patent 9102406 B2 for "Controlling unmanned aerial vehicles as a flock to synchronize flight in aerial displays." For the 2016 Christmas season, Disney and Intel launched drone light shows at Florida's Walt Disney World Resort, where no fewer than 300 super light Intel Shooting Star drones, made from Styrofoam and plastic, equipped with LED lights and weighing only 280 g (10 oz), created over four billion different color combinations, programmed and controlled from just one computer. Among the images: Christmas trees and angels.

In December 2016, Intel improved its Guinness World Record when, from a football field in Krailling, Bavaria, Germany, it simultaneously launched 500 Shooting Star drones to give a flawless ballet in the night sky, finishing by displaying the number "500" in the air. The entire swarm was controlled by one pilot and one PC. Each quadcopter's propellers are also protected by covered cages—all features designed to ensure the drone is safe to fly, is splash-proof and can fly in light rain.[3]

The Shooting Star fleet's next venue came in February during the halftime break at the 51st Super Bowl, the American football championship game between the New England Patriots and the Atlanta Falcons. Lady Gaga's singing was accompanied by the 300-strong fleet, recently over from Disney World, forming an image in red, white and blue of the American flag. Marking the first time drones had been used as part of a live television broadcast, the display was watched by an estimated 111.3 million people.

Was there a limit to the number of drones in the sky simultaneously? On the night of February 11, 2017, *one thousand* colored EHang drones flew up into the sky, alongside the landmark Canton Tower in Guangzhou, southern China, to celebrate the Lantern Festival, the last day of the Chinese New Year holiday. During a 15-minute performance, with an orchestra playing below, controlled by one computer, the thousand drones flew in six different formations of Chinese characters including "Blessing," "Lantern Festival," and the map of China. At the 2018 Winter Olympics, Intel again raised the record for the number of skyborne drones with no fewer than 1,218 Shooting Star drones flying in sync to create huge light-up images of Olympic sports and the iconic Olympic rings in the skies over Pyeongchang.

It is interesting to note that, to date, the greatest number of radio-controlled model aircraft airborne simultaneously is 179, achieved by on July 16, 2016, above Furey Field in Malvern, Ohio. This was organized by Howard Kaler of Plainfield, Illinois, in conjunction with the Flite Fest East family community. The previous record was 99 planes. Registrations for the event had been closed at 300 planes, with an additional 100 participants turned down. Although they did not collide with each other, unlike drones, these model aircraft were certainly not flying in formation. The same may be said for the average number of airplanes criss-crossing around the globe at any given time: over 16,000.[4]

On May 11, 2017, at the Xponential event held at the Kay Bailey Hutchinson Convention Center in Dallas, 10 quadcopters performed a synchronized routine on a fake bride, flashing LED lights that can create over 4 billion color combinations. On August 9, 2017, to celebrate Singapore's 52nd birthday celebration National Day Parade, a 300-strong drone display animated the night sky above Marina Bay with an outline of Singapore Island, a heart with a crescent and five stars, the NDP 2017 logo, a hashtag, the Merlion, children and an arrow shape, logos and even a map of the country. One month later, in Los Angeles, people looking skyward over Dodger Stadium witnessed an illuminated display celebrating a comic book superhero. But it was not the Bat Signal sending out a distress call for Batman—it was a fleet of 300 lit-up drones performing choreographed maneuvers to spell out Wonder Woman's trademark "W" symbol. The dynamic light show was produced by Warner Bros. in partnership with Intel's drone team, for the U.S. release of the film *Wonder Woman* on Blu-ray.

Drones will soon be flying around indoors at home. Back in September 2015, Tessie Hartjes and Lex Hoefsloot founded Blue Jay at Eindhoven University of Technology and Fontys University of Applied Sciences, southeast Netherlands. Their goal was to develop a domesticated drone which could navigate indoors using a Wi-Fi system with special lights produced in collaboration with Philips. These lights, mounted on the ceiling, each emit light at a different frequency. A camera on top the Blue Jay can then distinguish between the different frequencies. A laser that constantly measures the distance to the ground determines how high the Blue Jay is flying. R&D was carried out by a multidisciplinary team of 19 top students. In April 2016 the Blue Jay team set up the world's first drone café, featuring Blue Jays that could take orders and serve them. The drone café was part of the three-day Dream & Dare Festival marking the 60th anniversary of Eindhoven University's foundation.

Drones are also entering the classroom. Krishna Vedati, formerly with AT&T Interactive's Consumer Division, and his team at tynker.com in San Francisco create apps and curricula that teach children the basics of coding using games and real-world gadgets. Their downloads are used by some 60,000 schools in the USA (30 million kids). Tynker recently launched a new project—teaching coding through drone lessons. Schools typically buy between six and 12 drones via Tynker's partnership with drone maker Parrot and can then download Tynker's free set of drone lessons. Children learn to make drones do back-flips, as well as more complex ideas such as drones working together as a team.

In December 2015 Twitter patented a photo-sharing camera drone that is controlled by tweets; thus the selfie married the drone with the AirSelfie, as invented by Dylan Tx

Zhou. This is a pocket-sized flying camera that connected with smartphones to enable HD photos and videos of the owner and his friends. Its turbo fan propellers could thrust up to 20 meters (65 ft.) in altitude for a duration of three minutes. In October 2016 Amazon obtained a patent for a pocket-sized, voice-controlled camera drone that can be used to help police in chases, aid firefighters in tackling blazes, and even find lost children. In December 2016, Samsung, a maker of smartphones, cameras, and even a 360-degree VR-friendly camera, patented a completely circular drone, with a bulge at the bottom holding what seems to be a small camera.

For Christmas 2016, Casey Owen Neistat, an American YouTube personality, filmmaker, and blogger, conceived of an idea to release a video, *Human Flying Drone*, of himself using a giant drone to snowboard around Finland. Sponsored by Samsung, the 10-ft. (3-m)-diameter hexadecacopter was fitted with sixteen 31-inch (78-cm) carbon-fiber propellers powered by 16 individual electric motors. Neistat, wearing a Santa Claus costume for the feat, ski-jored (slalomed to and fro while towed by his drone) up and down the snowy slopes and at one point rose up into the sky. The YouTube video was watched by millions of viewers worldwide.

A remote-controlled flying-toys company has come up with possibly the best toy idea ever—drone toys of famous Star Wars ships such as Han Solo's *Millennium Falcon* and the rebels' X-Wing starfighter. Propel, which already sells a wide range of model helicopters as well as consumer UAVs, showcased its new Stars Wars Battling Quad drones at the Star Wars Celebrations convention in London, which took place over the weekend of July 15–17, 2016, at London's ExCeL Centre. Propel has shown off impressive miniature drone versions of famous Star Wars spaceships that can be used to play real airborne dogfights at speeds above 35 mph (60 kph) and engage friends and family in exciting multiplayer laser battles.

In 2016, Goitein Bezalel of Powerup Toys in Miami teamed up with Parrot to develop a paper drone. The "pilot," wearing a smartphone with a head-mounted display, can see what the drone sees and can control it with intuitive movements of his head. Once folded, the paper plane supports a frame that includes control electronics, two motors, a battery, and a tiny camera. There is also a microSD slot for recording the flights.

According to ABI Research, the drone industry is going to be worth $8.4 billion by 2019. This is not only from hardware sales, but mainly from the applications and services where most growth is expected.

Biomimicry Continues to Improve the Breed

From the fall of 2016, Professor David Lentink, an assistant professor of mechanical engineering at Stanford University in Palo Alto, California, and his team have benefited from a wind tunnel based on various measurement systems acquired with support from the Air Force, Navy, Army, Human Frontiers Science Program, and Stanford Bio-X program. The Stanford team's goal is to observe how tiny birds fly, then transfer the same skills to flying robots or drones. The tunnel is a 2-meter-long chamber that can blast gusts of up to 50 meters per second. The tunnel is the first of its kind able to create turbulence, using computer-controlled wind vanes.

Researchers at the National Centre of Competence in Research (NCCR) Robotics and the Laboratory of Intelligent Systems (LIS) at the École Polytechnique Fédérale in

Lausanne, Switzerland, have bio-mimicked the intricate folding patterns of rove beetle wings to develop a drone that opens up and takes to the skies in less than a second with a couple of swift movements, and can be carried in a backpack to remote and dangerous areas. The result is a drone that when folded up has only 43 percent of the wingspan and 26 percent of the surface area of when it is in operational mode, where its wings measure 200 × 500 × 16 mm (7.87 × 19 × 0.62 in). LIS's previous work includes a drone that uses its wings to crawl on land like a sea turtle and a grasshopper-inspired robot that can jump 27 times its body size.

Alongside such an approach, engineers at Airbus are looking at how future airplane shapes may mimic birds and dolphins. The most noticeable aspect of this approach is in the fuselage, which, instead of being wrapped in opaque steel, is composed of a web-like network of structural material that looks a bit like a skeleton.

On November 3, 2014, Nixie, a small camera-equipped drone that can be worn as a wrist band, competing against more than 500 other participants, won Intel's Make It Wearable competition, thus securing $500,000 in seed funding to develop Nixie into a product. It can be activated to unfold into a quadcopter, fly in one of its pre-programmed modes to take photos or a video, and then return to the user. Nixie, based in Palo Alto, California, was founded by Jelena Jovanovic, a Stanford physics researcher, and Christoph Kohstall, holder of a Ph.D. in experimental physics and former manager at Google, as well as technical program manager at OpenROV. Their goal is to develop their drone into the next generation of point-and-shoot cameras.

Laurent Eschenauer and Dimitri Arendt of Grâce-Hollogne, Belgium, have developed the Fleye (Flying Eye), a personal autonomous robot drone with a unique spherical design the size of a football, where all moving parts are fully shielded. Fleye has a powerful on-board computer, similar to the latest smartphones. It is a dual-core ARM A9, with hardware accelerated video encoding, two GPUs, 512MB of RAM, and it runs Linux. It also supports the popular Computer Vision library OpenCV. This means that Fleye can be programmed to execute missions autonomously, reacting to what it sees in its environment. On the functional side the Fleye's design takes a cue from industrial and defense UAVs, relying on a "ducted fan." Unlike most drones on the market that have multiple open rotors, the Fleye's lift is generated by a single shielded propeller with four control vanes providing directional control. The drone sports an array of sensors including an accelerometer, gyroscope, altimeter, GPS, sonar, optical flow, and a magnetometer, so right out of the box, the Fleye can track you, take 360-degree images, or fly other preplanned autonomous missions at the tap of an icon. Users need only switch on, open the app, choose select what they want Fleye to do, then toss it into the air, like a football.[5]

At Drone World Expo in November 2016, AeroVironment, still in the game after three decades, unveiled the VTOL Quantix, an industrial-strength fixed-wing drone for agriculture, energy, and transportation industries, among others. Quantix can map 40 acres (16 hectares) in about 45 minutes (its overall flight time is approximately an hour). About the size of their Raven, Quantix weighs 5 lbs. (2.3 kg) with a wingspan of 3.2 feet (1 m). But unlike the Raven, which is launched airplane-like, horizontally, Quantix will initially use four propellers to take off vertically like a consumer-brand quadcopter. Upon reaching cruising altitude of about 400 feet (120 m), Quantix can flip over and fly horizontally like an airplane at speeds up to 45 mph (70 kph). Landing will also be conducted vertically. Operators can fly Quantix easily using one-touch planning and launch via a dedicated Android tablet device. The AeroVironment Decision Support System (AV DSS™) is a full-service, cloud-based data analytics system.

Roman Luciani and Antoine Tournet of Toulouse, France, have established Airvada to develop the Diodon, an inflatable drone which, thanks to its patented inflatable structure, is at the same time easy to transport, waterproof, and rugged because of the flexible structure. In a patented system, CO_2 cartridges are used to inflate them in only 30 seconds, while packing away takes just 60. Thanks to the independent inflatable chambers, the Diodon is extremely reliable; up to two chambers can burst without compromising the flight.

At the very beginning of this history, we noted Leonardo da Vinci observing bird flight, then in 1620 we came across the "gansa" super-geese towing Domingo Gonzales into the sky and up to the moon. Three centuries later, Etienne Edmond Oehmichen published his work *Nos maîtres les oiseaux, étude sur le vol animal et la récupération de l'énergiedans les fluides* (*Our masters, the birds, a study on animal flight and recovering energy in fluids*). Now, one hundred years later, flying fauna still inspire more than ever.[6]

Bay Zoltan Nonprofit Ltd., a state-owned applied research institute of Logistics and Production Engineering in Budapest, Hungary, has developed a flying tricopter called Flike, its Li-Po batteries giving a potential 30–40 minute flight. The Flike achieved its first manned flight on March 7, 2015, at Miskolc Airfield in northeast Hungary, staying airborne for over a minute with a takeoff weight of 463 lb. (210 kg) and landing safely. The lift is generated by six rotors grouped in counter-rotating pairs on three axes, equally located around a circle. The rotation speed of individual rotors can be adjusted, and an onboard computer takes care of the craft's stability.[7]

Experimentally there is no limit to the number of copter-engines. To prove the point, in 2015 a British inventor created a "super drone" with 54 propellers, enough to keep a human airborne. The machine, dubbed "The Swarm," could only remain in the air for 10 minutes on a single battery charge and appeared to climb to a height of around 15 feet (7 m.). Two years later, a similar DIY one-off was YouTubed in the Swedish forest, where an engineer went up and down for eight minutes in his polycopter.

On August 29, 2016, the U.S. Federal Aviation Administration, following work with NASA, announced their regulations for the use of small drones, freeing organizations from having to request special permission from the federal government for any commercial drone endeavor—a waiver process that often took months. Under the new commercial-drone rules, operators must keep their drones within visual line of sight—that is, the person flying the drone must be able to see it with the naked eye—and can fly only during the day, though twilight flying is permitted if the drone has anti-collision lights. Drones cannot fly over people who are not directly participating in the operation or go higher than 400 feet above the ground. The maximum speed is 100 mph. Drones can carry packages as long as the combined weight of the drone and the load is less than 55 pounds. Dispensing with the requirement to have a pilot's license to fly a commercial drone, the regulations allow people over age 16 to take an aeronautical knowledge test at an FAA-approved facility and pass a background check to qualify for a remote pilot certificate. Real estate, aerial photography, construction and other industries that want to use drones for basic functions, such as taking a few photos or videos of a property, probably will benefit the most because their plans align more closely with the regulations, industry experts said. Although the new rules allow drones to carry loads, the visual line-of-sight rule and the weight restriction will keep more ambitious companies with plans for long-distance travel, such as Amazon and Google, from making significant deliveries that way. More than 3,000 businesses had already received a government exemption to fly, but given the estimation that there will be seven million small drones in operation by 2020, including 2.6 million aircraft for commercial use, NASA and the FAA will be further refining regulations.

The FAA launched its online drone-registration program in December 2016. It required all pilots, even hobbyists, to register their robots by February 19. It generally cost $5 per registration, but the FAA waived this cost through January 20. Information in the registry will be public record. In the program's first two days, the FAA collected 45,000 registrations.

In this dronomanicial adventure, growth of PAVS has been exponential.

In 2010, Alexander Zosel of Karlsruhe, Germany, teamed up with former Siemens IT expert Stephan Wolf and electrical engineer Thomas Senkel to design and build what they called a Volocopter, basically a scaled-up child's drone to carry a pilot. Between 2011 and 2014, coordinated by Professur Heinrich H Bülthoff of the Max Planck Institute für biologische Kybernetik, six research institutions across Europe studied the feasibility of the small commuter helicopters, with a $4.7 million grant from the European government: the University of Liverpool, the École Polytechnique Fédérale de Lausanne, the Eidgenössische Technische Hochschule Zürich, the Karlsruher Institut für Technologie, and the Deutsches Zentrum für Luft und Raumfahrt. The optimal solution would consist in creating a personal air transport system (PATS) that can overcome the environmental and financial costs.

On October 21, 2011, following elaborate simulations at Stuttgart University, the e-volo team had built the prototype 16-motor, all-electric VC1 "Volocopter," which Thomas Senkel test-flew for 90 seconds. The video of the flight achieved 1 million clicks on YouTube within a few days. Following the flight, the decision was made to decommission the VC1 and to merely use it as an exhibit. In 2012 a patent was issued, and the e-volo won the Lindbergh Innovation Prize for that year. From January 2013, the innovative concept of their electronic VTOL aircraft was able to so convince the German Federal Ministry of Transport that it resolved upon a trialing scheme spanning a period of several years for the creation of a new aviation class for the Volocopter. The DULV (The German Ultralight Association) was commissioned with drafting a manufacturing specification, work regulations and the training scheme for the future pilots in cooperation with e-volo. Following further refinements, the 2 kW VC-2 (or VC200), now fitted with 18 Czech-built MGM electric motors and nine batteries, was demonstrated unmanned in November 2013 at an enclosed arena in Karlsruhe, Germany. The following month, e-volo managed via the Seedmatch online crowd-funding platform to attract funding of €1,200,000.00 within just three days, a new European crowd-funding record. The first €500,000 of this sum was subscribed within just 2 hours and 35 minutes.

By August 2015, sophisticated intercommunicating electronic components had been manufactured and tested providing automatic height and position adjustment, enabling the pilot to take his hands off the joystick control, with the VC200 able to carry up to 200 kg including the pilot. That December, within the scope of the "SolutionsCOP21—Celebrate the Champions Night" at the Grand Palais on the Champs-Élysées in Paris, e-volo received Climate Champion COP 21, the award for their Volocopter.

In February 2016, the German Ultralight Flight Association granted e-volo a provisional certificate for its VC200 as an ultralight aircraft (certificate number D-MYVC).

2016: Volocopter founders Alexander Zosel (right) and Stephan Wolf (left) in front of the VC200 multicopter (Volocopter).

2016: The Volocopter VC200 200-5 watt power unit (Volocopter).

On March 30, 2016, Alexander Zosel was the pilot for the VC200's first three-minute manned flight over the dm-arena in Bruchsal, near Karlsruhe. Called *The White Lady* on account of its livery, it proudly carried the registration D-MYVC. Further encouraged, the Volocopter test flight program could continue. In June 2016, manned flights at speeds of up to 70 kph maximum and at higher altitudes were carried out at a special flight test area in Bavaria. At 50 kph, an autonomy of 27 km (17 mi) was achieved. Test flights within the third testing phase aim to validate the system at higher altitudes and in the full speed range of the VC200 up to 100 kph. A full aircraft emergency parachute was fitted, as well as multiple redundancy in all critical components such as propellers, motors, power source, electronics, flight control, displays, and highly reliable communication network between devices through a meshed polymer optic fiber network ("fly-by-light").

In January, Volocopter moved into its new headquarters in Bruchsal, bringing together the design office and the hangar/airfield at one site. At the 2017 AERO in Friedrichshafen in April, e-volo unveiled their two-seater Volocopter 2X optionally designed as manned taxis, remote-controlled and autonomous flights. With entirely new components, the 2X has been developed for approval as an ultralight aircraft and should receive multicopter-type certification that shall be created under the new German UL category in 2018, enabling anyone with a Sport Pilot License (SPL) for multicopter to fly it. For the future, e-volo is striving to obtain a commercial registration that allows for transportation of passengers as commercial taxi flights. The development of a 4-seater Volocopter with international approval (EASA/FAA) is one of the next planned steps. The e-volo team planned test flights of the 2X during the summer of 2017. In June, the Dubai government's Roads and Transport Authority (RTA) signed an agreement with e-volo regarding the regular test mode of autonomous air taxis in the Emirate. On September 26, 2017, watched by Dubai Crown Prince Sheikh Hamdan bin Mohammed, Volocopter A6-RTA made its first unmanned field test, near Jumeirah Beach. Trials between voloports will go on for five years. Primary reasons for choosing the Volocopter included the stringent German and international safety standards. Dubai plans to handle 25 percent of all of its passenger travel using autonomous transportation by as early as 2030. The following month, Volocopter agreed to a finance deal of over 25 million euros with the automobile firm Daimler from Stuttgart, the technology investor Lukasz Gadowski from Berlin, and further investors. Using this fresh capital, Volocopter could speed up the introduction process of the Volocopter serial model.

In late 2017, Brian Krzanich, the chief executive of Intel, became the first official passenger to ride in an air taxi when an 18-prop copter from Intel's partner, the German company Volocopter, lofted him within the confines of the company's hangar near Munich. In 2018, at the Consumer Electronics Show (CES) in Las Vegas, Krzanich indicated that from now on Volocopter and Intel would be working closely in the field of electric air taxis.

Another airplane type is the gyrocopter that uses an unpowered rotor in autorotation to develop lift, and an engine-powered propeller, similar to that of a fixed-wing aircraft, to provide thrust. The technology goes back to the 1920s, but it was not until June 2015 that an electric autogyro took to the skies. For two years, backed by Lower Saxony Aviation, AutoGyro GmbH of Hildesheim, Germany, had been refitting one of their two-seater Cavalon gas gyros with a Bosch SMG 180 (80kW / 200Nm), also used on the FIAT 500e electric Smart Car, and to help drive the rear wheels on the Peugeot 3008 Diesel HYbrid4. This was linked to an INVCON 2.3 electronic controller and a 16.2 Ah Li-Po battery. The Cavalon is only 4.7 meters (15.2 feet) long, 1.8 meters (5.9 feet) wide, and 2.8 meters (9.2 feet) high, topped by its 8.4 meter (27.5 feet) rotor. The 45-minute maiden flight took place from Hildesheim airfield on June 24, 2015.[8]

Sometimes concept aircraft remain concept; other times they are realized. NASA's Puffin VTOL concept has been taken up by a start-up called Lilium based in Gilching, Germany, in the form of a 100 percent electric short-haul private jet that may at last fulfill the promise of the flying car. "Lilium" comes from linking the name of 19th-century German gliding pioneer Otto Lilienthal with the lily flower and the lithium battery. The company was founded in 2015 by a quartet of engineers and doctoral students from the Technical University of Munich and nurtured in a European Space Agency-funded business incubator in Bavaria. "Our goal is to develop an aircraft for use in everyday life," says one of Lilium's founders, CEO Daniel Wiegand. As the Lilium team saw it, the problem with personal aviation is airports, which are expensive to operate and utilize, and usually sit well away from city centers, negating their use as commuter hubs. Lilium designed an airplane that could take off and land vertically and did not need the complex and expensive infrastructure of an airport. It would require an open space of just 225 m² (2,400 ft²)—about the size of a typical back yard—to take off and land. The Lilium Jet would cruise as far as 500 km (310 mi) at a very brisk 400 kph (248 mph), and reach an altitude of 3 km (9,900 ft). Overnight recharging could use a standard household outlet. In 2015 and 2016 Wiegand filed for patents using a VTOL pivotal aerofoil system. At the end of 2016, €10 million of finance for Lilium came from Niklas Zennstrom, Skype's cofounder and former CEO, through his venture capital firm Atomico. Lilium said it intended to use the new money to expand its existing team of 35 aviation specialists and product engineers. As head of recruiting, they appointed Meggy Sailer, former head of talent EMEA at Tesla, who oversaw Tesla's growth from 200 to 13,000 employees worldwide. In August 2017, Lilium received an additional $90 million worth of new investment from China's Tencent and several other investors including Obvious Ventures, co-founded by Ev Williams, who also helped create and run Twitter. This made it one of the best-funded electric aircraft projects in the world. Lilium also hired former Airbus and Rolls-Royce engineer Dirk Gebser to manage production and Dr. Remo Gerber as chief commercial officer. Due to Lilium's technology of moving from hover flight to forward flight, longer distances and higher speeds would give the German company the edge on other existing electric airplanes.

2018: Lilium Aviation, based in Gilching, Germany, has received €100 million for the development of their electric VTOL. From left to right: Daniel Wiegand, Matthias Meiner, Sebastian Born, and Patrick Nathen beside the two-seater Eagle prototype in Bavaria (Lilium Aviation).

In mid–April 2017, Lilium completed a series of short test flights with a pilotless full-scale prototype, announcing plans to run the Eagle, its first two-seater vehicle test, in 2019, with commercialization of a five-seater following in 2025. The short film of its trials, posted on the Internet, was watched by 10 million internauts. Lilium has been in discussion with the European Aviation Safety Agency (AESA), which has already carried out work for the insertion of drones.

In August 2016, Airbus announced that it was developing "an autonomous flying vehicle platform for individual passenger and cargo transport," the first test flight of which was slated for late 2017. The project name was Vahana, a name that stems from the Sanskrit word meaning "that which carries" (Sanskrit: वाहन). R&D by A^3, Airbus's innovation outpost in California's Silicon Valley, had been in progress since February. The project executive is Rodin Lyasoff. In 2002, Lyasoff was on a team that built the first aerobatic helicopter, the X-Cell 60, at MIT (where he also earned his undergraduate and master's degrees in aeronautics and astronautics). He spent several years at Athena Technologies designing flight software for a number of vehicles including the AAI Shadow, Alenia Sky-X, and the NASA Mars Flyer. Subsequently, Lyasoff led flight software at Zee.Aero. He then went to Airware, where he built the world's first hardware and software platform for commercial UAVs. Among his patents are "Variable geometry lift fan mechanism" (2015), a vertical takeoff and landing aircraft with rotors that provide vertical and horizontal thrust. During forward motion, the vertical lift system is inactive. A lift fan mechanism positions the fan blades of the aircraft in a collapsed configuration when the vertical lift system is inactive and positions the fan blades. In his LinkedIn entry, Lyasoff writes, "I love making improbable vehicles fly. I love building and growing engineering teams, and guiding product development to enhance customer amazement." Initial tests in January 2018 were tethered while Airbus hopes to have the Vahana certified and ready for use by 2020.

2017: A^3, Airbus's innovation outpost in California's Silicon Valley, is developing the Vahana, a name that stems from the Sanskrit word meaning "that which carries" (Sanskrit: वाहन) (courtesy Airbus).

The Vahana passenger drone will rely on obstacle detection and avoidance systems similar to those already seen in vehicles like the Mercedes-Benz E-Class. For this, Sanjiv Signh and his team at Near Earth Autonomy in Pittsburgh have developed a sensor system called Peregrine, named after the falcon. The system, mounted under the fuselage, contains lidar, Light Detection and Ranging, which uses lasers to measure air data parameters such as true airspeed, angle of attack, and outside air temperature; the craft also features inertial measurement, GPS sensors, and all kinds of processing power. When the aircraft is at an altitude lower than 65 feet, it laser-scans the ground in three dimensions, looking for objects bigger than 12 inches across, to determine whether the landing spot is clear and safe. If it does spot an impediment, it will suggest an alternative LZ and feed that back to Vahana's flight control computer. The company has designed the self-contained Peregrine sensor system as an easy retrofit to existing aircraft. A^3 planned to start flight tests of a prototype in November 2017. The flight tests were conducted from Pendleton Unmanned Aerial Systems Range in Oregon, where the company recently occupied a new 9,600-square-foot hangar, specifically configured to support the trials. Boeing's HorizonX division has also invested in Near Earth Autonomy's Peregrine system.

To work through this and allied vehicle efforts, the Skyways project involves Airbus Helicopters Deutschland GmbH and the Civil Aviation Authority of Singapore. Using a cross-flow fan patent developed by Sebastian Moresat Rotorcraft, development is underway of a multi-passenger VTOL, called CityAirbus. In October 2017, the CityAirbus team led by Marius Bebesel thoroughly checked the individual performance of the ducted propellers as well as the integration of the full-scale propulsion unit with two propellers, electric 100 KW Siemens motors, and all electrical systems. The full-scale demonstrator will be tested on the ground initially. In the first half of 2018 the development team expects to reach the "power on" milestone, meaning that all motors and electric systems will be switched on for the first time. The first flight is scheduled for the end of 2018. In the beginning, the test aircraft will be remotely piloted; later on, a test pilot will be on board.

CityAirbus and Vahana will then be tested on the campus of the National University of Singapore, and so help shape the regulatory framework for unmanned aircraft operations in that country. In the first phase, multiple drones will deliver parcels across the Singaporean university campus using defined aerial corridors. If successful, a second phase will extend deliveries to ships in the Port of Singapore. Airbus Helicopters has developed the "zenAirCity" business and mobility concept in which quiet, electrically powered aircraft such as Vahana and CityAirbus are integrated into the transport infrastructure of a megacity. The vision is a range of products and services from ride-booking and -sharing apps, through flying taxis and luggage services, to cybersecurity to protect the system. Although the prototype would be piloted, subsequent versions will be autonomous.[9] However, Neva Aerospace, a European consortium driving the development of key technologies for flying cars, such as their AirQuadOne, believes such fully autonomous flights remain a long way off. Another retro-style flying car concept has been inspired by a 1920s automobile design, complete with maroon and black coloring: the Hover Coupé.

Likewise, the Cormorant has been developed by Urban Aeronautics of Yavne, Israel, led by former Boeing airplane engineer Rafi Yoeli, also a reserve officer in the Israeli Air Force. Trials of the UA passenger-carrying Cormorant took place late in 2016 in Megiddo. The company claims that the Cormorant can fly between buildings and below power lines, attain speeds up to 115 mph (185 kph), stay aloft for an hour and carry up to 1,100 pounds. At present its power source is gas, which fuels a standard helicopter engine, with lift from two fans buried inside the fuselage, called a Fancraft. Of 47 U.S. and worldwide patents that have been applied for, 39 have already been granted. The civilian version, called the CityHawk (a play on Kitty Hawk, North Carolina, where the Wright Brothers made their first flights), would be built by Metro Skyways Ltd., a subsidiary of Urban Aeronautics. Although Israel is also working on boosting the energy density of batteries, Urban Aeronautics is also investigating running it on liquid hydrogen fuel and also 700-bar compressed hydrogen ... again, this plan depends on waiting for the infrastructure and technology to mature. It may even employ a system in which hydrogen is fed directly into a specially designed turboshaft engine, eliminating the need for fuel cells or electric motors.

In March 2017, at the International Motor Show in Palexpo, Geneva, Switzerland, Airbus and Italdesign-Giugiaro of Moncalieri near Turin premiered the Pop.Up, a 5-by-4.4-m autonomous passenger octocopter that can be docked with wheels to turn into a 2.6-m battery-electric carbon-fiber auto. Once passengers reach their destination, the air and ground modules with the capsule autonomously return to dedicated recharge stations to wait for their next customers. A fleet of Pop.Ups would be artificially intelligently managed to interact with trains and hyperloops to navigate around tomorrow's megacities. The Pop.Up is expected to take 7 to 10 years for realization.

The combination of an aircraft with a road vehicle has also been explored by Richard Glassock. Since 2008, Glassock, a Queensland University of Technology–based mechanical engineer, has specialized in the hybrid potential of UAVs made from off-the-shelf model aircraft components. In 2010, he proposed that another advantage of hybrid-electric aircraft would be their use for short-haul skydiving flights. Following a spell in Hungary, Glassock relocated to Nottingham University, UK. In 2017, he developed a highly innovative project whereby the range of an electric aircraft would be extended by a 50kW electrical generation power unit in the form of a gasoline-engined motorcycle slung underneath it, which could then be detached after landing and used on the roads. Glassock unveiled his RExLite and RExMoto in September 2017 at the International Conference on Innovation in European Aeronautics Research in Warsaw, Poland. The engine, generator, chassis and drive structure are claimed to have a novel layout and the whole unit weighs no more than 125 kg. Retractable wheels ensure that RExMoto can fit beneath the aircraft's fuselage or under a wing while minimizing drag in flight mode.

Another entrant came from Stephen S. Burns and Alan J. Arkus at Workhorse Group Inc. of Loveland, Ohio. Their planned SureFly quadcopter's fuselage and two fixed contra-rotating propellers on each of the four arms are made of carbon fiber–reinforced plastic (CFRP). The craft has a backup battery to drive the electric motors in the event of engine failure and a ballistic parachute that safely brings down the craft if needed. Formerly known as AMP Electric Vehicles, Workhorse has combined its experience in carbon-fiber drones and electric vehicles to design a two-seater with a 70-mile range. Early models will be pilot-operated. The goal is to introduce future autonomous models able to carry payloads of up to 181 kg. SureFly was unveiled at the Paris Air Show, then at EAA AirVenture in Oshkosh, then at CES 2018 in Las Vegas. Test flights are scheduled in 2018 for a Federal Aviation Administration certification in late 2019.

David Mayman and Nelson Tyler of Jetpack Aviation in Van Nuys, California, having developed the world's first jet turbine backpack, JB-9 and JB-10, progressed to an electric version capable of flying under battery power for five minutes.

In early 2017, Uber, whose rideshare app is available in over 66 countries and 507 cities worldwide, hired Mark Moore, who had worked as director of aviation for NASA for thirty years, including the PAV Puffin and the Maxwell X57. In late April, Moore ran their Elevate Network summit in Dallas, bringing together experts in Vertical Electric Take Off and Landing (VETOL). It announced that it would be teaming up with the governments of Dallas–Fort Worth and Dubai to test out its flying taxi network. The company would be working with Dallas real estate development firm Hillwood Properties to plan vertiports, sites where the aircraft would pick up and drop off passengers. To develop the vehicles, Bell Helicopter of Fort Worth would team up Embraer (of Melbourne, Florida), Pipistrel and Aurora Flight Sciences with their XV-24A X-plane program currently underway to develop airworthy vehicles. ChargePoint would develop charging stations. Uber also hired Celina Mikolajczak, a senior battery engineer who previously was in charge of battery cell quality and materials analysis while at Tesla. Uber planned to have its technology ready for demonstration by the World Expo in Dubai in 2020, with a fleet of 50 VETOLs to follow. Uber investigated the potential of Sydney and Melbourne as potential candidates for Elevate. Australia's Civil Aviation and Safety Authority confirmed it was ready to "meet challenges" involved in regulating air space for new flying vehicles. However, Australians will have to wait until at least 2023 to take their everyday commuters to the skies, and will have human pilots for the first five to 10 years while enough data is collected to convince regulators that sky taxis are safe.

At the CES in January 2018, Bell unveiled the cabin design of its hybrid-electric FCX-001 concept BellAirTaxi, with its rotorless anti-torque tail boom, extensive use of glass in the fuselage, gull-wing doors, and sustainable composite construction. It will have rotor blades capable of morphing to suit the need of the pilot. It would also be equipped to harvest, store, and distribute energy. The pilot would control the aircraft using the same augmented reality that the passengers can use to check the news, share a document, or watch a movie, among other things. At the show, visitors enjoyed simulated VR rides in the cockpit.

At the same summit, Mooney Aircraft announced partnership with Carter Aviation to produce another 4–6 seater VETOL, using Carter's patented Slowed Rotor Compound (SR/C) technology for efficient hover and efficient cruise at 175 mph. The General Aviation Manufacturers Association (GAMA) hosted the final day of the Summit along with the second training session about the U.S. government's Part 23/CS-23 rule rewrite for the design of small airplanes, which would enable such projects to be successful. Michael Hirschberg, executive director of the American Helicopter Society (AHS), considers that the next 50 years of vertical flight will see electric air taxis become as normal as automatic elevators.

Mike Tolkin, a techno-businessman behind such innovations as IMAXShift, a high-tech indoor cycling studio in which scenes are projected onto an extra-large screen, and Rooms.com, a home design website that offers virtual tours of designer rooms, is running for the Democratic nomination for mayor of New York. As part of his forward-thinking campaign, Tolkin has created the Smart Cities concept, in which electric Skybuses will enable faster and more efficient inter-borough transit.

Passenger Drone, led by Paul Delco and based in Zurich, has developed a 16 e-engined air taxi, which makes use of adaptive flight control, wireless fiber-optic internal communications, field-oriented motor control and encrypted communication channels. Although there is autonomous mode using LTE (4G) network, the pilot can take over at any time with touch-flight control or fly-by-wire joystick. Following a maiden flight in July 2017, Passenger Drone has started the certification process with the U.S. Federal Aviation Administration and the European Aviation Safety Agency in early 2018, and aims to make the drone commercially available in 2019. One thinks back to the sci-fictionalist Alberto Robida, who described a similar system back in 1893 (see Chapter One), although set in the 1950s.

The Russian approach to PAV is based on the Bartini Effect discovered by the Italian-Soviet aircraft designer Robert Bartin (aka Barone Rosso): the increase in thrust created by mounting co-axial counter-rotating thrusters within a nacelle. Since 2015, the startup Bartini, with its motto "Weird to think of—in 1985. Easy to hop on—in 2020," has been working on two VTOL variants: a 2-seater and 4-seater, with the intention of an air taxi service. Bartini, based at the Skolkovo Technopark, on the western outskirts of Moscow, has been funded by the company's eight founders, part of the Blockchain.aero consortium. The Blockchain consortium is the first crypto-currency platform for mass urban aviation, whose task is to automate the charging, maintenance and parking of flying cars. The Russian consortium plans to deploy urban community-driven aviation systems for passengers and city managers in 2020. First test flights are aimed for Dubai, Singapore or Sydney in 2018. Flight scheduling will be done via a mobile application and will offer different levels of service from fast to entertainment flight.

Indeed, in September 2017, Boeing put up U.S. $2 million in prize money to encourage bright ideas around building and designing an easy-to-use personal flying device. The GoFly prize followed a similar blueprint to the XPrize competitions and the Hyperloop Global Challenge, in that it tasks anyone who was willing and able to come up with technological concepts that would move humanity forward in a big way. The deadline for the first phase of the GoFly competition was April 4, 2018.

On June 5, 2017, Caihong (Rainbow), a solar UAV developed by Shi Wen and a team at the China Academy of Aerospace Aerodynamics (CAAA), a subsidiary of China Aerospace Science and Technology Corporation, made its first test flight to a height of 65,000 ft. Measuring 14 m long with a 45 m wingspan, its eight e-prop configuration recalling Helios, Caihong is designed to cruise at a speed of 150–200 kph and stay airborne for months at a time—a sort of Zephyr.

Higher still, in June 2017 Airbus launched the all-electric EUTELSAT 172B from Kourou, French Guiana, by Ariane 5, to provide enhanced telecommunications, in-flight broadband and broadcast services for the Asia-Pacific region. The 172B combines electric power of 13 kW with a launch mass of only around 3,500 kg, thanks to the latest EOR (Electric Orbit Raising) version of Airbus's highly reliable Eurostar E3000 platform. It uses full electric propulsion for initial orbit raising and all on-station maneuvers, with Xenon gas ejected at high speed using only electric power supplied by solar cells.

The Japanese Int-Ball spheroid camera drone, manufactured by JEM entirely using 3D printing, weighs 1 kg (2.2 lbs.), has a diameter of 15 cm, and has 12 propellers. In June 2017, it was delivered to the Japanese module *Kibo* on the International Space Station (ISS). Although the ISS flies at an altitude of between 330 and 435 km (205 and 270 mi), by July, remote-controlled from Earth by the JAXA Tsukuba Space Center, Int-Ball had been used on board to save crew members time by snapping pictures of experiments.

ARDN Technology, based in Kazan in southwest Russia, has developed the SKYF mega-drone capable of carrying a 400-pound (181-kg) payload and of flying for up to eight hours. Measuring 5.2 meters (17 feet) by 2.2 meters (7.2 feet), the SKYF has a maximum flight speed of 70 kph (43.5 miles per hour) at a maximum height of 3,000 meters (9,843 feet) and has a positional accuracy of 30 centimeters (11.8 inches). The drone uses its gasoline-powered engines for its two primary lift props, and uses all four sets of twin props with electric motors to help stabilize and steer it. Although it's fairly large in size, it can fold down so that two can fit into a 20-foot (6-meter) cargo container. In addition, it requires 10 minutes of setup before it can fly. ARDN has arranged for test flights at the Kurkachi Airfield (Kurkachi village, Republic of Tatarstan). They are planning to commence ARDN implementation in the Republic of Tatarstan, Krasnodar Region, Voronezh and Tambov Regions, and other subjects of the Russian Federation. Almost in parallel, Boeing's Horizon X is working on a similar heavy-lifter cargo drone with a planned carrying capacity of 400 lbs. for 15 minutes at 60 mph.

Although development is running parallel, there is one type of PAV which, once landed, can take to the roads thanks to its wheelbase. Flying cars have their own history.

TWELVE

Flying Cars

Although it is only recently that hybrid-electric flying cars are being developed, the flying car has a sci-fi history going back over 160 years.

In Andrew Jackson Davis's *The Penetralia: Being Harmonial Answers to Important Questions*, published in 1857 by Bela Marsh of Boston, the 30-year-old American clairvoyant predicted that "aerial cars will move through the sky from country to country." Interestingly enough, in addition to his foretelling the coming of both the airplane and the car, he predicted prefabricated concrete buildings. In more detail, he predicted the "internal combustion engine, carriages and traveling saloons on country roads—sans horses, sans steam, sans any visible power, moving with greater speed and safety than at present. Carriages will be moved by a strange and beautiful and simple admixture of aqueous and atmospheric gases—so easily condensed, so simply ignited and so imparted by a machine somewhat resembling our engines as to be entirely concealed and manageable between the forward wheels of these land-locomotives."

Forty years later, in Romania, Trajan Vuia, while still a young student, designed and built a scale model of a flying machine called a "winged automobile." This was a three-wheeled velocipede on which was assembled a metal frame set vertically, at its upper part being clamped a wing, direction rudder, engine and propeller. He endured scoffing and ridicule mixed with envy when he went to France in 1902, after earning his doctor's degree from the Budapest Polytechnic. Then on March 18, 1906, he tested his modular monoplane prototype Vuia I in Montesson, near Paris, taking off from an ordinary road, flying for 11 meters (36 feet) and landing. He later claimed a powered hop of 24 meters (79 feet).[1]

From 1910 to 1925, René Tampier built several different models of integrated roadable aircraft in his workshops at Boulogne-sur-Seine, a western suburb of Paris. On October 23, 1921, in Paris he drove his Avion-Automobile, and on November 7 he took to the air. This biplane, 3.7 m (12 ft.) tall and 7 meters (25 ft.) long, had four wheels conventionally placed, the front ones steerable and the rear axle equipped with a tiny differential. The fuselage and tail section remained rigid but the wings folded along each side. Cranks were used to fold them into a horizontal and longitudinal position. A second pair of rubber-tired wheels dropped into place. On the ground, the vehicle was powered by a small, 10-hp, four-cylinder, water-cooled, auxiliary gasoline engine, and in the air by a 300-bhp, Hispano-Suiza V-12 engine. Ten folding-wing units could fit where one regular aircraft was parked. Conversion took less than an hour, but problems emerged due to the vehicle's height and bulkiness. Air speed was 112 mph (180 kph) and ground speed

René Tampier's Avion-Automobile was the wonderment of Parisians in 1921.

15 mph (24 kph). Tampier tried unsuccessfully to interest the army in testing it for military uses.

Other than these two pioneers, the flying car remained in the realm of science fiction. In the 1920s and the 1930s, popular children's author Oliver B. Capelle wrote a series of stories about Uncle Nat Denny, Buster and Sally, featuring their adventures with the Magic Flying Auto. These appeared in *Children's Playmate* magazine. In the late 1940s, Texaco Oil included flying cars in advertisements for Sky Chief Brand gasoline. In the 1930s, both Victor Appleton and Victor Appleton II (pseudonyms) featured in their Tom

Swift series of books for boys, Tom's prolific vehicle intentions and exploits, including fantastic flying cars such as the *Triphibian Atomicar*.

In 1932, the Ambi-Budd plant in Berlin, which was making auto bodies for the Nazi Wehrmacht, produced an "Autoflugzeug" flying car for the Deutsche Luftfahrtausstellung (Aero Exhibition). With its overhead tricopter blades, the streamlined four-seat three-wheeler was given a registration to take to the road (IA-011032) and one to fly ((D-11032). A static exhibit, it never needed either, and was eventually destroyed during the wartime bombing of Berlin.

In 1945, designer Carl H. Renner at General Motors Special Body Development Studio painted his conception: the "Escacar" or "Unicycle Gyroscopic Rocket Car." This was shortly before General Motors design chief Harley Earl, inspired by World War II aircraft, particularly the twin-tailed P-38 Lightning, gave the Detroit company's latest automobiles tailfins, suggesting they might take to the air. Batman's Batmobile, the first one to be extremely stylized, was originally a one-of-a-kind concept car, specifically a 1955 Lincoln Futura with its huge outward-canted tailfins.

The Gernsback Airmobile of 1955 was envisioned by Hugo S. Gernsback, editor-publisher from the 1930s of many popular pulp magazines on science fiction and other similar topics. His vehicle was a narrow, two-wheeled gyro car, powered by atomic-electrical energy that used telescopic, retractable stabilizers (wings) and had a retractable tail. A counter-gravitational field was to be created around or below the Airmobile, which meant that the pilot could levitate it at will.

Among the sci-fi stories written by James Blish, in the four-volume *Cities in Flight* series published by Avon between 1950 and 1962, one discovers an autonomously controlled flying taxi cab called "Tin Cabby": "The cab came floating down out of the sky at the intersection and manoeuvred itself to rest at the curb next to them with a finicky precision. There was, of course, nobody in it; like everything else in the world requiring an IQ of less than 150, it was computer-controlled.... The cab was an egg-shaped bubble of light metals and plastics, painted with large red-and-white checkers, with a row of windows running all around it. Inside, there were two seats for four people, a speaker grille, and that was all: no controls and no instruments...."

Blish was not alone. The Aircab appears in H. Beam Piper's "Time Crime," published by *Astounding Science Fiction* in 1955, whilst John Weston creates a Helicab, a taxi cab that flies using helicopter rotors, for his "Heli-Cab Hack," published by *Amazing Stories* in 1950.

Robert Heinlein, in his *The Star Beast*, published by Charles Scribner & Sons in 1954, wrote: "They were half way home when a single flyer, hopping free in a copter harness, approached the little parade. The flier ignored the red warning light stabbing out from the police chief's car and slanted straight down at the huge star beast. John Thomas thought that he recognized Betty's slapdash style even before he could make out features; he was not mistaken. He caught her as she cut power."

So was it sci-fi or fact when, in 1947, the Bristol Development Board in southwest England published an advertisement titled "When Motor Cars Fly, Bristol Will Build Them"? At that time the Bristol Aeroplane Company was indeed manufacturing an automobile, the Bristol 400, at their factory in Filton Aerodrome, where over 5,000 Bristol Beaufighter airplanes had been built for the war effort, but with an avgas engine. In fact, it was not until 1960 that Bristol indeed became involved with a flying car. The British government invited proposals from several companies for the design of a flying car for

1947: When motor cars fly, Bristol will build them!

reconnaissance purposes. The wingless vehicle was to have a flight endurance of one hour at a speed of eighty miles per hour and have the land-borne cross-country performance of a Land Rover. The method of lifting had to be within the main dimensions of the vehicle. The first contender was Bristol-Siddeley, who proposed three powerful engines driving four gyro-controlled airjets which could raise the car up to 10,500 ft. in flight. Although a model was made, the full-scale craft was never built. Another proposal was put forward by Short Brothers & Harland. In the two-seater flying car, the road engine was a 90 hp Porsche, while a single 8,300-lb static-thrust Bristol-Siddeley would provide the nozzle thrust.

Was it therefore a coincidence that in 1964, English writer Ian Fleming, successful for his James Bond novels, wrote a children's story called *Chitty-Chitty-Bang-Bang: The Magical Car*, in which Commander Caractacus Pott pulls a switch that causes the vehicle to sprout wings and take flight over the stopped cars on the road? Commander Pott and his passengers fly to Goodwin Sands in the English Channel, where the family picnics, swims, and sleeps. The original Chitty-Chitty-Bang-Bang was an aero-engined motor car, built and raced at Brooklands by Count Louis Zborowski in the early 1920s, and maybe named after an early aeronautical engineer, Letitia Chitty.

In 1954, Scott C. Rethorst (Ph.D. at Caltech), President of the Vehicle Researching Corporation, South Pasadena, California, was issued U.S. Patent 2681773 for a roadable aircraft. Rethorst had also innovated a high-speed ship design, forerunner to the hovercraft. In 1964, Einar Einarsson of Farmingdale, New York, was awarded U.S. Patent 3090581 for a flying car. In his patent, Einarsson defines the purpose of the invention as to "provide a ground vehicle with propellers and wings, as well as wing flaps so that the vehicle may take off and fly in the air." Although the bird-like design is impressive to look at, this winged vehicle never quite made it to production, even though an advertisement asked, "Is it a bird? Is it a plane? No, it's the flying car. Your commute to work will never be the same! Beat your colleagues to work by simply flying over them!"

In the 1960s, American television viewers enjoyed the latest Hanna-Barbera animated sitcom, *The Jetsons*. George Jetson's workweek is typical of his era: an hour a day, two days a week, and to do this he leaves his Skypad apartment in Orbit City, hops into his bubble-top green aerocar, and flies off to his job at Spacely Space Sprockets. En route, George's children fly down to their schools, the Little Dipper Elementary and Orbit High School, and his wife Jane to the shopping center, each in their personal mini-drones. When George arrives, his aerocar folds itself up into a briefcase!

What we do not know is whether these vehicles benefited from silent electric propulsion. On the other hand, for his *Fantastic Voyage II: Destination Brain*, published by Spectra in 1987, Isaac Asimov creates a "Hushicopter": "In the glow of the car's headlights, Morrison made out a helicopter, its rotors turning slowly and its motor making only the slightest purr. It was one of the new kind, its sound waves suppressed, its smooth surface absorbing, rather than reflecting, radar beams. Its popular name was the "hushicopter."... The automobile stopped and the headlights went out. There was still the faint purr and a few dim violet lights, hardly visible, marked the spot where the hushicopter sat."

During the early 1960s, Moulton B. "Molt" Taylor of Longview, Washington, created a Lycoming-engined roadable airplane, the "Aerocar," that was operated by radio station KISN in Portland, Oregon, for traffic updates. Piloted by "Scotty Wright," it flew for "Operation Air Watch." Painted white with red hearts, it had the letters KISN on the top and bottom of the wings.

Flying cars became serious with the arrival of Paul Sandner Moller, born in 1936 in Fruitvale, British Columbia, Canada. Gaining diplomas in aircraft maintenance and aeronautical engineering at PITA, Moller obtained an MA in engineering and a Ph.D. in aerodynamics at McGill University. From the mid–1960s and for the next fifty years, the lone Canadian began work on his Moller Skycar, a prototype personal VTOL aircraft, powered by four pairs of in-tandem Wankel rotary engines. But despite ground effect hovering, Moller never quite achieved free flight. With the advent of lithium batteries, in 2007 Moller announced his all-electric M200G Autovolantor, powered by Altair nano batteries.

By this time, Carl Dietrich and a team of graduates of the Department of Aeronautics and Astronautics at the Massachusetts Institute of Technology and graduates of the MIT Sloan School of Management had developed the Terrafugia flying car to make personal aviation more accessible. Dietrich was well qualified for the task. In 1996, as a summer intern at the NASA Ames Research Center, he had developed C-code to assist with the transportation logistics of the Space Station Biological research project. In 2000, at MIT, after winning four design competitions and founding a student group to develop advanced aerospace technologies, Dietrich was formally recognized by the Aero/Astro Department at MIT as the youngest of sixteen exceptional graduates under the age of 35. He received his BS, MS and Ph.D. from the Department of Aeronautics and Astronautics at the Massachusetts Institute of Technology (MIT) shortly after receiving the prestigious Lemelson-MIT Student Prize for Innovation in 2006. In May of that year, with a team including Anna Mracek and Samuel Schweighart, Dietrich founded Terrafugia using the funds from his prize.

The Transition® was a Proof of Process for Terrafugia's longer-term vision for the future of personal transportation. This was followed by a flying prototype in 2009 and a second-generation prototype in 2012. The latter, TF-X™, is a four-place fixed wing aircraft with electrical assist for vertical takeoff and landing. The vehicle will have a cruising

2014 *Terrafugia TF-X* is a four-place fixed-wing aircraft with electrical assist for vertical takeoff and landing (VTOL). The vehicle has a cruising speed of 200 mph (322 km/h), along with a 500-mile (805 km) flight range. Thrust from a 300 hp engine will be provided by a ducted fan TF-X, and fold-out wings will have twin electric motors attached to each end (Terrafugia).

speed of 200 mph (322 km/h), along with a 500-mile (805-km) flight range. Thrust from a 300 hp engine will be provided by a ducted fan. TF-X will have fold-out wings with twin electric motors attached to each end. In 2016, the FAA finally gave the go-ahead to certify the Terrafugia Transition. In July 2017, Chinese automaker Zhejiang Geely, based in Hong Kong, acquired Terrafugia. Geely, which already controlled Swedish car maker Volvo and Lynk & Co. and has a major stake in Malaysia's Proton, had been looking to acquire the company since 2016. Units will be available by 2018.

Some projects are staying with gasoline engines, but with the intent to convert once the 400 Wh/k energy density barrier has been broken. Slovakia's roadable aircraft, the AeroMobil, has its own history. In the late 1980s, as students in Soviet-controlled Bratislava, Štefan Klein and Juraj Vaculik used to sit on the eastern bank of the Danube and stare longingly at Austria, the west and freedom. Vaculik, a drama student, found escapism in the theatre of the absurd. But Klein, an engineer then studying design, dreamt of a more practical solution, a flying car. He began this venture in his garage, at home in Nitra, Slovakia. In the early days, with the help of his family, he developed two prototypes—AeroMobil 1.0 during the early 1990s and AeroMobil 2.0 from 1995. But in 2010, things really took off when he joined forces with Juraj Vaculík to form the AeroMobil firm. In 2013, at the SAE Conference in Montreal, they unveiled the pre-prototype of AeroMobil 2.5, a sophisticated flying car, combining a luxury sports car and a light aircraft in a single vehicle. A year later, an experimental prototype of AeroMobil 3.0, its wings folding back like an insect, was developed under his lead with the team of 12 people, and presented at the Pioneers Festival in Vienna. On the road, AeroMobil is powered by a hybrid electric system. The generator is the same engine that powers the vehicle in the air; this in turn powers a pair of electric motors located in the front axle. In May 2015, the AeroMobil crashed at Nitra Airport during a test flight near Janíkovce (LZNI). The aircraft entered a spin and the ballistic parachute was deployed. The pilot, Stefan Klein, was sent to a hospital by ambulance complaining of back pain, but was later released. But the impact on the ground damaged the forward fuselage. Following a re-think and rebuild, on April 20, 2017, AeroMobil launched their 3.0 at the Top Marques Monaco, an exclusive supercar show, and announced that it would begin to take pre-orders for a "limited first edition" before the end of 2017.

Contrary to others, 51-year-old Pavel Brezina, an international para-motor builder and pilot based at Prerov-Bochor airport in the eastern Czech Republic, has created the hybrid-electric Gyrodrive. This is a mini-helicopter that can take to the road. Brezina's firm, Nirvana Systems, buys gyroplane kits from a German firm and then assembles and equips them with a system allowing the pilot-driver to switch between a gasoline engine propelling the rotors and an electric engine that drives the wheels. The Gyrodrive is the only airplane certified for the road. After landing, the pilot only has to fix the main rotor blades along the axis of the GyroDrive and pull out a built-in license plate to transform it into a road vehicle. For his first trip, Brezina flew some 140 miles (230 km) west to an airport on the outskirts of Prague, then drove downtown to have a cup of coffee in the Czech capital's central Wenceslas Square—and was stopped by the police on the way! Forty units are planned.

The Volkswagen hover car was a product of the "People's Car Project" in China, which called upon customers to contribute design ideas for Volkswagen's model of the future. The crowd-sourcing initiative debuted in China in 2011 and inspired 33 million website visitors to submit 119,000 ideas. The yoyo-shaped hovercraft would use electromagnetic levitation to float along its own grid above the regular road network; distance sensors would keep the craft from colliding with other vehicles. The disc-shaped pod would hold two people and could be controlled by a joystick that offered amazing maneuverability. The car could move both back-and-forth and side-to-side and could even spin on an axis. To top it off, the concept car produced zero emissions. Although a video was made, the hovercar was never built.

In March 2015, Umesh N. Gandhi and Taewoo Nam of Toyota Motor Engineering & Manufacturing North America, Inc., based in Erlanger, Kentucky, filed a patent called "Shape morphing fuselage for an aerocar." This biomimicry approach describes molding the body in tensile skin to keep wings hidden in interior space. The wings would fold up inside a compartment and unfurl through a hatch. The propeller is shown on the back bumper of the patent illustrations. The vehicle would be driven using a power system that includes a battery pack, internal combustion engine turbine, fuel cell or other energy conversion device. The patent was awarded in September 2015.

It was inevitable that the Japanese would take part in the growing development of flying cars. Cartivator was founded in 2012 by Tsubasa Nakamura, 32, an automobile expert, from Mikawa in Aichi Prefecture, to develop Skydrive, a flying car measuring only 9.5 ft. (2.9 m) by 4.3 ft. (1.3 m), with a projected top flight speed of 100 km/h (62 mph), while traveling up to 10 m above the ground. By 2014, Nakamura had assembled a group of 20 engineers and designers ranging in age from 26 to 35 from across Japan's auto industry, who donated their free time to work on the project. They also received some outside help from Masafumi Miwa, a drone expert and associate professor of mechanical engineering at Tokushima University, and Taizo Son, founder of GungHo Online Entertainment, a Japanese online video game developer. By August 2014, having conducted experiments at an abandoned elementary school in the mountains of the Aichi Prefecture, they had a ½-scale single-seat working model that combined electric drone and tricycle, one front wheel, two rear wheels, and a rotor in each of the four corners. Each rotor consisted of two propellers that allow the car to take off and land vertically. After presenting their prototype at an Ogaki Mini Maker Fair in Tokyo in 2014, the team set up a page on the Japanese crowd-funding website Zenmono, setting its fundraising goal at 1.8 million yen. By January 2015, it had raised almost 2.6 million yen (about $22,000 at the time). In May 2016, Japanese auto-making group Toyota announced its decision to give Cartivator 42.5 million yen (£274,000 or $392,642). Recently, the team has been working to reduce the weight of the vehicle, by replacing the 180 kg aluminum frame with a 100 kg frame made of carbon fiber-reinforced plastic. The team is also trying to improve the computer program that controls the rotation rate of the propellers. Cartivator plans to develop a manned prototype for a test flight by the end of 2018. To that end, it will work to develop technology to control propellers to stabilize the vehicle. The group hopes to commercialize the Skydrive by 2020 when Tokyo hosts the summer Olympic Games. The Cartivator team is aiming to run their vehicle on the track of the new National Stadium and fly it to the Olympic cauldron to light the flame at the opening ceremony. Mass production is planned for 2025, and by 2050 Cartivator could help make it possible for anyone to fly in the sky anytime.

In 2016, Elon Musk announced that his firm would develop a VTOL flying car version of their Tesla electric automobile, the Model F, which would be ready to ship in 2019. It would be built in collaboration with Volante Scherzo, an Italian roadable aircraft startup. After officially announcing the collaboration, Musk stated that the Model F would be able to reach a top speed of a staggering 482 km/h (300 mph) while in flight. But then on Friday, April 28, 2017, Elon Musk gave a forty-minute interview at the TED conference in Vancouver. Among his comments: "There is a challenge with flying cars in that they'll be quite noisy, the wind force generated will be very high. If something's flying over your head and there's a whole bunch of flying cars going all over the place, that is not an anxiety-reducing situation. You're thinking, 'Did they service their hubcap, or is it going to come off and guillotine me?'" Musk had rejected the idea of flying cars as too dangerous, and switched over to his high-speed Hyperloop transportation venture.

Prolific aircraft designer 73-year-old Burt Rutan's most recent design, BiPod, is a hybrid flying car or roadable aircraft. The twin-pod vehicle has a wingspan of 31 feet 10 inches (9.7 m); with the wings reconfigured (stowed between the pods), the car has a width of 7 feet 11 inches (2.4 m) and fits in a single-car garage. The design has two 450 cc four-cycle engines, one in each pod, which power a pair of generators that in turn power the electric motors used for propulsion. Lithium-ion batteries in the nose of each pod will provide power during takeoff and an emergency backup for landing. With a cruising speed of 100 miles per hour (160 km/h), Scaled Composites says the BiPod 367 would have a range of 760 miles (1,220 km). The prototype was built in a four-month period. Test hops have been performed with the prototype at Mojave Air and Space Port using propulsion from the wheels. The vehicle has been ground-tested up to 80 mph. No flight testing is planned.

Once the flying car marries the driverless car, then the four-wheeled drone becomes a reality. Of such quadcopter drones with wheels (car drones), drones with two functions, the most favored is the quadcopter with wheels or car-quadcopters: SY X25 Quadcopter Drone.

Among those investing in VTOL electric flying cars is billionaire Larry Page, co-founder of Google. In 2010, Page secretly funded a company called Zee.Aero, next to the GooglePlex headquarters in Mountain View, Silicon Valley, California. Zee.Aero was founded by Ilan Kroo, professor of aeronautics and astronautics at Stanford and former NASA researcher at Ames. He recruited a surprising number of students and colleagues from both organizations to launch his startup. In order to improve on the energy front, Zee assembled an in-house team of electrochemists and physicists to build a battery research laboratory and develop custom cells with established manufacturers. Chen Li, former battery research scientist at GM, now became senior electrochemical engineer at Zee.Aero. The startup also hired other experts such as electrochemical engineer Patrick Herring, who holds a Ph.D. in condensed matter and materials physics from dual advisors at MIT and Harvard University. The battery management system was designed by former SpaceX electronics engineer Drew Eldeen. Zee also hired a Boeing 787 autopilot engineer to lead the design. According to patents, Zee.Aero, with its four-bladed propellers, could fit in a standard shopping center parking space. Zee.Aero now employs close to 150 people. Its operations have expanded to an airport hangar in Hollister, about a 70-minute drive south from Mountain View, where a pair of prototype aircraft have made regular test flights. The company also has a manufacturing facility on NASA's Ames Research Center campus at the edge of Mountain View. Page has spent more than $100 million on Zee.Aero.

In 2015, a second Page-backed flying-car startup, Kitty Hawk, began operations and registered its headquarters to a two-story office building on the end of a tree-lined cul-de-sac about a half-mile away from Zee's offices. Kitty Hawk's staffers, sequestered from the Zee.Aero team, worked on a competing design. Its president, according to 2015 business filings, was Sebastian Thrun, the godfather of Google's self-driving car program and the founder of research division Google X. The first vehicle, *The Flyer* (tribute to the Wright Brothers), an octocopter whose twin pontoons enable it to take off and land on the water, was unveiled in 2017. Its controls are built into a set of handlebars and work similarly to the buttons and joysticks on a video game controller. *The Flyer* had been scheduled for demonstrations on Lake Winnebago at the EAA AirVenture seaplane base, though the demo was canceled because it was too windy.

Neva Aerospace is a European consortium in Brighton, England; Angers, France; and Vilnius, Lithuania, developing a heavy-duty EVTOL turbofan aircraft called the AirQuadOne. Basically a flying car, the AirQuadOne is designed to carry a single pilot at speeds up to 80 kph for up to 30 minutes at altitudes of up to 3,000 feet (900 m). The AirQuadOne is expected to weigh around 530 kg, including 150 kg of batteries for the full electrical version and 100 kg for the pilot. Neva expects the battery pack to be similar to or compatible with those of cars, with recharging at standard electrical stations via direct wire connection, induction or a battery pack switch. Neva is also looking at hybridization solutions for range extension. AirQuadOne was presented at the Paris Air Show in June 2017.

As part of the Goodwood Festival of Speed, held between June 29 and July 2, 2017, the Future of Speed Lab exhibit hosted the initial scale concept model of the NeoXcraft ducted-fan flying car. The vehicle was designed and created in Derby by VRCO, working with Institute for Innovation in Sustainable Engineering IISE, part of Derby University and home to the Rolls-Royce Innovation Centre. VRCO had recently signed a MOU with Astral LLC to build the world's only neuro-mechanical holoportation drone platform.

In July 2017, the Experimental Aircraft Association (EAA) held its 65th annual fly-in convention at Oshkosh, Wisconsin. The world's largest fly-in annually draws 10,000 airplanes to Wisconsin and a total attendance exceeding 500,000. India-born aeronautical engineer and entrepreneur Sanjay Dhall of Detroit Flying Cars exhibited his prototype with patented technologies that telescope, turn and lock wings that compress into the front and back of the two-seat vehicle when it's on the road. Dhall's flying car, featuring an electric engine for driving and an aviation engine for flying with a flight cruising speed of 125 mph (200 kph) and range of 400 miles (650 km), should make its maiden flight in 2018. Erik Lindbergh joined the VTOL hybrid-electric air taxi community with his Seattle-based VerdeGo (verde = green, vertigo) start-up.

In India, Naman Chopra, ex–Tesla Motors, now as chief product architect with Rexnamo Electro Pvt. Limited in Ghaziabad, U.P. India, co-launched his nation's first Electric Highway—a network of fast chargers spanning the national capital region in the north to the Himalayas (Uttarakhand). The Electric Highway consists of a total of 12 fast chargers, allowing a 30-minute recharge to about 80 percent capacity for most EVs. The chargers consist of a 43 kW Fast AC option (Type-Two, or Mennekes connector, suits the Rexnamo Super Cruiser Bike) and two 50 kW Fast DC options—SAE Combo for the BMW i3, and ChaDeMo for Japanese makes of car. A slow charger is also located at each station, able to deliver up to 7 kW. A humble 15 amp GPO is also available. Close on its heels, Chopra launched the Rexnamo Roadable Aircraft®, a street-legal vertical takeoff airplane that converts between flying and driving modes in under a minute. A working prototype could be ready by 2018.

Paul J. DeLorean of Bloomfield Hills, Michigan, nephew of automobile manufacturer

John DeLorean, formed DeLorean Aerospace in 2012 to design and build a flying car. Formerly a designer at Mattel, then General Motors, DeLorean began with a 30-inch (75-cm) scale model, then a one-third scale model to prove his concepts: incorporating a center-line twin vectoring propulsion system, stall-resistant canard wing at the front, the main wing at the back with small winglets underneath while two tandem seats in between hold the passengers. The full-scale DR-7, 20 feet (6 m) long and 18.5 feet (5.5 m) wide, will have wings that fold in so it can be parked in a large garage; its tests projected for 2018 will check out an autonomy approaching 200 km (125 mi) and whether its 1.21 gigawatt motor can give it a top speed of 88 mph (140 kph). The DR-7 will have an autonomous option. Paul DeLorean has been issued U.S. patent No. 9085355, with additional domestic and foreign patents pending. One recalls the DeLorean flying car in the movie *Back to the Future*.

If cars can fly, why not flying tricycles? The vision of pilot Captain Gary Lee Pylant and aerospace engineer Mark Rumsey of Spring Valley, Arizona, is the Fly-B, an electric-powered ultralight aircraft adapted to a modern lightweight recumbent tricycle frame, combining the best of both worlds: pedalectric bicycling and ultralight flying. With a flying speed of 62 mph (100 kph), the twin-prop Fly-B has a wing fitted with 24 ft² (2.2 m²) of solar paneling to recharge its battery while parked and folds up for easy storage. A heart-rate monitor measures the pilot's calorie burn.

The Hoversurf Scorpion 3, a quadcopter "motorcycle," is the brainchild of Alexandr Atamanov, IT businessman and aviation enthusiast of Moscow and Los Angeles. Its wooden rotors enable it to fly up to 4 meters. After two years of R&D, this third-generation e-hoverbike was publicly launched in December 2016. Scorpion 3, which was demonstrated at the Moscow Raceway, Oblast, 97 kilometers from the city, can carry 125 kg and is capable of reaching 60 kph. Its battery capacity allows it to stay in the air for 15 minutes.[2] In October 2017, it was announced that Dubai was planning to add Scorpion 3s to its police fleet. Also from Russia, weapons manufacturer Kalashnikov Concern, known for its AK-47 machine gun, the world's most used weapon, demonstrated their unnamed 8 double-prop prototype, calling it a "hovercycle." The batteries appear to be located under the rider linked to the rotors; the flying car has a seat and is maneuvered through the use of a joystick. Is this Russia's new cavalry? Indeed there is another project for a quadcopter with a submachine gun and a 100-round magazine controlled by a smart tablet.

Another hybrid-electric flying car has been developed by the U.S. Air Force Research Laboratory, sponsored by DARPA and contracted to Don Shaw and his team at Advanced Tactics of Torrance, California, as a Special Ops Transport Challenge. Built by Lockheed Martin, it was code-named the AT Black Knight Transformer to serve as both octocopter and truck, giving troops the flexibility to infiltrate enemy lines and evacuate wounded comrades from war zones. The first electric prototypes were flown in 2010, followed by a 2,000-lb gas-powered hybrid whose first test took place in March 2014 at the U.S. Army Telemedicine and Advanced Technology Research Center. The prototype went up for sale on eBay a few days before Halloween 2017 to help fund development of a flying car called the AT Transformer (minus Black Knight), whose underbelly pods can hold from three to six people, cargo for disaster relief or resupply missions, or even wounded people on stretchers. AT states that it will be ready for sale by 2018. Alongside this, AT has developed the Panther package delivery mini flying car.

According to a report published by the Frost and Sullivan research and consulting organization titled "Future of Flying Cars 2017–2035," over 10 companies are poised to launch flying cars by 2022, with OEMs and other major industry participants set to join them with prototypes in the following decade. At least eight companies have conducted flight tests in the past four years and several more have scheduled tests before the end of 2018. The report also identifies other uses for flying cars, including as air ambulances and for law and order, military and surveillance purposes.

It may not be long before flying cars enter into competition with each other.

For 2017, U.S. inventor Dezso Molnar announced a flying car race series, divided among three categories of vehicles: radio-controlled, electric, and unlimited flying cars. Radio-controlled vehicles are unmanned and guided by a human operator, while the others are both manned. Over the course, the vehicles must fly and drive 219 miles (352 km) from California's El Mirage Lake, a dry lakebed, to the planned El Dorado Droneport in Nevada, near Boulder City. Radio-controlled flying cars will be raced within visual range of a control area. Twenty-two teams have been invited. On May 20, 2017, Molnar gave a presentation of his GT Gyrocycle and a talk, "How to Make a Flying Car That You Can Race," at the Maker Faire Bay Area, San Mateo, California. Molnar's Street Wing concept is a fully electric, solar-supported roadable airplane, and the G2 Gyrocycle is a race-focused, 200 mph (320 kph) three-wheeler that's already rolling on the street, and nearly ready to fly.

With the aim of flying car races, over the past two years Matt Pearson and a team of five have been working in a Sydney warehouse in Australia to build the Alauda Mark 1 Airspeeder quadcopter with a top speed of 250 km/h (155 mph). With an aerospace aluminum frame and a carbon fiber composite body, it should have a net weight of 120 kg (265 lb.) and a power-to-weight ratio of 1.66. The plan is to make a test flight in 2018, followed by a head-to-head race between two of the single-seaters taking place in the south Australian desert late in the year. They'll be unmanned at first, as the team works on the car's safety systems. The first-ever Airspeeder World Championship, in which flying cars from different manufacturers race against one another, could subsequently be held in 2020.

The European Flying Car Association (EFCA) represents these national member associations on a pan-European level (51 independent countries, including the European Union Member States, the Accession Candidates, and Russia, Switzerland, Turkey, and Ukraine). The associations are also organizing racing competitions for roadable aircraft in Europe, the European Roadable Aircraft Prix (ERAP), mainly to increase awareness about this type of aircraft among a broader audience. EFCA members have launched the idea of organizing a European Grand Prix Competition for roadable aircraft (aka flying car). Because there are big differences in the technology used by the manufacturers, they believe there should be a competition per model or class of models. The races would, of course, have a driving part and a flight part. The weather will play an important role for these aircraft, so for safety reasons it may not be possible to plan the races exactly on a given date, unless the race tracks are designed in a way that allows the pilots to choose whether to finish the track only over road or also by air. Analysis should also be made of how filming is done with drones during drone competitions, like the World Drone Prix in Dubai. EFCA plans to organize the first races in 2018, if they manage to get all the necessary authorizations on time: ERAGP 2018 (European Roadable Aircraft Grand Prix 2018).

Eventually flying race cars may become driverless.

THIRTEEN

Into the Future

In my previously published books, I have sometimes put on my rose-tinted spectacles and dared to suggest what may happen in the years to come. In 1979, for my book *The Guinness Book of Motorboating Facts and Feats*, Chapter 16 was called "AD2000" and ran to 7 pages. Forty years later, most of the projections have still to be realized. My book *The Harwin Chronology of Inventions, Innovations and Discoveries*, published in 1987, had a section at the back titled "The Shape of Things to Come…," which included "Civil airliner (VTOL, composite plastic construction and hydrogen-fueled)." As this book goes to the printer, over a dozen projects and patents for the electric airplane have been announced, filed and obtained. In ten years' time, readers of this book will look back, as this historian has looked back, and say either "Nothing ever came of that," or "The progress made with that aircraft puts this book completely out of date."

Here are just some of those projects and concepts, updated to early 2018. In 2028 I shall be 78 years old and may have had the chance to update this book again. If not, then the blank page at the end (of the paper edition, of course) is therefore for those with private copies they may like to add to.

On July 26, 2016, Dr. Peter Harrop, Chairman of IDTechEx, gave a Webinar titled "Electric Aircraft Reach a Tipping Point": "About 20 companies make or will soon be making electric aircraft. Nearly all are pure electric and fixed wing, the motorized hang-glider and the self-launching sailplane being typical with one hour endurance. A bigger value market being addressed is training planes and bigger still will be hybrid fixed wing and vertical take-off aircraft, hybrid and pure electric, with the pure electric ones only managing 30 minutes. In this webinar we discuss possible uses, improvements and other types too." It is now clear that the manned electric aircraft (MEA) business will be around $24 billion as early as 2020, but the new analysis by IDTechEx sees truly hybrid and pure electric aircraft being a $24 billion business in 2031. Half of that will be relatively low-priced craft such as leisure and small work aircraft, and the high-priced half will be a mix of such things as helicopters, military aircraft, and feeder aircraft, according to IDTechEx projections, with large airliners not quite there. "Manned Electric Aircraft 2016–2031" reveals how much of this will no longer be a reworking of land-based technology, but will be based on such things as superconducting power distribution and traction motors with at least four times the kW/kg and Distributed Electric Propulsion (DEP) along the full length of the wing. However, new concepts being progressed first on land, such as supercapacitor bodywork and some other structural electronics, may have a place in these new ultra-lightweight aircraft.

Here is my own review of e-aircraft power systems and design configurations, although the two are inextricably interlinked.

Power Systems for the Future...

The concept of a supersonic commercial airliner powered by superconducting hybrid electric propulsion is championed by English-born Richard Lugg with his *HyperMach* (Mach 4 cruise) *SonicStar*. Lugg's Patent 8636241, filed in 2006, describes a fixed-wing VTOL aircraft, featuring an array of electric lift fans distributed over the surface of the aircraft. A generator is (selectively) coupled to the gas turbine engine of the aircraft. During VTOL operation of the aircraft, the engine drives the generator to generate electricity to power the lifting fans. Power to the lifting fans is reduced as the aircraft gains forward speed and is increasingly supported by the wings. The airliner also has supersonic double delta laminar flow, electromagnetic drag reduction technology and boom reduction. Lugg followed his father into the U.S. aerospace industry, via a detour into medical research that led him to work with NASA on how the impact of deep-space flight could affect the design of ships. But perhaps his most important experience was his involvement in the early stages of the NASA Hyper-X program in the mid–1990s, which would eventually produce a world-record-beating unmanned aircraft—the scramjet-powered X-43A—which flew at almost 10 times the speed of sound for 11 seconds in late 2004. Making some improvements in *SonicStar*'s turbine to increase the maximum capacity from 24 to 32 passengers, and bringing the cruising speed up to around 3,431 miles per hour (5,522 km/h), meant that the original entry date was pushed back from 2021 to June 2024, depending on the $220 million financial commitment from aircraft manufacturers. *SonicStar* will be able to fly from New York to London in all of 71 minutes.

Leik N. Myrabo with his *Lightcraft* sees laser flight carrying people around the globe and into space by 2020. Ground-based lasers called LightPorts would provide the energy needed to propel the crafts.

On June 30, 2015, Boeing patented a laser nuclear fusion jet engine, fitted with one or more free-electron lasers for providing pulsed laser beams to vaporize pellets comprising the propellant (deuterium and tritium). As a result of the compression of the deuterium and tritium, the gas mixture reaches sufficiently high temperatures to cause a release of energy beyond the "break-even" level, so increasing the overall thrust and exhaust velocity: a specific impulse of 100,000–250,000 seconds may be provided. No one is building this yet, but the patent lays the groundwork for a new kind of jet engine which could change the way jet engines are designed in the future.

In September 2015, Boeing also filed a patent for generating electricity from airport noise. Chin H. Toh, an inventor at Boeing, has worked out a way to harness the noise emanating from an aircraft to generate electricity. His invention achieves this by installing the acoustic electricity generating system on sides of a runway. The acoustic electricity generator of Boeing consists of four parts: acoustic wave collectors, an acoustic converter, a turbine and a generator. The device collects acoustic waves from the noise produced by aircraft on a runway, and directs the collected waves to an acoustic converter. The acoustic converter receives these waves to produce output airflow. For producing output airflow, a vibrating drum is mounted within the converter assembly. The drum moves up and down when excited by the incoming acoustic waves. As the drum vibrates, it acts as an air pump to draw the air in, and then pushes the drawn air down to form an output air flow. This output air flow is then directed to the turbine chamber, where it rotates the turbine shafts. The turbine shafts are further coupled to the generator, which generates electricity.[1]

The European Aeronautic Defence and Space Company (EADS) also announced its long-term plans to develop their ZEHST, a zero-emission hypersonic airliner, that could be whisking passengers from Tokyo to London in under 2.5 hours by the year 2050. In the short term they would be working on the VoltAir, a proposed all-electric airliner that could be flying within 25 years. Two next-generation lithium-air batteries would power two highly efficient superconducting electric motors, which would in turn drive two co-axial, counter-rotating shrouded propellers at the rear of the aircraft. An advanced carbon fiber composite airframe design, aerodynamics and low weight would make the airliner as easy to push through the air as possible. The batteries would be housed in the lower front section of the VoltAir, where they could be removed and installed just like baggage, at the airport. Recharging would take place when the batteries were out of the aircraft, so planes would simply land, swap out their depleted batteries for charged ones, and take off again. Not only would this arrangement make turnaround times similar to those of conventional refueling, but it would also reduce the weight and technical complexity of the aircraft. With advances currently being made in the field of high-temperature superconducting (HTS) materials, however, EADS saw a potential solution on its way. While certain materials are able to achieve superconductivity—an electrical resistance of almost zero—at very cold temperatures, others can achieve it at higher (but still cold) temperatures. These are the HTS materials. In the VoltAir's electric motors, HTS wiring would take the place of conventional copper coils, and would be cooled to the necessary temperature with liquid nitrogen. This would result in an almost lossless electrical current, and emissions that would consist of nothing but harmless nitrogen gas. EADS anticipated that as the technology is developed, high-density superconducting electric motors will actually exceed the power-to-weight ratio of existing gas turbine engines. The streamlined VoltAir's rear-mounted propellers would be able to "ingest" the wake from the fuselage, while the wings are able to remain streamlined and engine-free.[2]

Among those who reserved their judgment about the future of electric airplanes was Pratt & Whitney of Connecticut, the U.S. aero-engine manufacturer since 1925. In May 2014, Alan Epstein, P&W's vice-president of technology and environment, stated that three technological "miracles" must occur before electric flight could go mainstream. First, battery technology must improve by 50 to 100 times, with a commercial aircraft like a Boeing 737 requiring about 10MW of energy during cruise. Battery-powered aircraft could be viable with current technology only. Epstein continued:

> P&W had also spoken with engineers at the Massachusetts Institute of Technology about developing an electric engine capable of powering large aircraft. Such powerplants could be built, they have determined, but would require new, complex superconductivity technology. Also, engineers would need to remove the engine's magnetic shielding to reduce its weight. But without magnetic shielding the engine could kill the people sitting next to the motors. Three miracles are about two-and-a-half too [many] for an industrial organization and one-and-a-half [too many] for most companies. I don't see major commercial [electric aircraft without innovations] that have yet to be invented. Still, P&W's parent company United Technologies will be at the forefront of electric-aircraft design when, and if, the technology becomes viable.[3]

In his paper "Are electric airplanes possible in the future?" published in *Quora* in December 2014, Joseph Guindi, a thermodynamics engineer for airliners such as the A350, pointed out:

The energy density of jet fuel far exceeds that of any battery. That means, you would need to carry far more weight to have an electric aircraft. More weight means exponentially more fuel. Even supposing the energy density were the same, jet fuel has an additional very useful property: once it is consumed, it is no longer on the aircraft, resulting in weight loss, resulting in reduced fuel consumption over a flight. This would be impossible with batteries. Additionally to all that, batteries require a long time to be recharged; the turnaround time for an Airbus A320 is as little as half an hour. At present, no commercially available battery offers that degree of performance. Any new technology to replace jet engines and jet fuel with electric motors and batteries would significantly alter the entire structure of an aircraft. Before that is allowed to be done, rigorous testing, lasting at least a decade, would have to be done before it is approved for civilian commercial use and proving that it is as safe and as reliable as jet engines. Since none of the milestones laid out above have occurred, *I wouldn't expect an electric commercial aircraft in the next quarter century at a minimum, if at all.*

Tesla CEO Elon Musk has said that once batteries are capable of producing 400 watt-hours per kilogram, with a ratio of power cell to overall mass of between 0.7 and 0.8, an electrical transcontinental aircraft becomes "compelling." Indeed, Tesla engineers continue to work towards this goal. They have developed a new battery cell in partnership with Panasonic now in production at the Gigafactory in Nevada. The new 2170 format cell features a new battery chemistry slightly different from the current 18,650 cells used by Tesla in its vehicles and energy storage products. Powerpack 2.0 has a higher energy density and a capacity of 200 kWh. In Musk's terms, they are only halfway there.

In 2014, NASA launched the University Design Challenge: All Electric Aviation Vehicle competition to design a four-seat, all-electric aircraft capable of entering service by 2020. More than 20 universities entered the competition, including Georgia Tech, University of California Davis, and Virginia Tech. The winner was Tom Neuman, a 24-year-old engineer at Toyota's Technical Center in Michigan, who had completed co-ops at Boeing, Sikorsky, and Rolls-Royce while studying aerospace engineering at the Georgia Institute of Technology in Atlanta. In 2008, aged 17, Neuman had already worked on another electric plane—one with a 9-foot (3-meter) wingspan, and designed for a remote-controlled aircraft competition. For the NASA challenge, Neumann incorporated the technology of the fuel-cell-powered Toyota Mirai automobile, which delivers 800 Wh/kg at 55 percent efficiency. That was certainly better than 400 Wh/kg for the best lithium-based batteries. He then installed it into a Cirrus SR22 composite aircraft. He ran yet another analysis and found "a sweet spot in efficiency" using two rather large propellers attached to a pair of motors. Instead of mounting them conventionally, on the wing or fuselage, he put them atop the plane's V-shaped tail, where the airflow is cleaner. That is when Neuman dubbed his V-tailed, hydrogen-powered design Vapor. Neuman's Vapor could theoretically carry at least 400 pounds (180 kg) of extra cargo, fly at least 575 miles (1,000 km) during a single flight, cruise at a speed of at least 150 mph (250 kph), and be able to take off in less than 3,000 feet (900 m) under normal conditions.[4]

To further encourage the DEP, NASA's University Aeronautics Design Challenge 2015–2016 took in their concept of an electric-powered, commuter-sized airplane featuring "green aviation" technologies. Sponsored by NASA's Aeronautics Research Mission Directorate, the engineering design contest specifically asked students to incorporate distributed electric propulsion, or DEP, in their concept for an airplane that could enter service by 2025. The proposed aircraft design also had to be able to carry 19 passengers, cruise at 250 mph (400 kph), fly as high as 28,000 feet (8,500 m), take off or land on a runway no longer than 3,000 feet (900 m), and be able to fly in all types of weather, including icing conditions.

First place was taken by a team of 13 students from the University of Virginia in Charlottesville, with BLItz, a distributed electric propulsion commuter aircraft with eight fans partially embedded in a smaller wing that has a turboelectric generator at each wingtip. The fans' placement would duct air over the wing to improve lift, and also enable boundary layer ingestion, which takes advantage of the airflow over the aircraft to make the fans operate more efficiently and reduce drag. Second place went to a seven-member team at Virginia Tech in Blacksburg. Their concept was for an aircraft called Ion, which would use ten electric motors covering 90 percent of the wing's surface, with six of those folding away during cruise. Third place went to another seven-member team at Virginia Tech for their concept called Partior Q-1, which incorporated on its wing leading edge eight propellers for takeoff, six of which would not be needed during cruise. In early November 2016, the winners presented their papers at a student conference at NASA's Langley Research Center in Virginia. Plans for the University Aeronautics Design Challenge for the 2016–2017 academic year included two technical areas: a supersonic challenge and a low-noise subsonic challenge.[5]

Another line of R&D, taken by Philip J. Masson of the Department of Mechanical Engineering at the University of Houston, with Cesar A. Luongo, senior superconducting magnet engineer at Jefferson Lab, Newport News, Virginia, concerns superconducting motors. They believe these could solve problems of weight and autonomy by using liquid hydrogen to run an electric fuel cell. Liquid hydrogen is cold enough to make the superconducting magnets work, but also has four times as much energy, weight-for-weight, as aviation fuel.

In their paper "HTS machines as enabling technology for all-electric airborne vehicle," published in 2007 in *Superconductor Science and Technology*, Volume 20, Number 8, they concluded:

> Electric propulsion for aircraft would require the development of high power density electric propulsion motors, generators, power management and distribution systems. The requirements in terms of weight and volume of these components cannot be achieved with conventional technologies; however, the use of superconductors associated with hydrogen-based power plants makes possible the design of a reasonably light power system and would therefore enable the development of all-electric aerovehicles. A system sizing has been performed both for actuators and for primary propulsion. Many advantages would come from electrical propulsion such as better controllability of the propulsion, higher efficiency, higher availability and less maintenance needs. Superconducting machines may very well be the enabling technology for all-electric aircraft development.[6]

Funded by the USAF Research Laboratory and NASA, a team coordinated by Masson and Luongo has developed design concepts for a revolutionary aircraft using distributed propulsion whereby the very high specific power required for the airborne generators and motors can be achieved by using superconductors. Analytical 2D sizing models were created and showed very promising results. The next step was 3D modeling, where the magnetic flux distribution is calculated using Biot-Savart's law coupled with the magnetic moment method for the back iron. The code also includes thermal and mechanical models, allowing for a full and accurate design. Masson adds that the team is now looking for an industrial partner to build a prototype of the superconducting "turbofan." "The technology is there," he says, "it is a matter of finding a source of funding."

In 2009, researchers at the U.S. Navy Research Laboratory located in southwest Washington, D.C., developed the *Ion Tiger*, a UAV whose fuel cell was powered by gaseous hydrogen stored at 5,000 psi. This enabled it to stay aloft for 26 hr and 2 min. The team continued to work on their prototype so that by April 2017 *Ion Tiger* had been refitted with liquid hydrogen fuel in a lighter weight, cryogenic fuel storage tank and delivery system. This enabled *Ion Tiger* to stay aloft for 48 hours and 1 minute. The NRL LH_2 flight capability is being developed by NRL's Tactical Electronic Warfare and Chemistry Divisions, and is sponsored by the Office of Naval Research.

Rod Badcock and researchers at the Robinson Research Institute, Victoria University, New Zealand, are collaborating with NASA's Electric Aircraft Technology Roadmap in Wisconsin on superconductivity for electric airliners.

Alongside the X57, NASA is looking other projects for the coming thirty years. The SUGAR (Subsonic Ultra Green Aircraft Research) Volt, a hybrid aircraft being developed by Boeing Research & Technology, uses both jet fuel and batteries, longer wings for lift, and open-rotor engines. It plugs in at the airport, charges its batteries up, and flies its mission. To enable portions of flight with low or zero emissions, electricity is used as a supplement or replacement. Dual-turbine engines would be powered by traditional jet fuel, and at cruising altitude, the system could turn over to electrical power. The wings can fold when landed to accommodate airport gate space. Boeing is looking at a 2030 to 2050 time frame for the SUGAR Volt.

Then there is a fully turboelectric, superconducting aircraft called the N3-X, with its hybrid or blended-wing body. Two wingtip-mounted superconducting electric generators would drive the distributed fans to lower the fuel burn, emissions, and noise. Building on the R&D from the N3-X is the partially turboelectric STARC-ABL, a subsonic commercial aircraft concept with conventional underwing gas turbofan engines powering a ducted aft boundary layer propulsor in the tailcone. Attached to each turbofan is a generator that extracts mechanical power from the fan shaft and converts it to electrical power. Electrical wires send it to a rear-mounted boundary layer ingesting, electrically powered fan. In October NASA awarded Aurora Flight Sciences a contract to perform a comprehensive evaluation of the STARC-ABL design. The turbofans provide 80 percent of thrust during takeoff and 55 percent at the top-of-climb. The rest is done via the aft-fan propulsor located at the end of the aircraft below the tail.

In May 2016, Blaine K. Rawdon and Aaron J. Kutzmann, engineers at Boeing Phantom Works in San Pedro, California, filed a patent application for an electric-powered plane with solar cells covering its wings, including winglets that stick up from the ends. The winglets can be angled more directly into the sunlight, even when the sun is at a low angle, so extending the range.

In 2011, Miguel Sánchez of Madrid, an aeronautics engineer and private pilot, formerly the project engineer and project manager at Airbus, and his friend, electronics engineer Daniel Cristóbal, discovered that the aeronautical industry had still not delivered the answer to preventing accidents caused by gas engine failure. Together the Spaniards created AXTER, and working with Andrés Barrado and researchers at Carlos III University in Madrid, developed a hybrid propulsion system that adds up to 30 kW of combustion engine extras or turbo-boost for short takeoffs, high ascent ratios or compensation of the main engine power loss during the hottest months. By December 2013, AXTER carried out its first flight of an ultralight airplane EC-ZEL with the AX-40S parallel hybrid propulsion system. "Our airfield has a very small runway and the weight of our aircraft is 600kg. During the summer (40°C) it is very difficult to take off. Thanks to our hybrid system we reduced the runway by 70 meters. We launched our product on the market at AERO 2015, by flying from Madrid across the Pyrenees to Friedrichshafen."

Since then the Spanish system has been installed in a Tecnam P2004 Bravo (EC-ZRE) and a 4-seater Australian Jabiru light aircraft from Casarrubios Airfield, near Toledo. In the future, this new hybrid electric system could prevent 600 accidents, 70 deaths and 24 million euros in losses recorded per year.[7]

Funded by DARPA and NASA, Seth Kessler and a team at Metis Design Corporation (MDC), a small technical consulting firm in Boston, has developed a microturboalternator which drives a lightweight permanent magnet generator and consumes only one pound of fuel per hour per kilowatt output. This gives it a power source 12–15 times the energy density of Li-Po. The MDC design places the power turbine driving the bypass flow at the front of the combustor rather than at the engine exhaust. A compact recuperator recovers heat from the exhaust and decreases fuel consumption. Given company figures, a 20-kilowatt microturboalternator would weigh under 12 lb. (5 kg) and consume 20 pounds (three gallons) of fuel per hour at full output. With electrical output ranging from under 10 kW to over 100 kW, the microturboaltenator would be ideally suited to the next generation of hybrid light-sport aircraft and UAVs, as well as hybrid road vehicles and stationary power generation.

Sergio Bortoluz of Konner Helicopters in Amaro, Italy, is also equipping his 90 hp TK-150 gas turbine helicopters with a hybrid electric back-up. The hybrid version, called TG-250HY, integrates in the complex turboshaft an electric motor that can deliver 50hp in case of need.

There is retrofitting. Kevin Noertker and a team at Ampaire, housed in the Los Angeles Cleantech Incubator, are developing a process for retrofitting a standard turboprop airplane into one that would operate solely on electric power. Ampaire's team originates from top institutions including Caltech, Stanford, Penn, USC, Northrop Grumman and SpaceX. A retrofitted six-passenger regional airplane with a range of up to 100 miles would be used for both passenger and cargo flights. The company has also designed an all-electric aircraft called TailWind that will come in two models—the all-electric TailWind-E and the TailWind-H hybrid electric, which is designed for longer-range flights.

Millionaire Martine Rothblatt founded United Therapeutics Corporation in an effort to commercialize a pill for her daughter's disease, a rare life-threatening form of pulmonary hypertension. She is the founder of Sirius satellite radio and vehicle navigation company GeoStar, launching several satellite communications companies including the first nationwide vehicle location system (GeoStar, 1983), the first private international spacecom project (PanAmSat, 1984), and the first global satellite radio network (World-Space, 1990). Owner of 7 Tesla automobiles, Rothblatt decided to electrify a helicopter for the purpose of delivering transplantable organs much more cleanly and quietly. Rothblatt is experimenting with growth of kidneys, hearts and ultimately lungs in laboratories for transplant, in an attempt to solve the problems of supply and rejection. In September 2016, Lung Biotechnology Tier 1, the red Robinson R44 helicopter, retrofitted by Glen Dromgoole of Tier 1 Aviation, with its 200-lb payload, made a five-minute flight at 400 m altitude above Los Alamitos Army Airfield. Five months later, with Rothblatt and Dromgoole at the controls, it increased its duration to 30 minutes at 800 feet and flying at peak speed of 80 knots with 8 percent battery state-of-charge remaining after a safe hover landing. On March 4, 2017, Rothblatt and Ric Webb set a world speed record for electric helicopters of 100 knots at Los Alamitos Army Airfield under an FAA experimental permit for tail number N3115T.

According to another IDTechEx report, "Electric Vehicle Energy Harvesting/ Regeneration 2017–2037," electric vehicles will make their own electrical energy using daylight, wind, waves and more: energy harvesting for regeneration (EH/R). The report presents several technologies for this: electrodynamic, GaAs photovoltaics, triboelectric, dielectric elastomer nano-generators, thermoelectric and piezoelectric. One example would be solar cloth for electric airships.

Dale Martin Walter-Robinson of Guildford, Connecticut, has invented an in-flight energy-cell regenerative system for electric airplanes. When Walter-Robinson found that the cooling fan in the nose of his electric car was not efficient, he improved it and then tested it in a Cessna 152 for eventual application in electric airplanes. His system works by capturing propeller-blasted air rushing beneath the cowling in flight and converting it to electrical power to recharge on-board batteries, using a special alternator and centrifugal fan that is rounded to reduce drag. This continuously produces 2,000 watts, fed to a series of batteries. The system automatically switches away from fully-charged batteries to feed others, supplying a continuing source of power for the aircraft's propulsion and on-board electrical systems including a cooling device. It is similar to the Ram Air Turbine (RAT) designed by Alexander Lippisch for the rocket-powered Messerschmitt Me163 Komet in early July 1944. In December 2015, Walter-Robison was awarded World Patent 2015195856 for his system and trademarked the words Eviation and ElectronAir.

Another approach has been taken by a team of 35 scientists and engineers, working at four NASA centers and led by Patricia Loyselle of NASA's Glenn Research Center in Cleveland, Ohio. The project is called M-SHELLS (for Multifunctional Structures for High-Energy Lightweight Load-bearing Storage), and the quest is for a material that is as strong as today's aircraft-construction materials, can store large amounts of energy, and can both charge and dispense that energy rapidly.

QinetiX of Memmingen, Germany, is developing their QPD-40 unit in which the controller is integrated in the casing along with the 40 kW engine, with a total weight of 12 kg (26 lb.), so facilitating installation.

In October 2016, France further stepped up its commitment to hybrid-electric airplanes. Led by Xavier Roboam, HARTEC (Hybrid Aircraft, Academic Research on Thermal and Electrical Components) is a €1.5 million, five-year project, involving 6 theses and 2 postdoctorates, whose aim is to identify promising technologies and to develop new tools to reduce the fuel consumption of aircraft and reduce nuisance factors (CO_2 and noise on the ground). Part of the European Commission for the Aeronautics Industry under the Clean Sky 2 program, this research will be carried out by the Toulouse laboratory consortium of CIRIMAT, specializing in materials and batteries, by the LAPLACE Laboratory, involved in the field of electrical energy conversion, working together with the Poitevin-based Prime Institute, dedicated to thermal management. The goal is, in particular, to double the power/mass ratio of the machines and their power supply, which will reduce the on-board weight of the aircraft by approximately 1.8 tons, thereby reducing fuel consumption. For a regional flight, fuel consumption would decrease by 3.5 percent.

In 2017, the Netherlands Aerospace Centre (NLR) and Delft University of Technology launched a project called NOVAIR (Novel Aircraft Configurations and Scaled Flight Testing Instrumentation) to design aircraft configurations with hybrid propulsion as part of Europe's Clean Sky 2 joint undertaking. With this system, a gas turbine motor generating electricity to power an aircraft by electric motors has the potential to reduce fuel consumption by approximately ten percent. As the e-propulsion will be detached from a gas turbine motor, it is possible to position the propulsion on the wing or fuselage in a way that improves the aerodynamic properties of an aircraft. A hybrid aircraft configuration will thus make an extra contribution to reducing aircraft fuel consumption and emissions. The first demonstration model of such an aircraft concept is expected to be tested in 2021.

In southeast Queensland, Australia, Dr. Jason Chaffey and his team at magniX, a subsidiary of Singapore-based Heron Energy, have developed an electric aero engine with a power density of more than 5 kW/kg—more than twice as high as the best conventional motors. MagniX believes that it could eventually achieve power densities of 25 kW/kg—three times higher than modern aircraft engines. Located in Arundel, magniX received a A$2.5 million (U.S. $1.9 million) grant from the Australian government as part of a A$12 million (U.S. $9.1 million) three-year collaborative project—indicative of the significance of the breakthrough. The "magniXmix" combines permanent magnet and superconducting motors and generators, all based on precision placement of the magnetic field. The aim is to maximize the field strength where it is useful and to minimize it where it is not, thus boosting power densities. This in turn reduces the amount of steel in the rotor, and thus its weight, resulting in fast transient responses and the ability either to reverse direction quickly or to switch rapidly from motoring to generating. The project, also involving both the University of Queensland Composites Group and Ferra Engineering in Tingalpa, aims to optimize the design, thermal management and materials needed for the high-power-density motors. In March 2017, magniX began by testing its 50 kg, 250-kilowatt prototype simulating aircraft takeoff and landing procedures to test speed capacity and reliability. Further funding will be needed to commercialize the technology when magniX plans to offer two commercial versions of its motors with lower power densities. One, weighing 210 kg (463 lb.), delivers 150 kW and 1,500 Nm continuous (200 kW and 2,500 Nm peak), with an efficiency of more than 97 percent. The 480 mm/19 in-diameter and 450 mm/17.7 in-long motor has an operating speed of 900–1,000 rpm (with a maximum of more than 2,700 rpm). The other version, aimed at high-speed applications around 4,000 rpm, weighs 220 kg (485 lb.) and can deliver 200 kW and 485 Nm, with an efficiency above 96 percent.[8]

On February 1 and 2, 2017, an international More Electric Aircraft conference "vers des aéronefs plus électriques" was held at the Palais des Congrès in Bordeaux. The MEA colloquium gained its technological and international credibility in a particularly strong competitive context (German and American competition). This credibility is due to the success of the three previous editions: Toulouse in 2009, Bordeaux in 2012, and Toulouse again in 2015. Their diffusion can only have a positive effect.

At ONERA, the French Aerospace Lab based in Lille, Jean Hermetz is leading a team to develop a DEP aircraft named Ampère. Thirty-two small motors have been mounted on the 2.9 m (10-ft) wing of a ⅕ scale model which was wind tunnel tested between September 2016 and February 2017. The full-scale airplane will be propelled by 40 motors powered by ten fuel cells, one for every four motors. A 4- to 6-passenger flying taxi version of Ampère with a range of 300 miles (500 km) should be ready by 2030. The Ampère model was exhibited at the Paris Air Show in June 2017.

Design Configurations

In May 2016, Blaine K. Rawdon and Aaron J. Kutzmann, engineers at Boeing Phantom Works in San Pedro, California, filed a patent application for "Solar Power Airplane" with nine propellers along its staple-shaped frame, which is essentially one giant wing with upturned edges. There is no cockpit and no pilot on board. Instead, the plane is controlled remotely from the ground. Like the Pathfinder (see Chapter Ten), Boeing's plane is designed with solar panels on its upper side to soak up and store energy for unusually long flights. But the Pathfinder's relatively flat design made it difficult to absorb sunlight that came in at low angles. To overcome this challenge, Boeing has designed large winglets at each end of the main wing, which should allow the plane to maintain a stable flight path and high altitude while absorbing sunlight even when the sun is low on the horizon. If the design works as Boeing has planned, the plane may be able to collect and store enough energy during the day to sustain flight at night. It may never need to land.[9]

Another wing format would be based on the Custer Channel Wing, developed from the late 1920s by Willard R. Custer, whereby eight semicircular ducts channeling airflow around the rotating propeller enable extremely short takeoff and landing (STOL) airplanes. Hop Flyt, led by Rob Winston, a U.S. Naval aviator and former lead engineer at Helix Aero, and his wife Lucille, based at Chesapeake Ranch Airport, Maryland, is building a short-range (200 miles) commuter airplane electric version of the Custer Wing, using eight DEP motors. In 2005, Winston designed a Surface Independent Extremely Short Takeoff and Landing (XSTOL) Aircraft (patent 7,487,935) based on the advanced concepts of the combination of an aero-morphing variable incidence wing with an air-cushion landing system.

Norway, with its vast fjords, could well benefit from an electric amphibian. Inspired by the designs of Guenter Poeschel, in 2008 Tomas Broedreskift of EAN (Equator Aircraft Norway) in Oslo, an industrial designer with a passion for gliding, teamed up with Oeyvind Berven to start work on the new EQP2 Xcursion. A 100 kW (approx. 130 hp) Engiro DMG60 generator will power output of the prop, and a Wankel Super Tec (WST) KKM 352 engine running on bio-diesel fuels will produce 57–60 kW of power charging the Kokam batteries, which in turn drive the tail-mounted motor with its custom DUC propeller. All of this is controlled by a single lever in the cockpit, a kind of hybrid FADEC. The P2 Xcursion aircraft is the first aircraft design made by the cooperative EAN (Equator Aircraft Norway SA). The wing uses a NACA laminar flow profile, with a 1.4 deg blended twist. The span is about 10 m dependent on the type of winglets chosen; they are modular and can be exchanged simply. With the flaps extended the aircraft should reach a stall speed of 45 kt, and on cruise should be able to fly comfortably at 130 kt.

The Synergy airplane, its double boxtail configuration designed and patented (U.S. 8657226; 9545993) by John McGinnis of Kalispell, Montana, is a proposed five-seat, single electric engine, kit aircraft. In 1995, McGinnis commercialized the world's first thermoplastic carbon fiber snowboards and invented several other profitable, high-volume composite production processes. He founded Synergy Aircraft in 2010. The double boxtail arrangement allows a simpler, stronger wing of the minimum induced drag for a given wingspan loading. The two horizontal tails intentionally push down, moving air upward at the wingtip in opposition to wake vortex. To date, a working model of the Synergy has been built and flown. It is planned that the full-scale airplane would use aluminum-air batteries.

Another configuration may be a triplane with a cross-box tail and interconnected winglets. In the 1980s, Trevor Cloughley had developed ASVEC to design and build innovative UAVs. Now his son Neil, based near Bristol, has started up Faradair (after British scientist Faraday) to develop the 5-passenger BEHA (Bio-Electric Hybrid Aircraft). Power will come from a biofuel-burning 200 hp Hybraero H600, designed and built by Prodrive Motorsports, coupled to two electric motors for takeoff and landing. BEHA's three wings are positively staggered (top wing foremost, lowest wing rearmost) with ample space for solar cells. Annular shrouding of the pusher propeller at the stern enables vectoring thrust sideways or up and down. Faradair's Cloughley plans to have the BEHA in service by 2020 to revive Britain's regional air services so popular during the interwar years.

Vision has been conceived by English inventor and pilot Michael Waters. Waters has built a number of different types of aircraft and also flown over 200 different aircraft including general aviation, experimental, low-wing, high-wing, mid-wing, bush-planes, seaplanes, hang-gliders, ultralights, paragliders, microlights, biplanes, aerobatic, tail-draggers, trikes, twins, autogyros, and helicopters. He was also a competition sailplane pilot for over 12 years. The Vision's fuselage is off-the-shelf, the controls are off-the-shelf technology, the motors are off-the-shelf, and now that energy storage has reached a certain milestone, Waters claims that he has what he needs to put all these things together. The Vision would use 8 electric motors, 4 for thrust and 4 for rotor tilt, which allows maximum performance and flexibility. There are no control surfaces or conventional wings, even though the ducted fan shrouds are, in fact, highly efficient circular wings.

Following experience with over 600 remote-controlled extremely maneuverable working models in the range of 25 to 50 feet (7 to 15 m), a team led by former schoolteacher Daniel Geery of Salt Lake City has developed the ultra-efficient Hyperblimp, using various envelope materials, propulsion from Li-Po batteries, solar panels, and eventually fuel cells (separately or in combination). Always on the lookout for advanced materials, these ships already work as short range UAV platforms. The Hyperblimp's rear prop, mounted to move in all directions, propels the airship with great responsiveness, including vertical takeoffs and landings. Though presently on hold, Geery's next large goal is a 24/7 high-speed solar version to fly by remote control around the world. The technology is all there, as this company works on more funding and locating knowledgeable, interested personnel.

For his final year project for Transport Design BA (Hons) in 2013 at the University of Huddersfield, England, Mac Byres conceived of a luxury airship called the Aether with Vertical Electric Take Off and Landing propulsion. Dining areas feature giant windows, and the inside has the mixed feel of a modern office complex and an upscale hotel bar. Large bedrooms feature their own sitting areas and the windows extend to right behind the pillows, so passengers can wake up from their falling dreams and find themselves face to face with clouds. Byres has also designed an airship called Hemera, designed to land solely on water, which he aims to unveil at Expo 2020 in Dubai.

Andrew Winch Designs of Barnes, London, which works on private jets and superyachts, has conceived of the Halo, a residential airship with a living area the size of four soccer fields comprising 20 bedrooms, a spa, a cinema and a ballroom. A 266-ft/81-m-long prototype made a test flight four years ago and received certification by the U.S. Federal Aviation Administration. It is slated to be capable of vertical takeoff and landing, and be eight times more fuel efficient than a jet plane with a range of 6,000 miles (10,000 km).

If an airplane can morph its wings like a bird, then it will need less electrical power. In 1905, Orville Wright steered the brothers' pioneering airplane by lying prone in a saddle and twisting the tips of the plane's fabric-and-wood wings with a sway of his hips.

A century later, from 1996 to 2005, the U.S. Air Force had been collaborating with NASA to develop an Active Aeroelastic Wing, which used the power of the airstream to twist itself for better roll control during high-speed maneuvers. But that technology was intended only for fighter jets, and the program eventually lost support. From 2010, two research projects continued to investigate this potential. With €51 million in support from the European Union, Smart Aircraft Structures (Saristu) is coordinated by Piet Christof Woelcken of Airbus and bringing together 64 partners from 16 European countries. This resulted in the design of a morphing flap for an imaginary 90-seat airliner of the future in which just the final 50 cm (20 in) of the flap is adaptive—and performed extensive wind tunnel tests on a 4.9-m/16-ft-long section of it in a massive Moscow wind tunnel. The morphing part had 10 electric motor-driven actuators that were used to change the profile of the flap's adaptive section in different flight conditions. In the USA, the aerospace firm Flexsys of Ann Arbor, Michigan, worked with NASA and the Air Force Research Laboratory (AFRL) and came up with a wing surface called FlexFoil that is able to shift shape in midflight, thanks to seamless bendable, twistable materials. The testing has involved 22 research flights that have been completed over the last six months at the Armstrong Flight Research Center in California. The advanced lightweight materials used to build the flexible wings will not only reduce the weight of wing structures, but allow engineers to tailor them to improve fuel economy. The technology can be retrofitted to existing airplanes.

NASA is taking its investigation of aeroservoelasticity and flutter suppression into the sky with the X-56A, nicknamed "Buckeye," a 7.5-foot drone with a 28-foot wingspan, with highly flexible, lightweight wings built by Lockheed Martin. From November 2017, test flights have taken place from the Armstrong facility.

Going one further, NASA's Mission Adaptive Digital Composite Aerostructure Technologies (MADCAT)—comprised of researchers and students from MIT; University of California, Davis; University of California, Berkeley; University of California, Santa Cruz; and Cornell University—has developed a morphing aircraft wing. This is made of a lattice of tiny, lightweight subunits that robots could assemble. The subunits are covered by overlapping parts reminiscent of scales or feathers. The wing components are made from advanced carbon-fiber composite materials. Computers and motors can help change the shape of the wing for better efficiency even while an aircraft is flying. The new wings could also be manufactured using much simpler and more streamlined processes. In 2016, following tests in a 12-foot tunnel at NASA Langley, a demonstrator was successfully demonstrated at NASA Crows Landing facility in Modesto, California. The MADCAT demonstrator utilizes a wing-twist actuation mechanism that generates a linear spanwise wing-morphing capability, thereby producing both lateral and longitudinal directional control authority. In addition, the aerodynamic lift/drag can be modulated by varying wing-tip twist oscillation frequency. During the flight test, the pilot reported that, overall, MADCAT flew quite easily with sufficient control authority, and did not seem to fly any differently from conventional aircraft with ailerons. Additional flight tests are planned with instrumentation of an advanced onboard video camera and sensing devices.[10]

Professor Bharath Ganapathisubramani and Southampton University's Aerodynamics and Flight Mechanics Group, working with Dr. Rafael Palacios and Imperial College London's Department of Aeronautics, have developed a 50-cm (19.7-inch)-wide working seaplane drone with its wings working like artificial muscles, changing their shape in response to physical forces they experience. They use electroactive material that changes shape when an electric current is passed through the polymers. The drone is powered by rotor engines on the front, but as the voltage alters the shape of the wing membranes, the aerodynamic characteristics can be altered as the drone flies. The design biomimics the wing movements of bats.[11]

Another biomimic approach has been taken with the monocopter, first tested by Papin and Rouilly in 1913. In 2009, almost a century later, students from the University of Maryland's Clark School of Engineering unveiled their samara (maple seed)-inspired micro air vehicle, which was billed as "the world's first controllable robotic samara monocopter." In August 2011, at the Association for Unmanned Vehicle Systems international conference in Washington, D.C., Lockheed Martin performed the first public flight of its Samarai Flyer, a disc-like unit that contains its battery and electronics, joined to a single wing with a propeller mounted at the far end. The Samarai can take off from and land on the ground, or be launched by being thrown into the air like a boomerang. It is 16 inches (40.6 cm) long, weighs less than half a pound (around 227 grams), and has only two moving parts, so it lends itself to being stuffed in a backpack, then pulled out for use. A team of students from the Singapore University of Technology and Design, again inspired by the samara or maple seed, have developed their Transformable HOvering Rotorcraft (THOR) or two-winged monocopter. The THOR's opposing wings are mounted at right angles to each other and rotate into alignment when making the transition from helicopter-style hover to fixed-wing-style cruising. The students have also created a passive system to shuffle weight around based on flight mode. When the craft switches from hovering vehicle to fixed-wing aircraft, or vice versa, the centrifugal force involved in the switch is used to move the ballast into a position to keep the aircraft balanced.

Boeing's 777X airliner, forthcoming in 2020, will not be a morphing plane per se, but it will change shape on landing. The 777X will have a 3.5 m long (11.5-ft) wingtip that is folded up vertically when the plane is on the stand or taxiing, but is locked down for flight.

In England, in March 2016, Airport Parking and Hotels Ltd. unveiled plans for a new hybrid-electric commercial aircraft, which it said would carry as many passengers as a jumbo jet. The idea is the brainchild of Adam Omar—an aircraft design Ph.D. student at Imperial College, London. The proposed hybrid-electric aircraft with its blended-wing body (BWB) design would use small biofuel engines turning six electric fans, positioned in clusters along the back of the craft rather than mounted on the wings. This positioning would help to suck in the thick layer of air around the aircraft body to reduce drag. Omar's proposal also takes advantage of superconductivity—a phenomenon of zero electrical resistance that occurs when certain materials are cooled below a critical temperature. This would mean that no electricity is lost due to friction when passing through the plane's power system, leading to a reduction in fuel requirements. Boeing Microlattice—a metal that is considered to be the world's lightest material—would be used across large parts of seating, flooring and walls to contribute to even greater fuel efficiency.

Oscar Viñals, a nuclear engineer and CGI designer from Barcelona, has projected his concepts beyond the year 2030. The AWWA-QG Progress Eagle design uses six hydrogen fuel engines—one to drive a central screw-type engine at the rear to achieve the thrust needed to take off. Once it reaches the right altitude, the central engine turns off and starts to generate electricity from the air flowing through it, powering five superconductive engines. It also has solar panels on the wings and a rear engine that doubles as a wind turbine. The three-deck airplane with its 314-ft (96 m) wingspan would have the capacity for more than 800 passengers along with beds and offices for crew. On landing, wing sections could be folded to make the aircraft easier to maneuver in airports. Excess energy stored in the aircraft's batteries could be recovered by special electrical storage trucks on the ground when it lands. According to Viñals, the aircraft would be made from lightweight materials such carbon fiber, aluminum, titanium and ceramics. The Spanish engineer has even gone beyond this bold concept with his Flash Falcon, capable of carrying 250 passengers at Mach 3, in an airframe more than 130 ft. (39 m) longer than a Concorde and with a wingspan twice as wide. Its engines would even be able to tilt up to 20 degrees to help the aircraft take off and land like a helicopter. Its propulsion system would use a nuclear fusion reactor pumping energy to its six electric engines.

Of this second concept, Simon Weeks of the Aerospace Technology Institute warns there are some major issues that come with putting a fission reactor on an airplane. Not only would it require a "closed loop system"—a reactor that reuses the waste fuel—but it would also need large amounts of heavy shielding. Nuclear fission produces a lot of neutrons which can be very harmful.[12]

Inspired by airships and paragliding sports, the Sky Voyage concept, designed by Jet Shao of Yanko Design, aims to expand traffic networks to the sky by utilizing airspace to relieve urban traffic congestion on the ground. The hybrid glider/airship would take off vertically by inflating the gasbag in an upright position. Once airborne, the craft can be maneuvered through the wind with assistance from a hydrogen fuel cell–powered turbine engine.

Daphnis Fournier, a Paris designer, has developed what he calls the Ecologic Aircraft, an electric passenger plane using an inflatable structure above the main cabin that contains flexible photovoltaic panels to collect solar energy while flying above the clouds. For efficiency during takeoff (when the most power is lost) the "balloon" remains flat, inflating only after reaching a maximum altitude. Considering most of energy of a plane flight is spent on taking off, its balloon does not consume any energy of fuel for taking off. Fournier's study suggests that this 65-m (210-ft) plane can transport up to 300 passengers. It is reminiscent of a design presented in the 19th century.

There are some free spirits whose thinking goes beyond conventional battery technology. Consultant Luke Workman, who had already come up with innovative lithium technology for Zero Motorcycles in Scotts Valley, California, is one of these. He has reasoned that instead of packaging cells with heavy lithium and aluminum, the wings of a large airliner could become the battery itself, using a composite sandwich of aluminum and copper. In this way, the entire cells are basically in contact with one another, conducting through their whole surface plane, instead of out through tabs. Workman calculates that using such a system would obtain around 13,300 amp hours per 0.2 mm of thickness for each foil layer. Nine hundred layers would then supply 3.3 kV nominal and around 44 megawatt hours of battery storage. The aircraft would have a total weight of approximately 104,000 kg, with an extraordinarily high percentage of that mass being active material; hence lower conduction losses would give 423 watt hours per kilo and a serious flight range. This is provided there are 300 m^2 (3,200 ft^2) of wing area, with a foil core about 20 cm (8 in.) thick and 1 cm current conductor plates on the top and bottom. The bigger the battery, the more efficient it becomes.

Arthur Léopold Léger and a team at Elixir Aircraft in the Industrial Zone of Périgny in La Rochelle, France, are adapting the use of boatbuilding techniques to airplane construction. The carbon wing with a span of 7.8 m (25.6 ft) has been made as a single piece and its rudders and fuselage are also one shot, eliminating riveting and reducing overall weight. Partners include Dassault Systèmes for electronics and software, C3 Technologies for composite structures and parts, and Simair (aeronautical equipment manufacturer) for metal parts.

Also in France, the Space Agency CNES is investigating the use of solar sails positioned between the nacelle of a stratospheric balloon to study the radiation of our galaxy from an altitude of 45,000 km (28,000 mi) where wind and oxygen are weak. The sails would use a patented high-efficiency ultralight flexible photovoltaic film developed by sailmaker Alain Janet of Solar Cloth System in Mandelieu-la-Napoule, and the orientation of the balloon will be made by two small electric motors.

One of the weak points of a solar-powered airplane might be a lightning strike during a storm. In 2016, Bertrand Rives and his team at Airbus Group SAS obtained a patent whereby the airplane fuselage, nacelles and wings would be covered by a layer of flexible polymer, a photovoltaic film, and a protective skin.

The ghost of pioneer electro-modeler Fred Militky (see Chapter Four) has been walking abroad in the form of the twin pusher-engined Hy-Fly he designed and that was marketed by Graupner in 1973. In 2014, forty years later, Gérard Risbourg, a veteran aeromodeler and multiple RC glider champion at the Aero Club Saint Remy of Provence based at the Aerodrome of Romanin les Alpilles, was invited to work with the European Association for the Development of Gliding (AEDEVV), Dassault Aviation and research lecturers and students at the ISAE Group (Higher Institute of Aeronautics and Space Group made up of SUPAERO, ENSMA, ESTACA and the Air School). Their joint goal has been to produce a French two-seater electric glider for pilot training. On February 21, 2016, the 8.80-meter-wide RC Euroglider scale model was loaded with four Mobius ActionCam cameras for varying angles. After ten minutes of flying above the Alpilles, the ASH 25 configuration completed its presentation flight without hindrance. Working groups are now dedicated to the preparation of the BEV (flying test bench); this will be used during the second half of 2018 to test the experimental class for the twin-engine configuration which will be presented to the DGAC, as well as for the validation of the onboard energy chain and its flight control systems. The Euroglider will be classified CS-22 and not ULM. At the same time, the AEDEVV, in accordance with its plan of action, will undertake contacts with potential partners for industrialization. With the Euroglider, the association wants to generate a production line as far as possible in France. The objective is to fly a first prototype in 2020.

If an airplane can be made feather-light and ultra-strong, it will need less energy for propulsion. In 2004, Konstantin Novoselov and team isolated the 2-dimensional material graphene, identified many of its extraordinary properties, and subsequently described other 2-dimensional materials. Their work is of such importance that both were awarded the 2010 Nobel Prize in Physics, knighted by the Queen and by the King of the Netherlands, and over the past decade have been showered with numerous honors and awards. A decade later, researchers at MIT have designed and tested Porous 36D forms of graphene, made by compressing and fusing flakes of graphene, a two-dimensional form of carbon. The new material, a sponge-like configuration with a density of just 5 percent, can have a strength 10 times that of steel. The team was led by Markus Buehler, the head of MIT's Department of Civil and Environmental Engineering (CEE), and the McAfee Professor of Engineering, Zhao Qin, a CEE research scientist.[13]

UCLA researchers have developed a new grapheme-based material, holey graphene, enabling production of a capacitor that has unparalleled energy density, 10 times that of currently available supercapacitors. Holey graphene features superior electrical conductivity, exceptional mechanical flexibility, and unique hierarchical porosity, making it ideal for use as a cathode in electrochemical capacitors and batteries. These characteristics may give some credence to claims that a 10,000 Farad ultra-supercapacitor, smaller than a paperback book, has been produced—a major development if true and one with revolutionary potential for electric propulsion. Over the following decade, this is likely to have two far-reaching effects: firstly, it will be possible to store large amounts of electricity more efficiently generated from solar energy by more efficient solar panels, and secondly, the range, recharge time and reliability of electric vehicles (EVs) of all descriptions will be vastly improved and much cheaper.

The miniaturization of digital electronics over the past half-century has followed a similar exponential trend, with the size of transistor gates, used in computers, reducing from approximately 1,000 nanometers in 1970 to 23 nanometers today. With the advent of transistors made of graphene showing great promise, this is expected to fall further to about 7 nanometers by 2025—approximately the size of a human red blood cell. The increase in computational power and decrease in circuit size, when combined with progress made with 3D printing, will mean that tiny integrated computers powerful enough to control an aircraft will be possible in the next decade. Systems using a biologically inspired digital "nervous system" with receptors arranged over the aircraft to sense forces, temperatures, and airflow states are expected to drastically improve the energy efficiency of an airplane. In the future, they could even be paired with software and hardware mechanisms to change the shape of the aircraft to make it extra efficient.

As for Drones...

In April 2016, Amazon Technologies Inc. obtained a patent for an "airborne fulfillment center" (AFC) or warehouse and drone airport, hanging from a blimp. It will hover at 45,000 feet (14,000 m), so that its delivery drones will glide down with Amazon packages, delivering goods within minutes of when they are ordered. Instead of having the empty drones fly all the way back up to the AFC, they instead would meet up with smaller restocking blimps for the flight back up. Such AFCs would be located above festivals or sporting events and smaller airships could act as shuttles taking drones, people and supplies to the warehouse.[14]

On December 29, 2016, Gregory Karl Lisso and the team at Amazon Technologies Inc. obtained another patent for a "Collective Unmanned Aerial Vehicle" (CUAV) configuration, capable of lifting virtually any size, weight, or quantity of items, and flying greater distances than smaller drones. Individual modules could detach from the collective drone body once they were no longer required, and operate independently to deliver smaller burdens, and would also be able to travel longer distances. Another Amazon patent, received December 20, 2016, addresses countermeasures to protect airborne drones against threats from hackers and from "malicious persons" armed with bows and arrows.[15]

The same month, Japanese mobile phone giant NTT DoCoMo, Inc., in a continuing quest to create new business, unveiled the "world's first spherical drone display," a drone inside an omnidirectional spherical frame, with 8 curved LED strips that spin rapidly to create the illusion of a spherical screen. The "resolution" is 144 pixels high and 136 pixels around the circumference with a maximum diameter of 88 cm. The highly maneuverable 4 kg drone can be operated virtually anywhere, including venues such as concert halls or arenas where it can fly around as part of a performance or deliver advertising messages and event information. Following a demonstration of the display at the Digital Content Expo in October 2017 at the Miraikan in Odaiba, DoCoMo aims to commercialize its product from March 2019.

In July 2016, Airbus filed a patent for a type of carrier aircraft, known as a high-altitude platform system, that could transport heavier solar-powered planes with more equipment up above 60,000 feet (18,000 m) and deposit them there. The advantage to this strategy is that the aircraft deposited by the carrier platform would not need the thrust capabilities required to reach the stratosphere on its own, allowing it to carry more science and communications payloads, which could remain airborne for months or even years.

Defense specialist BAE Systems is planning to chemically "grow" drones in large vats. The UK-based defense company is working with Leroy "Lee" Cronin, Regius Chair of Chemistry in the School of Chemistry at the University of Glasgow, one of the developers of the Chemputer. While a 3D printer physically makes the parts for a machine, the "chemputer" speeds up the chemical reactions from the molecular level. This would artificially emulate a life form, such as bumblebee, using a gel. Once grown, the drone would most likely combine a fuel-cell with trickle recharge solar energy. So far Cronin and his team are working in centimeter lengths, but later grown drones of several meters could be created from chemical compounds in weeks, rather than years. British warplanes are already flying with parts made from a 3D printer. The date for the first chemdrones is far into the future.[16]

In December 2016, Warren East, the CEO of Rolls-Royce, stated, "There's a lot of chatter about hybrid electric flight, not just little airplanes but regional airplanes. I'm convinced we will see these things happen sooner rather than later. There is a race on. We need to be ready by 2020 because people are talking about entry into service by 2030."[17]

Pierpaolo Lazzarini of the design company Jet Capsule in Naples has conceived a UFO-looking two-seater electric ocotocopter taxi called the IFO, or Identified Flying Object. On land, the carbon-fiber IFO, with its 4.70-m (15-ft) diameter, would stand 10 feet (3 m) high on six fold-in extendable legs. The doors of the cockpit capsule swing upwards, enabling passengers to climb in and out using detachable footbridges; or they can enter from underneath through an elevator that descends from the vehicle's main body. IFO would have a battery life of 70 minutes and could hit a top speed of 120 mph (190 kph). It is equipped with an emergency parachute. Lazzarini has also designed an enclosed 740bhp diesel engined 13 passenger waterjet ski boat he calls a mini-yacht, with an all-electric option. One recalls Nikola Tesla's circular electric airplane concept of 1911, while Lazzarini Design's motto is "Think about the future, never forget the past."

Then there is the lighter-than-air drone, one of which is the PLIMP. As identical twins growing up in Washington in the 1970s, James C. and Joel D. Egan flew model gliders and threw weighted balloons. While James became a prominent attorney in Seattle, the twins continued their passion for aviation. In 2013 they applied for a patent for a "plummet-proof" plane-blimp hybrid aircraft they trademarked as PLIMP and started up Egan Airships to promote it, first as a drone and later as a passenger aircraft. Weighing under 55 lb. (25 kg), the 28-ft. (8.5-m) prototype PLIMP is an airplane body attached to a helium-filled blimp, electrically powered by two housed lateral propellers that can rotate and travel in any direction. It can deliver forward speeds of more than 40 mph (64 kph) with at least an hour of flight time and offers an unpowered descent speed of only 9.5 mph (15 kph) should engines fail, as well as smooth flight and acceleration for nearly stable platform filming. The PLIMP was presented at Interdrone 2017 in Las Vegas. It is strangely reminiscent of ideas conceived in France in the 1890s.

Electricity Harvesting Aircraft

Alongside aircraft using electricity, there are also aircraft that will be used to regularly generate electricity. It was in 1978 that Miles L. Loyd at the Lawrence Livermore National Laboratory, Livermore, California, applied for a patent (U.S. 4251040) for large-scale wind power production by means of aerodynamically efficient kites. Based on aircraft construction, these kites would fly transverse to the wind at high speed. The lift produced at this speed is sufficient to both support the kite and generate power. This would come to be known as crosswind kite power.

Twenty-eight years later, in 2006, Australian inventor Saul Griffith and kite designer Don Montague teamed up to build a similar type of generator, naming their company after the Hawaiian word for wind, Makani. Funded by DARPA, they built a 20-kilowatt turbine-carrying glider and flew it in 2009; the higher the altitudes, where the winds are stronger and more reliable, the more electrical energy is harvested. By 2011 Makani was testing developed models from the tarmac of the former Alameda Naval Air Station. In 2013 Google bought Makani. Although facing significant regulatory obstacles including wildlife preservation issues as well as the technological challenges, they were eventually able to produce their eighth generation prototype designed by Damon Vander Lind, a 600-kW carbon-fiber energy kite with eight rotors, each 7.5 ft (2.3 m) in diameter, and the 85-ft (26-m) wingspan of a small jet airliner. The turbine-driven generators would also function as motor-driven propellers in a powered flight mode, which could be used for vertical takeoff and landing. A perch adapted to facilitate the takeoff and landing would pivot such that the pilot is oriented towards the tension direction of the tether. On May 18, 2017, the Makani 600-kW kite produced power for the first time.

Alongside this is the European Ampyx tethered glider, the 250 kW AP3 powerplane, with its 12-m (40-ft) wingspan, developed since 2006 by a team led by Richard Ruiterkamp of TU Delft, Netherlands. Funded by the European Commission and E.ON, with their control center near Winchelsea, England, the team is conducting tests up to an altitude of 450 m (1,476 ft) in Mayo County, Ireland. Ampyx announced that a 2 mW commercial version would be available by 2020.

In southern Norway at Lista airfield, since 2008, Kitemill has been testing their Spark prototype airplane to harvest 30 to 100 kW of airborne wind energy at altitudes up 1,500 m (1 mi). Among the team are electrical power engineer and yachtsman Olav Aleksander Bu, and Jon Gjerde, a brewer and former world champion in acrobatic gliding.

Another approach is with a tethered aerostat floating at a great height above the earth without moving, but remaining stationed where the strong winds passing through its rotors generate electricity. In late 1979 Charles M. Fry and Henry W. Hise conceived and patented such a device (U.S. 4165468 A). Again the principle has been taken up thirty years later by aeronautical and astronautical engineer Ben Glass while conducting research on compact, efficient turbomachinery at the Massachusetts Institute of Technology's Gas Turbine Lab. Glass conceived and patented a ring-like shroud with an airfoil cross-section filled with lighter-than-air helium gas to use a lightweight wind-harvesting electroturbine. Glass co-founded a start-up, Altaeros Energies, at Greentown Labs, a leading clean technology incubator located in Somerville, Massachusetts. They have built and tested BAT (Buoyant Airborne Turbine), or "Super Tower," in partnership with the Alaska Energy Authority; it resembled the stern end of a conventional airship. The Alaska Project will deploy the BAT at a height of 1,000 feet (305 m) above ground, over the community of Fairbanks, Alaska, for 18 months, a height that will break the world record for the highest wind turbine in the world and also send down consistent, low-cost energy for the remote power and micro-grid market. Such aircraft have the potential harvest of 100kW of electricity, while others may be beaming the Internet to 4.5 billion people still waiting for it.

Although this book covers more than a century of electric aircraft, current development is happening at a fast pace.

The cruise wingtip motors that will power NASA's first fully electric X57 *Maxwell* airplane are being tested on the ground. Passing through various modifications, the fifth motor will undergo testing on Airvolt at its full operational capability and will then be taken apart to have its components inspected as part of what's called a "destructive inspection." The state of the bearings, rotor and magnets will be observed and analyzed to see how healthy they are.

Boeing HorizonX, working with Zunum, has taken a stake in battery start-up Cuberg, based in Berkeley, California. Cuberg's high energy, dense lithium-metal anode, high-energy cathode battery components are lighter and less flammable compared to current battery technology. It has also invested in Near Earth Autonomy of Pittsburgh, a company that focuses on technologies that enable reliable autonomous flight. Boeing is also continuing to test more than 30 new technologies aboard *ecoDemonstrator*, a FedEx-owned 777 freighter, equipped with Safran Electrical & Power's more electric technology.

The famous 1919 England to Australia Great Air Race (30 days/18,000 km) is to be re-created but this time for electric aircraft. The race to be held in September 2019 will begin at Biggin Hill Airport in south East London and end in Darwin.

EViation of Israel has entered into a battery supply deal with Kokam to provide energy for their nine-passenger *Alice* aircraft. The 900 kWh battery pack should give *Alice* an autonomy of 650 miles (1,047 miles). The battery will have 9,400 cells distributed throughout the aircraft including the ceiling, floor and wings, weighing 3.8 tons, or 60 percent of the maximum takeoff weight. EViation aims to begin service in 2021.

Israel Aerospace Industries (IAI) is to develop an all-electric aircraft, an initial prototype of which will fly in around three years. Range will be about 500 nm (926 km).

Daan Moreels and the team at Magnax at Kortrijk, Belgium, working in close collaboration with Ghent University on next-generation axial flux direct-drive electric motor and generator technology for aircraft and drones, has won European funding in the ultra-competitive Horizon 2020 SME instrument program.

A report by MarketsandMarkets projects that the aircraft battery market will grow from an estimated U.S. $475 million in 2017 to U.S. $667.8 million by 2022.

S. Yokoyama, pilot for Aircraft Olympos, Ltd., has been test flying a conventional solar-powered airplane weighing 86 kg (189 lbs.) on only 2.2 kW above Takikawa SkyPark on the large island of Hokaido, Japan.

In March Pipistrel completed delivery of another four Pipistrel ALPHA Electro aircraft to Calstart in Fresno, California. Following certification, these aircraft are entering service as flight training aircraft in the local areas of Reedley and Mendota, making the first electric aircraft training base in the world.

Following a demonstration of its eAircaft in Chicago on March 27, 2018, Siemens USA CEO Lisa Davis projected that e-propulsion will be "the standard solution for *all* aircraft segments" by 2050.

California-based Joby, developing its 16-prop VTOL flying taxi, has received $100 million (£70 million) in funding from a group of investors led by Toyota, JetBlue and Intel. On January 31, 2018, Airbus's A3 Vahana prototype made its maiden 53-second, five-meter pilotless flight from Pendleton in northern California. André Borschberg's Hangar 55 (H55) has been awarded first-round funding by U.S. venture capital firm Nano-Dimension to develop an electric air taxi.

The EHANG 184 quadcopter has achieved a series of manned flight tests in Canton, carrying one and two passengers, including many Guangzhou government officials. In all, about forty people have been carried, not counting the thousands of pilotless test flights. German automobile maker Porsche, already involved with the Mission-E Cross Turismo sportscar with 310 miles on a charge and 800-volt DC fast-chargers, is planning a flying passenger drone. Audi has further invested in the Airbus/Italdesign air taxi program, *Pop.Up Next*.

Google co-founder Larry Page's autonomous all-electric two-seater air taxi prototype, developed by a team led by Fred Reid at Zephyr Airworks *Cora* (N301XZ), has been making test flights over Christchurch, New Zealand.

Bye Aerospace has raised more than $5 million in a Series C financing round led by Galileo Global Securities and Ashanti Capital. On April 10, 2018, John Penney took *Sun Flyer 2* on its maiden flight from the Centennial Airport in Englewood, Colorado. Sun Flyer is powered by EPS equipment including battery modules, battery management unit, and power distribution unit.

China's *RX1E-A* light electric monoplane has been testing its two-hour flight range, while preparing its larger four-seater version.

UK-based Samad Aerospace, led by Seyed Mohseni, is developing the *Starling Jet*, the world's first hybrid-electric business aircraft capable of vertical takeoff and landing, and of transporting ten passengers at 460 mph between 900 and 1,500 miles. With provisional orders placed, first deliveries are scheduled for 2024.

Electric aircraft startup Wright Electric has teamed up with Jetex, a flight support company from Dubai. The proposed design they are working on with Wright Electric aims for a range of 540 km or 333 miles, which would enable passengers to fly from Dubai to Muscat or Malaga to Casablanca on a single charge.

Thales Alenia Space is using the digital integration of the most advanced manufacturing technologies, including augmented and virtual reality and the robotic assembly of panels to prepare the *Stratobus* demonstrator for late 2018.

DJI's 300gm *Spark*, its tiniest drone yet, is the first that can be controlled by hand gestures. If a user simply frames her face with her fingers, the hovering Spark will snap a 12-megapixel selfie.

In March, 180 U.S. Marines at Camp Pendleton, California, were training with prototype 7.5-lb. *Drone Killer* lasers with a two-mile range depending on visibility. They are built by IXI Technology at Yorba Linda.

Yates Electrospace has patented *Silent Arrow® ER-700* (Electric, Reusable, 700 lbs. payload) cargo delivery drone product line that allows persistent operations from improvised airstrips as well as via airdrop from a variety of fixed and rotary-wing aircraft. YEC has been contracted by the U.S. government to build and fly ten of its tandem-wing Silent Arrows.

Following Hurricane Maria, workers from Duke Energy in North Carolina have been using five AceCore's *Zoe* quadcopters to locate and repair fallen power lines across Puerto Rico.

To become the world's first 100 percent solar-powered airport, Cochin International Airport Limited (CIAL), southwest India, started with a 12-megawatt project which expanded to 15.5 megawatts. The airport will eventually increase its capacity to 40 megawatts, including a solar-powered carport. CIAL is set to help Ghana take this clean energy path as well at its Kotoka International Airport at Accra, Kumasi International Airport, and the Navrongo Domestic Airport.

The state-run Norwegian aviation firm Avinor—which runs 45 airports in Norway—is planning to embrace electric aircraft as soon as they hit the market. In Oslo, Norwegian firms Haptic Architects and Nordic Office of Architecture have aimed higher, seeking to design Aerotropolis, the world's first energy-positive airport city, for the Norwegian capital for both electric planes and driverless cars. Construction of the Oslo Airport City is expected to begin in 2019, with the first buildings completed in 2022.

Full Circle?: Out of the Ancient Past and Into the Future!

Could reports in ancient documents provide the solution for future electric aircraft? For some concepts we need to tread very carefully. Telekinesis is moving objects while levitation is making the human body hover, both through mind power alone. The "Vimāna," a flying palace or chariot as described in Hindu Yogic texts and Sanskrit epics, was controlled by the mind. From the ancient accounts found in the Sanskrit epic *The Mahabharata*, we read that a Vimāna measured twelve cubits in circumference, with four strong wheels. Their method of propulsion was "anti-gravitational." It was based upon a system analogous to that of "laghima," the unknown power of the ego existing in man's physiological makeup, "a centrifugal force strong enough to counteract all gravitational pull."

In 1895, on Chowpatty beach near the city of Mumbai (Bombay, Maharashtra, India), Shivkar Bapuji Talpade, a 30-year-old technical instructor in the art and craft department of Sir Jamsetjee Jeejeebhoy School of Art, with a passion for ancient Sanskrit manuscripts, is reported to have proved that heavier-than-air flight was indeed possible. While carefully researching the descriptions of Vimāna as recorded in ancient Indian scripts, the *Rigvedādic Bhāshya Bhumikā* and *Rigved and Yajurveda Bhāshya*, Talpade also read newspaper reports about the unsuccessful attempts of aviation pioneers such as Thomas Alva Edison or Hiram Maxim's captive steam-driven aircraft as described in Chapter Two of

E2 Lange's experimental fuel-cell craft heads into the future (courtesy Large Aviation GmbH).

this book. Taking the guidance of Vedic Acharyas from Karnataka, he built his own cylindrical-shaped aircraft, or Vimāna, which he called *Marutsakhā*, derived from the Sanskrit *Marut* ("air" or "stream") and *sakhā* ("friend"), which together mean "Friend of wind," and then test-flew it from the beach in Mumbai, reaching an estimated altitude of 1,500 ft. (457 m) before dropping back down to the sand. According to K.R.N. Swamy, "a curious scholarly audience headed by a famous Indian judge and a nationalist, Mahadeva Govinda Ranade and H. H. Sayaji Rao Gaekwad witnessed the event."[18] After Chowpatty, Talpade tried to raise funds to build another aircraft. He unsuccessfully asked for funds from the then Maharaja of Baroda, and there is even one recorded instance of his appealing to a group of businessmen in Ahmedabad. He died in 1917. Kindly analysis of Talpade's device suggest that he was using a "Vedic Ion design," apparently a concept similar to electric propulsion. More recently NASA has researched the Shuttle's ability to use solar electric propulsion—solar power combined with mercury bombardment thrusters to deliver 600 kW.

In the 1970s, Israeli Uri Geller became the world's best-known psychic and made millions traveling the world demonstrating his claimed psychokinetic abilities, including bending spoons with his mind. One test was performed as Geller flew across the USA in an aircraft, reportedly using his mind to mend broken watches on the ground.

In 2009 a "telepathic" microchip that enables paraplegics to control computers was developed by Dr. Jon Spratley, a British scientist who assembled it while studying for a

Conception of Lilium parking in 2020.

Ph.D. at Birmingham University. The chip is implanted onto the surface of the brain, where it monitors electronic "thought" pulses. It means paraplegics, amputees or those with motor neurone disease, such as the late Stephen Hawking, could be able to operate light switches, PCs and even cars by the power of thought alone.[19] In 2013, Bin He and researchers at the University of Minnesota revealed a drone that can be controlled merely by thought. Published in the *Journal of Neuro Engineering*, the report of the project has implications in everything from unmanned vehicles to paraplegic mobility. To control the basic Parrot AR Drone, the "pilot" wears an electrode cap and controls takeoff, turning and landing.[20] In 2016, Panagiotis Artemiadis and a team at the Human-Oriented Robotics and Control Lab at Arizona State University had progressed to mind-controlling small swarms of robotic drones using the human brain.[21]

Parallel researches had been made for full-scale aircraft. In May 2014, a team of researchers from the Technische Universität München and the TU Berlin in Germany, led by Professor Florian Holzapfel, developed the technology to fly full-scale aircraft with thoughts alone. Seven subjects wearing a cap connected to EEG electrodes (one with no cockpit experience at all) used a flight simulator; all of them navigated the virtual skies with enough accuracy to pass a flying license test. In November 2016, *Wired*'s Jack Stewart, wearing an electrode cap developed by Santosh Mathan, an engineer with Honeywell Aerospace, used his mind to fly a Beechcraft King Air C90 above Seattle. With what Stewart describes as "a tiny amount of practice in a simulator," he could actually make the twin turboprop climb, descend and turn simply by focusing on certain areas of a tablet screen. After the flight, Stewart said the system followed his planned command about 90 percent of the time.[22] The next stage would be for a pilot or passenger to wear a helmet to control his electric flying car or Vahana—but then, why not call it Vimāna? Perhaps vibration-free electrical propulsion will render mental communication more responsive.

Which brings us back to our question: when mythical Icarus moved his arms to flap his waxed-feathered wings, was he not also sending electrical messages from his brain?

Whatever the case, where the sky's the limit, it's up in the air.

APPENDIX A

The Birth of Drone Racing

FPV, standing for first-person view or first-person video, was the catalyst which led to the birth of drone racing, in which participants could control drones equipped with cameras while wearing head-mounted displays showing the live-stream camera feed from the drones. In 1999, an Australian called Thomas, aka "Mr. RC-CAM," started an MSN group of drone-flying enthusiasts who desired to fly beyond the line of sight, first using baby monitors and later security cameras and small video transmitters. But it was not until 2012 that Team Drop Bear, a group from Melbourne, who had been FPV flying multi-rotor craft for a number of years, decided to do a proper race. Among them was a Queensland park ranger called Justin Welander aka "Juz70," who had recorded and posted countless breathtaking FPV videos with his Lumenier QAV multi-rotor airframes, able to do crisp flips and 4-point rolls. Somebody calling himself "Blackout," also from Queensland, came up with the Mini H Quad frame, almost half the size of the average drone. Using social media, "Blackout" started posting some amazing videos with very aggressive, sporty-style FPV flights on his YouTube channel, and reporting and writing about his newly designed frame, which would eventually become the world standard for FPV flight.

In November 2014, an event specialist called Terence Boyton of Gold Coast Mc, Queensland, set up Drone Racing International Pty. Ltd. to organize FPV meetings in abandoned warehouses and parking lots. By 2015 Australia held its first state-level championship in a six-race series. With events in Melbourne, Sydney and Brisbane, with hundreds of people taking part in and watching races among quadcopters achieving speeds of over 100 mph (160 kph) through a course of illuminated checkpoints. Like all enjoyable new sports, FPV drone racing soon crossed the seas. After a successful event held at Matamata, New Zealand, alongside the MFNZ Nationals, Matthew Wellington founded Rotorcross NZ (inspired by Motocross). This has become an organized racing league, with around 8 regional championship races held throughout New Zealand every year.

By this time, FPV was giving way to full Projected Virtual Reality (PVR), a sport in which people would eventually add 360° spherical cameras to drones. However, the wireless bandwidth demands (not to mention weight and battery constraints) for live streaming such footage currently preclude this capability.

The fledgling sport's organization in the USA came in 2015, when the first annual U.S. National Drone Racing Championship was held in a stadium at the California State Fair. Ironically, the winner was Chad Nowak of Queensland, Australia!

Justin Haggerty founded a company on April 3, 2015, to promote and build drone racing into a world renowned sport. The International Drone Racing Association, Inc. (IDRA), is a professional racing organization that sanctions and governs multiple drone racing events. IDRA's major events and series include the 2016 World Drone Prix, the 2016 North America Cup, the 2016 Asia Cup, and the 2016 California Cup.

In 2016, drone racing received a big boost when real estate mogul and Miami Dolphins owner Stephen Ross invested $1 million in the New York–based startup Drone Racing League, described as the nation's largest and most authentic drone racing championship in the history of the sport. The organization's vast network of drone racing chapters around the United States provided a platform to hold the nation's most competitive series based on pilot skill, not popularity. The organization strategically divided the U.S. into 15 regions that will hold over 55 qualifying events convenient to pilots' hometowns, making the sport easily accessible and approachable to the masses.

The next phase was a World Championship, but where? And by whom? IDRA began working with the United Arab Emirates government and Aerial Grand Prix in October 2015 to co-found and host the World Drone Prix in Dubai. This was held under the patronage of Sheikh Hamdan bin Mohammed bin Rashid Al Maktoum, Crown Prince of Dubai, on March 11 and 12, 2016, at Skydive Dubai. The $250,000 prize was won by 15-year-old British competitor Luke Bannister for his team Tornado X-Blades, beating 150 teams from all over the world, including Dubai and Russia. This followed a series of pre-qualifying events held in Los Angeles, Seoul, Berlin and cities across China.

The event gave the United Arab Emirates city the opportunity to announce the start of the "World Future Sports Games," which was expected to commence in December 2017. Sky Sports Mix Channel has invested $1 million in the Drone Racing League (DRL). On January 26, 2017, IDRA announced its professional 2017 Drone Racing Series on social media platforms. The Drone Racing Series, composed of 6 international races, started in China and ended with finals in the Netherlands.

But another series, the World Drone Racing Championships, was scheduled at Kualoa Ranch, Hawaii, for October 2016. Pilots from over 30 countries around the world competed on four world-class courses, and six separate racing events all day, every day.

The 2017 DRL Allianz World Championship saw sixteen racers competing at venues in Europe and North America, including NFL stadiums, abandoned factories, and historical landmarks. The drones had been designed and built by the League, which is a sports and media company, and are crafted differently for every race. Each model was worth from $500 to $1000, and could travel from 80 to 90 mph (130 to 140 kph). In order to prevent racers from cheating, standardized equipment was used, and DRL did not even let the participants touch the drones during the competition. The DRL Allianz World Championship Final took at Alexandra Palace, London, on June 13, 2017, as part of London Tech Week, Europe's largest festival of technology. The winner was the DRL 2016 World Champion, 25-year-old Jordan "Jet" Temkin from Fort Collins, Colorado, in a fierce final race against eight other pilots during a round-robin format, hitting speeds of over 80 mph and zooming through the first-ever FPV Power Loop, a series of light gates covering 180 degrees, which forced the drones into an inverted dive at full speed. The final was seen by millions of fans in 75 countries, divided up into broadcast views and views of its digital content.

The 2017 Drone Champions League races were held in six colorful venues including the Champs-Elysées in Paris, the ruins of Schlosskopf fortress in Austria, the caves of the Turda Salt Mine in Romania, and Berlin Station.

In sharp contrast, in October 2017, at the World Drone Expo, Attollo Engineering, makers of the Sabrewing Rapier's advanced sense-and-avoid system iRobotics, launched the Pacific Drone Challenge™: the first to fly the 4,500 miles (7,300 km) from the Pacific coasts of Japan to Silicon Valley, without stopping to refuel—a distance beyond the reach of current non-military drones. One of the teams already signed up for the challenge, Silicon Valley's Sabrewing Aircraft Company, is building the SWA-3 Rapier, a fixed-wing hybrid-electric quad-rotor drone, with a 30 ft., 1 in (9 m) wingspan designed to take off from a standard runway and cover as much as 8,800 km (5,500 mi). It will be controlled via satellite, and in constant communication with both the Launch and Recover Element ("LRE," located at the launch point in Japan) and Mission Control Element ("MCE," located at the destination landing point of Moffett Field, Mountain View, California). Monitoring will be by two pilots on the ground, in constant contact with air traffic control. The trip is expected to last about 45 hours. Sabrewing Aircraft will be pitting its technology against that of Japanese company iRobotics.

Who might have thought that in less than twenty years, a hobby would become an exciting international sport?

APPENDIX B

The Adventure of *Solar Impulse*

The round-the-globe flight of a solar-powered airplane, watched by the world, was a dream made reality by two men: Bertrand Piccard and André Borschberg.

Bertrand Piccard was born in Lausanne, Switzerland, on March 1, 1958, into a family of explorers. His grandfather, Auguste Antoine Piccard, was a physicist, inventor and explorer. As a physics professor at the Federal Polytechnic School in Zurich, the elder Piccard was friend of both Albert Einstein and Marie Curie, who discovered Uranium 235; he was also passionate about the protection of nature. On May 27, 1931, Auguste and Paul Kipfer took off in their balloon from Augsburg, Germany, and reached a world record altitude of 15,781 m (51,775 ft.). During this flight, Piccard was able to gather substantial data on the upper atmosphere, as well as measure cosmic rays. On August 18, 1932, launched from Dübendorf, Switzerland, Piccard and Max Cosyns made a second record-breaking ascent to 16,201 m (53,153 ft.). Auguste ultimately made a total of twenty-seven balloon flights, setting a final record of 23,000 m (75,459 ft.). He then turned his attention to the bottom of the sea. By 1937, he had designed the bathyscaphe *Trieste*, a small steel gondola built to withstand great external pressure. In 1953, the year it was launched, Auguste and his son—and co-designer—Jacques set a new world record by taking it down to a depth of 3,150 meters (10,330 ft.) in the Tyrrhenian Sea, with the aim of exploring deep marine life.

In 1960 Jacques Piccard set a record that still stands nearly 50 years later: together with Lt. Don Walsh of the U.S. Navy, he went down to a depth of 10,916 meters in the Mariana Trench, located in the western North Pacific Ocean. Auguste and Jacques Piccard subsequently worked together on the mesoscaphe *Auguste Piccard*, designed by the father and built by the son. It turned out to be one of the main attractions at the 1964 national exhibition in Lausanne. It was the first and largest-ever submarine built for pleasure trips, and took some 33,000 passengers down to the bottom of Lake Geneva and back.

Bertrand Piccard grew up in this world of ballooning and undersea exploration. It is scarcely surprising that his eyes turned skywards. As a child, when his father was working in the United States, he met such legends as the flying pioneer Charles Lindbergh; Hermann Geiger, who developed the art of landing on glaciers; and rocket designer Werner von Braun, inventor of the Apollo rockets, who invited him and his family to witness the launch of several space flights from Cape Canaveral. From an early age he was also fascinated by the study of human behavior in extreme situations. He received a degree from the University of Lausanne in psychiatry, specializing in hypnosis. His love of flight led him to obtain licenses to fly balloons, gliders, and motorized gliders. In

Europe, he was one of the pioneers of hang-gliding and microlight flying, becoming the European hang-glider aerobatics champion in 1985.

On March 1, 1999, Bertrand Piccard and Brian Jones set off in the balloon *Breitling Orbiter 3*, a bright red, carbon-composite, egg-shaped craft measuring 16 ft. (4.9 m) long and 7 ft. (2 m) in diameter, from Château d'Oex in Switzerland on the first successful nonstop balloon circumnavigation of the globe—the first circumnavigation requiring no fuel for forward motion. Piccard and Jones, in close cooperation with a team of meteorologists on the ground, caught rides in a series of jet streams that carried them 25,361 miles to land in the Egyptian desert after a 45,755 km (28,431 mi) flight lasting 19 days, 21 hours, and 47 minutes. Following this success, Bertrand Piccard was decorated with the Légion d'Honneur, the Olympic Order, the Fédération Aéronautique Internationale (FAI) gold medal, the Harmon Trophy, and the Charles Green Salver, and was honored by the National Geographic Society and the Explorers Club. An honorary professor and an honorary doctor in science and letters, he also received the Grand Prix of the Académie des Sciences Morales et Politiques.

Indeed, it was just after his balloon touched down that Piccard conceived of an idea to fly around the world again, but this time in a heavier-than-air machine and without using a drop of fuel. Back in Switzerland, encouraged by his wife Michèle, Bertrand's plans were soon underway. Then he met André Borschberg and persuaded him to join his impossible quest. Like Piccard, André Borschberg had been fascinated by aviation from his childhood. He trained as a fighter pilot in the Swiss air force, flying first Venoms and then Hunters and Tigers for over 20 years and specializing in aerobatics. Training as a mechanical and thermodynamics engineer, he also graduated at the MIT Sloan School of Management, becoming a consultant to the McKinsey firm. He then launched on his own account with two start-ups, and co-founded a company specializing in microprocessors:

> In 1999 I had decided to make a break in my professional activities. I wanted to open up to new ideas and new people. It was in 2003 at the École Polytechnique Fédérale de Lausanne (EPFL), the Switzerland's MIT that I met Bertrand Piccard. I already knew about his family and his world balloon flight so when he told me about his plan to fly around the world in a solar-powered airplane, I felt that here was somebody with whom I could associate. He asked me to lead a feasibility study and put together the technical team, while Bertrand found the partners to help back the project.[1]

It would take some 15 years for Piccard to raise the $170 million needed from companies including Omega, Solvay, Schindler, Nestlé, and ABB. In November 2003, following a feasibility study, they announced their project, in cooperation with the École Polytechnique Fédérale de Lausanne (EPFL), for a solar airplane they had decided to call *Solar Impulse*. "We did consider Solar Spirit, but in the end we chose Impulse as in French this gives the idea of starting something, even if in English impulsive means doing something on the impulse, without thinking, which was far from our case."[2]

The company Solar Impulse SA was officially founded on June 29, 2004, by Bertrand Piccard, André Borschberg, Brian Jones and sports marketing specialist Luiggino Torrigiani. A core technology team was then put together and the first scientific partnership agreements signed with the Ecole Polytechnique Fédérale de Lausanne, the European Space Agency and Dassault Aviation. As a mechanical engineer, co-founder André Borschberg would direct the construction of each aircraft and oversee the preparation of the flight missions. By 2009, he had assembled a multi-disciplinary team of fifty specialists from six countries, assisted by approximately one hundred outside advisers.

While this was underway, Piccard realized that he would have to obtain a license to fly an airplane. He would have to train for this over six years, accumulating hundreds of hours before he was even allowed to fly a prototype. "We even went to see Paul MacCready of AeroVironment, the doyen of solar flight, who told us of his challenge, some twenty years before, of building and flying a solar airplane, Solar Challenger, across the English Channel. With him we were able to share our philosophy of making the impossible possible."[3] Eight years before, AeroVironmment's unmanned *Pathfinder* had been equipped with a 99-ft wingspan (slightly longer than a Boeing 737) covered with 8 kW maximum of solar panels and a power source of six 1.5 kW electric motors. On September 11, 1995, *Pathfinder* had set an unofficial altitude record for solar-powered aircraft of 50,000 feet (15,000 m) during a 12-hour flight from NASA Dryden (see Chapter Ten). So the initial thinking of the EPFL team was for a similar, very large wingspan incorporating extremely efficient aerodynamics.

In early 2004, more detailed studies led to the first prototype version with its engines placed forward beyond the leading edges, in order to balance the thrust of the aircraft against the aerodynamic forces. The cockpit in this version would be in a separate underwing nacelle. Finally, a third version was born from the work of the engineers who decided to opt for a first prototype with a non-pressurized cabin. It took boldness and confidence in the projections to freeze the design and start construction. In November 2007, after four years of research, complex calculations and simulations, Piccard and Borschberg presented the final design with a wingspan of 63 meters (206 ft.) and weighing 1,600 kg (3,527 lb.), registration HB-SIA.

In May 2008, the flight simulator, developed in collaboration with Dassault and EPFL, enabled both pilots to "fly" HB-SIA for the first time for up to 25 hours, equipped and harnessed as they would be for real flights: helmet, safety harness, parachute, and oxygen mask, and with food and accessories for their natural needs. Five projection screens arranged 210° around the cockpit gave the pilots the impression of live flight.

For over a decade André Borschberg had been studying and practicing yoga, recently with a sangi in India: "When you practice yoga and its body postures, you learn to become an observer about yourself and to release yourself from stressful situations. This would be of immense help during the marathon flight sessions."[4]

By September 2008, after assembly of the cockpit and the tail boom of *Solar Impulse HB-SIA*, construction of the wing could begin at Dübendorf. Three rectangular carbon-fiber and honeycomb sandwich beams, the longest over 20 meters (65 ft.) in length, made by Décision SA, were placed end-to-end to form the central wing spar, the backbone of the wing, with its total span of 63.4 m (208 ft.). The *Solar Impulse* project then went through two further important test phases. In December, the Deutsches Zentrum für Luft und Raumfahrt (DLR) in Göttingen, an institute that specializes in aero elasticity calculations, carried out a week of vibration tests. These were aimed at assessing the risks of flutter, identifying the aircraft's specific frequencies, and verifying by physical experimentation the match between the engineers' theoretical models and the technical characteristics of the actual aircraft. The fuselage, the wing spar, and the horizontal and vertical stabilizers were assembled for the first time.

With the assistance of Eric Raymond, experienced solar airplane builder and pilot, 11,628 photovoltaic cells were mounted on the upper wing surface and the horizontal stabilizer to generate electricity during the day to power the electric motors and to charge the batteries allowing flight at night, theoretically enabling the single-seat plane to stay in the air indefinitely. Raymond also worked out propeller spinners and wing ribs.

In mid–February 2009, a series of more robust tests put the wing spar through its paces. These were designed to directly test its resistance to high loads. Fully conclusive, these impressive load tests were conducted in an almost religious silence in order to be able to hear the slightest crack, and also to permit the intense concentration and the essential coordination of all involved. The electrical generating set and propulsion system also underwent detailed testing: the four engines each developing 10 hp (7.5 kW), the 3.5-meter diameter (11.5 ft.) twin-bladed propellers, the lithium-polymer batteries, the optimization and control circuits, the cabling, and power controls. The combined efforts of each team member, the courage to undertake a completely new project, and the trust of all the partners, contributed to the construction of the aircraft, which was completed in June 2009. After its presentation to the public and the media, the integration work was finalized in the fall of the same year.

HB-SIA Solar Impulse 1 carried out its first tests in November, and on December 3 succeeded with its first "ground hop" at Dübendorf airfield with professional Daimler-Chrysler test pilot Markus Scherdel at the controls in the 3.8-cubic-meter cockpit. On April 7, 2010, watched by several thousand spectators, the plane, again piloted by Scherdel, made an 87-minute test flight from Payerne, a military base of the Swiss Air Force, located approximately halfway between Lausanne and Bern. Scherdel spent the time familiarizing himself with the prototype's flight behavior and performing the initial flight exercises before making the first landing on the Vaudois tarmac. This flight reached an altitude of 1,200 m (3,937 ft.). On May 28, 2010, the aircraft made its first flight powered entirely by solar energy, charging its batteries in flight. Two weeks later, on May 24, André Borschberg took the controls of *Solar Impulse 1* for the first time: "Coming from the world of jet fighter planes, I had to get used to the fact that in this airplane, everything happens extremely slowly. When you alter the flight controls, you have to wait a couple of seconds. We call this pilot-induced oscillations due to over-correcting. It was a question of learning new techniques, even for Bertrand Piccard who came from another world of balloons and gliders."[5]

After three "fast taxi" trials along the Payerne runway, the man who had been in charge of building the solar airplane was at last able to get airborne in it himself. Watching was Prince Albert II of Monaco, one of the project's patrons. In July 2010, *Solar Impulse 1* achieved the world's first manned 26-hour solar-powered flight, again flown by Borschberg. It took off at 6:51 a.m. Central European Summer Time (UTC+2) on July 7 from Payerne Air Base, Switzerland. Having flown through the night, it returned for a landing the following morning at 9:00 a.m. local time. During the flight, the plane reached a maximum altitude of 8,700 m (28,500 ft.). At the time, the flight was the longest and highest ever flown by a manned solar-powered aircraft; these records were officially recognized by the Fédération Aéronautique Internationale (FAI) in October 2010.

The next target was to fly outside Switzerland, making "European Solar Flights" to Brussels and then Paris. It was only natural that in late May 2011, *Solar Impulse 1*, which had obtained the patronage of the European Commission even before it had been built, should select Brussels as its first international destination, where her pilots met the leaders of the European institutions and used it as a tool to help raise public—as well as official—awareness about environmental matters.

On May 13, 2011, at 9:30 p.m. local time, the plane landed at Brussels Airport, after completing a 13-hour flight from its home base in Switzerland. It was the first international flight by the *Solar Impulse 1*, which flew at an average altitude of 6,000 ft. (1,800 m) for a distance of 630 km (391 mi.), with an average speed of 50 km/h (31 mph). The aircraft's slow cruising speed required operating at a mid-altitude, allowing much faster air traffic to be routed around it. Bertrand Piccard said in an interview after the landing: "Our goal is to create a revolution in the minds of people … to promote solar energies—not necessarily a revolution in aviation."

Next stop was the Paris International Air Show at Le Bourget as a "Special Guest." On June 14, 2011, after a 16-hour flight, Borschberg successfully landed the aircraft at Bourget Field, Paris. Visitors to the Air Show—over 200,000 people—were able to approach very close to *Solar Impulse 1* in the static display and every morning, weather permitting, to marvel at in-flight demonstrations of its silent electric motors. On June 5, 2012, the *Solar Impulse 1*, originally conceived as a one-seater, now adapted to allow both Piccard and Borschberg to co-pilot, successfully completed its first intercontinental flight, a 19-hour trip from Madrid to Rabat, Morocco. During the first leg of the flight from Payerne Air Base to Madrid, the aircraft broke several further records for solar flight, including the longest solar-powered flight between pre-declared waypoints (1,099.3 km [683 mi.]) and along a course (1,116 km [693 mi.]). Cruise speed: 70 kph (43 mph). Endurance: approximately 36 hours. It also flew an entire diurnal solar cycle, including nearly nine hours of night flying, in a 26-hour flight.

Soon after, Piccard nearly lost his co-pilot, when Borschberg was involved in a helicopter crash but emerged without a scratch.

In 2013, *Solar Impulse 1* was taken across the Atlantic, where in May, Piccard and Borschberg made a cross-U.S. flight. They took off from Moffett Field in Mountain View, California, and flew to Phoenix Goodyear Airport in Arizona. Successive legs of the flight lasted between 14 and 22 hours: Dallas–Fort Worth Airport, Lambert–St. Louis International Airport, Cincinnati Municipal Lunken Airport to change pilots and avoid strong winds, and Washington's Dulles International Airport. The aircraft's second leg of its trip on May 23 to Dallas–Fort Worth covered 1,541 kilometers (958 mi) and set several new world distance records in solar aviation. On July 6, 2013, following a lengthy layover in Washington, *Solar Impulse 1* completed its cross-country journey, landing at New York City's JFK International Airport at 11:09 p.m. EDT. The landing occurred three hours earlier than originally intended, because a planned flyby of the Statue of Liberty was canceled as a result of damage to the covering on the left wing. But air traffic control had blocked all other flights for the landing; *Solar Impulse 1* was placed on public display. In August 2013, it was disassembled and transported back to Dübendorf Air Base, where it was placed in storage in a hangar. During the American flights, a new partner had joined them: Google.

Solar Impulse 1 had always been considered as a prototype for a round-the-world airplane. No. 2 was designed and built following lessons learned during the above flights. It would use 140 carbon-fiber ribs with a gap of 50 cm (20 in.) intervals, in order to provide wingspan rigidity. With a length of 22.4 m (73.5 ft.) and a wingspan of 71.9 m (236 ft.), and a weight of only 5,000 pounds (2,268 kg), the solar airplane carried 17,248 monocrystalline silicon solar cells, 135μ thick and mounted on the wings, fuselage and horizontal tail plane. She had more powerful brushless, sensorless engines, 13 kW (17.4 hp). Her lithium-polymer batteries, manufactured by Kokam, had been optimized to have a density of 260 Wh/kg, capable of storing 164,580 watts of power. With a total mass of 633 kg (2,077 lb.), the batteries would store power needed to allow the aircraft to fly during nights.

Construction had begun in 2011, with completion initially planned for 2013, with a

25-day circumnavigation of the globe planned for 2014. But a structural failure occurred on the aircraft's main spar during static tests in July 2012, and it exploded when they reached 100 percent of the load. This led to delays in the flight testing schedule to allow repairs. *Solar Impulse 2HB-SIB* was first publicly displayed on April 9, 2014, at Payerne Air Base, while her first flight, with Markus Scherdel at the controls, took place on June 2, 2014. The aircraft climbed to a randomly orbiting path within a 20-mile (32 km.) compass of the airfield to the southwest, Lake Neuchatel to the north, and Belleville to the northeast. It averaged a ground speed of 30 knots (56 km/h) on the 2-hour flight, and reached an altitude of 5,500 feet (1,670 m).

The first night flight was completed on October 26, 2014, and the aircraft reached its maximum altitude during a flight on October 28, 2014. Flight testing was completed in 2014 and the aircraft was delivered to Masdar, Abu Dhabi, by a cargo plane for the World Future Energy Summit in late January 2015. Prince Albert of Monaco, His Excellency Dr. Sultan Al Jaber of the UAE, and Richard Branson joined Bertrand Piccard and André Borschberg to launch the FUTURE IS CLEAN movement one day before takeoff.

On March 9, 2015, *Solar Impulse 2*, with Borschberg at the controls, took off on its circumnavigation of the Earth, with the return to Abu Dhabi in August 2015. The route followed was entirely in the Northern Hemisphere. Seventeen stops were planned along the route; at each stop, the crew would have to wait for good weather conditions along the next leg of the route. A mission control center for the circumnavigation, directed by Raymond Clerc, was set up in Monaco, using satellite links to gather real-time flight telemetry and remain in constant contact with the aircraft and the support team.

2015: André Borschberg (left) and Bertrand Piccard discuss flying strategy in front of the *Solar Impulse 2* **(© Solar Impulse).**

Departure was from Al Bateen Executive Airport Abu Dhabi in the United Arab Emirates. Destination: Muscat (Oman). This first 772-km (480-mi) leg took 13 hours 1 minute. In Leg 2 (1,593 km/990 mi), leaving the Sheikh Zayed Grand Mosque below and heading out across the Arabian Sea to Ahmedabad (India), Piccard took 15 hours 20 minutes, completing the longest distance ever flown by a solar airplane in aviation history. On March 18, *Solar Impulse 2* took off from Ahmedabad's Sardar Vallabhbhai Patel International Airport, flying 1,170 km (728 mi.) over the Ganges River to Varanasi (India), touching down after 13 hours 15 minutes. Onwards to Mandalay and its temples (Myanmar), another 1,536 km, which took 13 hours 29 minutes. On March 29 the flight from India over the Yangtze River and down to Chongqing International Airport (China) a distance of 1,636 km (1016 mi.), took 20 hours 29 minutes. It was vital to switch on the automatic pilot and take 20-minute naps or use yogic breathing techniques or other exercises to promote blood flow and maintain alertness. Flying above the mountainous provinces of Yunman and Sichuan required Piccard to perform a steep climb at the beginning of the journey, and due to continuous flying at high altitude, he had to wear an oxygen mask and to face temperatures of -20°C. Crosswinds in China caused weeks of delays. Following the 1,384-km (860-mi) flight from Chongqing (China) to Nanjing (China), taking 17 hours 22 minutes, Chinese President Xi Jinping gave the team a great welcome.

Borschberg has described the challenge of Leg 7:

> This was the moment of truth for our entire project: the first time that I must fly solo for more than twenty-four hours. To fly Si2, you need very calm air and the air is never calm so you do not know if you are flying in updraft or downdraft; five days was the minimum duration and the weather forecast five or six days ahead is only 30% reliable. So what would the weather be like when I arrived over Hawaii, one single point in the middle of the ocean? It could also be the longest solo flight in the history of aviation—Steve Fossett had made a three-day flight, so we had always wanted to double his record. So I took off from China. As we got close to Japan, the weather worsened, forcing me to curtail the flight and land at Nagoya-Komaki airport until conditions improved. Despite this, by remaining airborne for 3 consecutive days and nights, Si2 had broken all distance and duration world records for solar aviation. As it happened, call it coincidence, on closer inspection in Japan, we discovered that one of the main electrical systems which provided power for the cockpit had a fault. Potentially this would have stopped working in the middle of the Pacific Ocean had I decided to press on. As Japan had not been on our schedule, it was difficult to find a hangar for our wingspan. We were grounded for one month, waiting day after day for the weather to clear. When eventually I did take off again, there were technical problems. Due to insufficient air cooling, our lithium batteries were overheating due to high climb rate and an over insulation of the gondolas. The engineers asked me to return to Japan where they could be fixed. When I announced that I was not intending to do so, this created a huge emotional crisis. I had to have full conferences from the cockpit because people were threatening to resign. It was incredibly fraught. It was at times like this that yoga was vital and cat naps proved vital.[6]

Borschberg ran the risk. After 8,924 km (5,545 mi) across the Pacific, a grueling 4 days 21 hours 52 minutes in the air, the Swiss co-pilot touched down at Kalaeola Airport, Oahu, Hawaii. Here it was decided that, until the battery problem had been solved, they would winter on the volcanic island. This meant they would miss their "weather window" to cross the Atlantic before the end of 2015. The team also faced financial troubles in 2015 after raising €20 million from sponsors. While they were waiting, in November 2015, Piccard and Borschberg attended the COP 21 in Paris, where they met U.S. President Barack Obama, confidently inviting him to see *Solar Impulse 2* in the USA the following year. Bertrand Piccard was also designated the United Nations Environment Program (UNEP) Goodwill Ambassador. Meanwhile, in Hawaii, the team had replaced the batteries

2014: *Solar Impulse 2*, in flight (© Solar Impulse).

and installed a new integrated cooling system. Between February and mid–April 2016, some thirteen test flights were carried out to ensure proper functioning, as well as training flights for Piccard and Borschberg. In addition, to give them more flexibility for route planning, they lined up four potential destinations: Phoenix, San Francisco, Los Angeles and Vancouver.

Flights resumed on April 21, 2016. Piccard left Hawaii behind him and, after 2 days 17 hours 29 minutes in the air, covering 4,523 km (2,810 mi), he arrived in the USA.

On April 22, while flying over the Pacific, he spoke directly from the cockpit of *Si2* to United Nations Secretary-General Ban Ki-Moon at the UN in New York, where 175 nations had just signed the Paris Agreement on Climate Change. Piccard commented, "If an airplane like Solar Impulse 2 can fly day and night without fuel, the world can be much cleaner." There was another iconic moment when the solar airplane flew over Golden Gate Bridge, at what appears to be about 3,000 feet (900 m). It was dark by the time the experimental craft made it down the peninsula to Mountain View, touching down at Moffett Airfield, operated by Google, one of the *Si2*'s sponsors. Google's Planetary Ventures are nearby, as well as NASA's Ames Research Center.

All this time, "internauts" had continued to watch the flight "live" via the website solarimpulse.com and followed the experience in real time thanks to 5 onboard cameras, as well as one showing Mission Control. For most of its time airborne, *Solar Impulse 2* would cruise at a ground speed of between 50 and 100 kph (30 and 60 mph), usually at the slower end of that range at night to save power.

From the West Coast, the European sun-plane now took 15 hours 52 minutes to cover the 1,199 km (745 mi) to Phoenix Goodyear Airport, then on to Tulsa, passing over

April 23, 2016: After its flight from Hawaii, *Solar Impulse 2* flies over the Golden Gate Bridge (©Solar Impulse).

Missouri, Illinois, and Indiana, and then to land, as gracefully as usual, after dark, its sixteen headlights lighting the runway at Dayton International Airport, Ohio, home of the Wright Brothers, another flight of over one thousand kilometers taking over sixteen hours. After landing, pilot André Borschberg was met by two relatives of the Wright Brothers, neither of whom ever married: great-grandnephew Stephen Wright and great-grandniece Amanda Wright Lane. They gave models of the Wright Flyer to both Borschberg and Piccard, "the Solar Brothers." Piccard noted that Dayton had served as the base for Orville and Wilbur Wright's airplane-building operation more than a century ago. He told a crowd of onlookers in Dayton, "People told the Wright Brothers, and us, what we wanted to achieve was impossible. They were wrong. If everything goes well, we'll land in Abu Dhabi in July this year and in 10 years' time, there will probably be electric airplanes."

On May 25, *Solar Impulse 2* continued its silent and globally followed flight across the United States, reaching Lehigh Valley, Pennsylvania, and then on to New York, a shorter hop of 230 km (142 mi) taking just 4 hours 41 minutes. Borschberg ended this leg by circling the Statue of Liberty and flying in front of the skyline of Manhattan, touching down at JFK at 7:59 a.m. During *Si2*'s stay, it was visited by Ban Ki-Moon, sponsors and their personnel, and 500 children.

On June 20, 2016, Bertrand Piccard took off from New York City and headed out over the Atlantic Ocean on one of the toughest stages of their attempt to fly around the globe using solar energy. Meditation, hypnosis and short naps would again play a vital role during a flight lasting almost four days. In fact, thanks to nine days of meticulous work from the Mission Control Center in Monaco and their weather specialists, they were able to identify a narrow window, bypassing a cold front that was situated in the middle of the Atlantic; the 6,765-km (4,203-mi) flight took only 2 days 23 hours 8 minutes. During the flight, Piccard announced the intention to create an NGO, the International Committee for Clean Technologies (ICCT). Its goal: to continue the legacy *Solar Impulse* started, promoting concrete energy efficient solutions in order to solve many of the challenges facing society today.

Touchdown was at 5:38 a.m. in Seville, Spain, and as he prepared to land, Piccard was welcomed by the Spanish Patrulla Águila, and then, once on the tarmac, by Borschberg. "Charles Lindbergh completed the first transatlantic flight in 1927 from New York to Paris, and now Bertrand has brought aviation to the next level with clean technology. I bet you Lindbergh never imaged a solar airplane could do the same!"

Now it was the 64-year-old Borschberg's turn. His final flight, to Cairo, was his chance to break the late Steve Fossett's 2006 absolute world record for "speed around the world, non-refueled." He took off on July 11 in the pre-dawn darkness from Seville, crossed the Mediterranean through the airspaces of Tunisia, Algeria, Malta, Italy, and Greece, taking 2 days 50 minutes to cover the 3,745 km (2,327 mi). As *Si2* approached Cairo, Borschberg flew over the pyramids at sunrise. The ancient pyramids poked through the morning city mist, beautifully juxtaposing the solar-powered airplane that flew above. Borschberg: "To fly over the country of the great Pyramids reminded me that there was a time when the sun was worshiped as a god. These pyramids had been built for eternal life or sustainability which is also what we are trying to promote three thousand years later."

For the final leg of this marathon circumnavigation, Leg 17, back to Abu Dhabi, Piccard was to take the controls. On July 15, the weather conditions were favorable, the plane was ready, but then 58-year-old pilot Piccard fell ill, forcing the historic flight to be postponed, because he could not possibly fly for 48 hours in that state. Mission Control identified a weather window that could allow *Si2* to overcome the high temperatures across Saudi Arabia. Conditions were tricky with winds challenging *Si2*'s limitations on the runway. With Piccard recovered, the flight began across the Red Sea and past the Persian Gulf. To avoid the severe turbulence above the hot Saudi desert, he had to fly at high altitude: "There were moments in the last night that I could not rest at all, I just had to fight with my flight controls."

But on Tuesday, July 26, 2016, *Solar Impulse 2* touched down at Al Bateen Executive Airport, after 2 days 47 minutes of flying covering 2,694 km (1,673 mi). Climbing out to cheers and applause from team and spectators, Bertrand Piccard commented, "The future is clean. The future is you. The future is now. Let's take it further." Co-pilot Borschberg commented, "Now that *Solar Impulse*'s technology has demonstrated that unlimited endurance is possible for an airplane, I am keen to start developing future applications such as drones...."

By landing back in Abu Dhabi after a total of 23 days of flight and 43,041 km (26,744 mi) traveled in a 17-leg journey, with 4 continents, 3 seas, and 2 oceans crossed, *Si2* had proven that clean technologies can achieve the impossible. Nineteen official FIA aviation records had been set during the global adventure, in particular the crossing of the Pacific and the Atlantic.

Ban Ki-Moon, the UN Secretary-General, said: "*Solar Impulse* has flown more than 40,000 kilometers without fuel, but with an inexhaustible supply of energy and inspiration. This is a historic day for Captain Piccard and the *Solar Impulse* team, but it is also a historic day for humanity."

July 26, 2016, the victorious *Solar Impulse* team at Abu Dhabi's Al Bateen Executive Airport (©Solar Impulse).

The Swiss post office released a special 100-franc stamp to mark the achievement.

On November 11, 2016, the Solar Impulse Foundation launched the World Alliance for Clean Technologies during COP22 at Marrakech, as a legacy to the first-ever solar flight around the world. Its goal is to federate the main actors in the field of clean technologies, in order to create synergies, promote profitable solutions to the world's most pressing environmental and health challenges, and give credible advice to governments. During COP23 in Bonn, the WACT announced its goal of selecting 1,000 solutions that can protect the environment in a profitable way, and bring them to decision makers at COP24 to encourage them to adopt more ambitious environmental targets and energy policies.

One can only wonder what the ghosts of electric airplane pioneers, such the Tissandier brothers in the 1880s, Fred Militky in the 1970s, Paul MacCready in the 1980s, and other brave spirits might have thought about a flight which had attracted the attention of the whole world.[7]

In February 2017, the 460-page *Objectif Soleil: L'aventure Solar Impulse* (*Objective Sun: The Solar Impulse Adventure*), with Piccard and Borschberg as contributors, was published by Stock. On June 17, as part of the National Geographic Explorers Festival, Piccard and Borschberg were awarded the 2017 Further Award in recognition of their pioneering solar-powered flight around the world.

André Borschberg has commented:

> The Solar Impulse airplane is not ready for retirement after our round-the-world mission. It has been designed to fly a total of 2,000 hours. By the end of the mission, we will have only flown 700 hours,

and therefore still have 1300 hours remaining. We are, therefore, considering using the plane for further testing on solar technologies with a test pilot to learn how to make the plane fully autonomous. The mistake made by people in the past was to take an existing airplane, get rid of the combustion engine replacing it with electric motors and test it. It makes less noise but you see the limits of the technology instead of the potential because the mileage is very small. Electric propulsion has a fantastic future, because it is highly efficient, quiet, reliable and easy to control. What you have to do is to build an airplane around this technology. I am starting a new VTOL project whose goal is to revolutionize air transport in cities. We are at a time when a lot of things are converging – what's happening in cars, in battery technology, software solution, new materials is going to change things very soon.

■ ■ ■

While *Solar Impulse* used solar panels for marathon flights, Jean-Luc Soullier, a former French airline pilot and holder of the world record in microlight electric propulsion (see Appendix C), plans to use fuel cells for his *Etlantic* microlight. In this, he has been helped by Roman Marcinowski, head of AéroSkyLux, a research laboratory for non-polluting energies based in Gap (Hautes-Alpes). Their fourth prototype, designed by Grégory Cole, is made of carbon, aluminum and titanium with a 50-ft (15-m) wingspan and will use a "de-turbulator" which reduces the energy consumption of flying craft today by 20 percent. The power supply for the 15 kW Rotex engine is supplied by 100 ft^2 (10 m^2) of solar panels and fuel cells that supply Li-Po batteries and then lithium-thionyl chloride batteries to achieve a speed of 110 mph (180 kph). Its 35 hours of autonomy of flight would allow Soullier to cover the New York–to–Paris transatlantic distance of 3624 miles (5,831 km). The Frenchman then plans to fly from London to Darwin, 8,602 miles (13,843 km), in 2018. One is reminded of the plans presented in the late 1880s by another Frenchman, Arthur DeBausset, as chronicled in Chapter One.

Also in France, with initial sponsorship from Paris Airports, Raphael Dinelli of Les Sables d'Olonne, an experienced offshore sail racer, has planned a translatlantic flight from New York to Paris Le Bourget in his solar-powered biofuelled *Eraole*. Dinelli and his team at the Océan Vital Laboratory have spent two years researching the best biofuel for the task, settling finally for an oil made from micro-algae, cultivated specifically for this purpose. The *Eraole*'s average speed of around 100 kilometers per hour means Dinelli will be stuck in its tiny cockpit for nearly 60 hours.[8]

Transoceans.fr, based at the Dôme Airpark, Le Champ Pres Froges, southeastern France, has also been planning a series of airship records. Firstly, a speed record with the transoceans *Lelio*, for a target 120 kph (75 mph). For *Lelio*'s design, transoceans resorted to services by the Laboratoire Aérodynamique Eiffel (Eiffel Aerodynamics Laboratory) and investigated Eiffel's research work discussing airship profiles. They also used their partner Airstar's wind tunnel for minor profile studies. For example, the pilot is positioned *inside* the envelope, leading to eliminating the nacelle. Following this, a second record is envisaged: a world record for the airship crossing of the Mediterranean Sea, and ultimately an Atlantic crossing by *Stream Continental 1*, this time using fuel cell propulsion. Cruising from 80 to 100 km/h (50 to 62 mph) it would take about sixty hours to cross the Atlantic Ocean. At the time of writing, transoceans has been held back due to lack of finance.

APPENDIX C

Speed and Altitude Records

Speed records for gas airplanes go back a long way. In February 1912, Parisian Jules Védrines, piloting the Deperdussin Monocoque, with its 140 hp (100 kW) Gnome double Lambda engine, broke the 100-mph barrier when he clocked 100.92 mph (162.5 kph) above Pau, southwest France. It was to be another decade, interrupted by war, before Joseph Sadi-Lecointe, test pilot for Nieuport-Delage, took their Sesquiplane, powered by a 12-cylinder 447 kW (600 hp) Hispano 12Hb engine, over the 200 mph (320 kph) barrier. The location was Ville Sauvage and the date September 1921. It would be another seven years before the 300 mph mark was broken by Mario de Bernardi in the Macchi M.52bis seaplane with a speed of 318.62 mph (512.77 kph).

In October 1883, the Tissandier electric airship had cruised above Paris at a gentle speed of 10 kph (6 mph.). In 1973, although the main goal was the duration and not the speed, the *MBE1* was timed at approximately 120 kph (75 mph). In late 1980, Paul MacCready's *Solar Challenger*'s regular flight speed, as measured on her airspeed indicator, was 42 mph (68 kph), although at higher altitudes, such as 12,000 ft (3,700 m), her true airspeed limit would have been 20 percent higher, or about 50 mph. But there was never an official FIA speed or altitude record attempt made, or homologated.[1]

This was the same for the electric airspeed record established by an Italian astronaut, Maurizio Cheli. Born in Zocco near Modena, Italy, on May 4, 1959, Cheli attended the Marco Minghetti Classic College of Bologna and then entered the Pozzuoli Aeronautical Academy, coming second in the Corso Urano III, in 1978. In 1982, he finished his studies at the University Federico II of Naples, obtaining a degree in aeronautical science. In 1983 he obtained operational assignment as the reconnaissance pilot for the famous F-104G. In 1988 he trained as a test pilot at the Empire Test Pilots' School in Boscombe Down, England, winning the McKenna Trophy as the best student in his class, the Sir Alan Cobham Award for the best results in the flight activity, and the Hawker Hunter Trophy for the student preparing the best Test Preliminary Report.

In 1992 he joined the European Space Agency (ESA) and was sent to NASA's Johnson Space Center in Houston, where he obtained a master's degree in aerospace engineering from the University of Houston. In 1996, having trained with the USAF, boarding the Space Shuttle *Columbia*, Cheli took part in the mission STS-75 Tethered Satellite with the role of mission specialist, passing more than 380 hours in orbit. That same year he joined Alenia Aeronautica and two years later he became chief test pilot for combat aircraft. His last test program was for the Eurofighter Typhoon with its top speed of 2,495 kph (1,550 mph/Mach 2).

In 2008, Cheli took part in a conference with a big audience at the prestigious Science Festival of Genoa, organized by journalist Giorgio Pacifici. Cheli described his experience on the Shuttle and spoke about Hyský. This was a new prototype 1 kW fuel cell electric plane developed by Paolo Pari at the Department of Aeronautics and Space Engineering of Turin Polytechnic, which had reached an altitude of 5,000 meters (16,000 ft). Soon after, Cheli founded DigiSky with the goal of setting a world air speed record for electric airplanes. In this he was accompanied by the team at Turin Polytechnic led by Professor Paolo Maggiore. The aircraft chosen was a Pioneer 300, offered by Alpi Aviation of Pordenone as a testbed. Its wooden frame made it lightweight and easily adaptable to the testing of new installations. Other technical partners had soon joined the team. The SKP-VAL2 motor was a 75 kW (101 hp) brushless unit fully designed by DigiSky and the Polytechnic of Turin and built by SICME; named "Valentino," it derived its energy from 102 7.5 kWh Li-Po cells. Between March and October 2008, ground tests were carried out with the integration of the technical systems by January 2009. On January 28, the aircraft, called the SkySpark, was presented at a press conference of the Piedmont Region, with the announcement that it would attempt world speed and endurance records. The plane was then modified by adding a new inverter, normally used in F1 racing automobiles, and new control software developed by Magneti Marelli. Maurizio Cheli made his first test flight on June 10, 2009, as part of the World Air Games in Turin. Two days later, SkySpark achieved a silent speed of 252 kph (155 mph). Although not homologated by the FAI as a record-setter, Cheli thus remains at the top of an honorable list that will grow with the passage of time.

On September 5, 2010, pilot Hugues Duval established a speed record for electric aircraft in the MC15E CriCri "E-Cristaline." The aircraft was equipped with French Electravia engines, controllers, batteries and propellers. During the Pontoise Air Show, a top speed of 262 kph (141 kt) was recorded by Aero Club de France organizers. Then, on June 25, 2011, during the official flight presentation at the Paris Air Show (Salon du Bourget), Duval established a new world record of 283 kph (175.46 mph), although it too did not qualify under the FAI Section 2 sports aircraft rule.

Another electric airplane speed record was established on April 13, 2011, by Jean-Luc Soullier at the Friedrichshafen Air Show. Flying his single-seater Luciole MC30E ULM, Soullier achieved a speed record of 136.4 kph (84.7 mph), although his record was not ratified because he had not filled in the claim form correctly. It was an achievement which had taken this 54-year-old very experienced French commercial airline pilot and instructor some five years. In 2006, when Soullier had begun to examine the potential of a lightweight electric airplane, the first candidate was a Rutan Quickie, a lightweight single-seat taildragger airplane of composite construction, configured with tandem wings. By 2008 Soullier had progressed to an MC15 "CriCri," as designed in the 1970s by Michel Colomban. To electrify, Soullier replaced the twin JPX gasoline engines with German Plettenberg Predator 37/6s electric units with energy from 4 packs of Kokam batteries yielding up to 6 kWh. On September 8, 2009, Soullier made a short flight in the MC15E from runway 23 at Maubeuge Elesmes, northeast France; the flight might have lasted longer had it not been for the controllers burning out simultaneously. Soon after, Soullier turned to Michel Colomban's latest Firefly MC30, which could be classified as either airplane or ULM. Soullier recalls:

> Working with Colomban, and Werner Eck, the designer of the Geiger 15 kW Flytec brushless motor, we rolled up our sleeves and less than a year later, on 1st August 2010, the MC30E took off from track

10 in Brienne-le-Chateau, north-central France, for a short and successful flight. In subsequent flights we improved the internal cooling of the propulsion chain, and also, with Richard Kruger of Helix Carbon, designed and built the blades for Eck's variable pitch propeller. Our battery was the same 4 Kokam packs as the CriCri totalizing 6 Kwh. We planned to establish some medium-term FIA world records in the RAL1E class. At the beginning of 2011, we signed an agreement with the Albert II Foundation of Monaco to act in its name and in its colors during the Aero 2011 aviation fair in Friedrischafen. Although unpaid, the prestige of such an action could not be overlooked, especially since very few world records in electric propulsion had been registered at this time, so the advertising effect would be maximum and coordinated with the issue of a stamp bearing the image of the machine by the Monegasque postal services. As the stakes were high, a maximum of precautions were to ensure a credible result at a fixed date and time. Indeed, the Prince wanted the result of the flight to be announced at the end of a meal he chaired to gather all that the world counts in the matter of green aviation. On the 13th of April 2011 at exactly 1 p.m., I took off from Runway 25 at Friedrichshafen. Watched by several hundred plus local TV, and in a gentle breeze and clear sky, I made my attempt with our untested craft. The speed I registered on a 15 km round trip was successfully completed at an average speed of 135 kph. Mission accomplished, even if eight months later the FAI would reject the file for defect of form while the flight was perfectly eligible, I then felt no feeling of injustice because I am one of those who think that the regulation is published for everyone that guarantees its fairness. On the other hand I had a formidable desire to start again.

Soullier decided to replace his Flytec with a 15 kW Agni pancake motor developed by Cedric Lynch in England and the variable pitch propeller with a fixed one. Despite recurring controller problems, on February 27, 2012, the Frenchman climbed to an altitude of 743 meters (8,000 ft.) without problems. Seven months later, on September 29, 2012, Soullier took off from Koksyjde, West Flanders, in Belgium and, timed by FAI judge Fred Van Aersen, increased his speed record to 189.87 kph (117.98 mph), duly homologated. Soullier recalls: "I succeeded despite a cloudy sky, rough atmosphere and adverse wind component on both legs—autumn is not a friendly season along the North Sea coast and this deprived me of the honour of being the first electric airplane to reg-

July 2012: Jean-Luc Soullier in the *MC30E* ultralight aircraft prior to establishing the 117.97 mph (189.87 kph) record over Koksyjde, Belgium. Note the deturbulator devices set on the wings and behind the canopy (courtesy Jean-Luc Soullier).

ister a speed of more than 200 kph. However it was the True Air Speed during the flight, but the FAI only registered Ground Speed, a pity, but then again, five years later, the performance still holds."[2]

Although dogged by technical bad luck, at the time of writing, Soullier remains determined to perfect his MC30E to further increase his record in the ULM RAL1E (FAI sporting code section 10) and Aircraft C1a/0 G6 (FAI sporting code section 2) categories.

Once a record is established, somebody else wants to break it. Towards the end of 2011, William Morrison Yates III, aka Chip Yates, very recent holder of several electric motorcycle records at 200 mph, became more than interested in electric aviation. Yates was born on February 11, 1971, in Portsmouth, Virginia, spending his early years in Pittsburgh, where he displayed an early interest in mechanics. By the time he was thirteen years old, he could disassemble and reassemble complete motorcycles. At age fourteen, Yates was sent to Culver Military Academy, a co-ed boarding school in Indiana where he received his high school education. He went on to receive a master's degree in business entrepreneurship from the University of Southern California, where he was later hired as adjunct faculty. In 1997, Yates replaced automotive designer Chip Foose at ASHA Corporation, where he invented and patented a series of hydraulic control valves for the 1999 Jeep Grand Cherokee. He also launched a start-up company called "SWIGZ®" to market his patented dual-chambered fitness bottle concept. From 2001 to 2004 Yates served as a technology-marketing executive of the Boeing Company, and then Honeywell Aerospace from 2004 to 2015.

Meantime he went in for auto racing. He competed in the SCCA Club Rally and Pro-Rally Series driving a 1989 Toyota MR2 that he built with a 1.6 liter supercharged engine. In 2001, Yates won the SCCA Southern Pacific (SOPAC) Group 5 (2-wheel drive class) Rally Championship. In January 2007, at age 36, Yates switched to motorcycles and entered a beginner's motorcycle track riding course at Auto Club Speedway near Los Angeles. He became drawn to motorcycle racing, earning enough points during the 2007–2008 amateur road-racing seasons to turn professional within nineteen months of his first track experience. In 2009, Yates competed in the AMA Pro Daytona SportBike class in televised professional races at Auto Club Speedway, Infineon Raceway, Laguna Seca, and Heartland Park, before his season ended prematurely with a broken pelvis sustained in a high-speed racing crash during AMA competition.

While unable to race due to his broken pelvis, Yates recruited two volunteer aerospace engineers, Ben Ingram and Robert Ussery, to develop an electric racing motorcycle capable of meeting his goal of equaling gasoline-powered motorcycle lap times. Yates announced plans to ride the hand-built prototype in the newly formed TTXGP and FIM e-Power electric motorcycle race series. To accomplish gasoline performance parity, Yates and his team developed and filed patents on several new electric vehicle technologies including a kinetic energy recovery system ("KERS") designed to capture braking energy from the front wheel of the motorcycle. In 2011, Yates rode his 258 hp electric superbike at over 200 mph (322 kph) to eight official world land speed records on the Bonneville Salt Flats, 4 AMA National Championship Records, the Pikes Peak International Hill Climb record, and the *Guinness Book of World Records* title of "World's Fastest Electric Motorcycle."

To prove his stated mission to show that electric vehicles do not have to be slow and boring, Yates and his team next leveraged the technology from the electric superbike including the 193 kW (~258 hp) UQM electric motor to assemble an all-electric airplane.

Purchased by Yates's venture Flight of the Century (FOTC) in April 2012 as an R&D plane for development of the company's patented midair recharging technology, the Rutan Long-EZ underwent a complete restoration and conversion from gasoline power to all electric power in just two months at FOTC headquarters.

While this was progressing, on May 22, 2012, Yates announced plans to build a 100-foot wingspan custom electric airplane that he intended to fly along Charles Lindbergh's 3,600-mile transatlantic route. Using a patent-pending midair recharging concept, the aircraft would receive battery recharges from a series of five unmanned recharging aircraft or drones en route, with the goal of matching or exceeding Lindbergh's average speed. The challenge of midair refueling is not new. In 1923, two Airco DH-4B biplanes flew in formation with a hose run down from a hand-held fuel tank on one aircraft and placed into the usual fuel filler of the other. In this way an endurance record of 37 hours was set by three DH-4Bs, a receiver and two tankers, involving nine midair refuelings.

By July 2012 all was ready for the speed attempt. Yates having received his private pilot's license on July 12, 2012, after two months of training, the unprecedented test program then moved to FOTC's Inyokern Airport facility, in Kern County, California. Yates made his first taxi test on July 14, his first runway test July 16, and conducted the first flight on July 18. Then on July 19, on the second test flight, he achieved a speed of 202.6 mph (326 km/h), beating Frenchman Hugues Duval's record by 27 mph. The flight ended in an emergency dead-stick landing following an in-flight lithium-ion battery problem, the same pack he lifted from his record-setting electric motorcycle. On-board video footage shows Yates barely making the runway at Inyokern Airport after the flight. FOTC was engaged in a cooperative relationship with the Naval Air Weapons Station China Lake, who deployed high-speed telemetry, radar and tracking cameras to capture Yates's historic flight adjacent to their restricted airspace. After the flight, officials from China Lake visited the FOTC hangar at Inyokern Airport to corroborate the flight data.

During the summer of 2013, Yates and team installed a brand-new battery pack from EnerDel. The new 450-volt pack could produce a continuous 600 amps, and most importantly, it had not been abused setting world records on a motorcycle. Weighing 525 pounds, the new battery takes up the entire back seat of the airplane, leaving just enough space for the pilot. Renaming his e-airplane Long-ESA ("Electric Speed and Altitude"), he set out to create records in other categories. Many of the records for electric airplanes had yet to be established, and Yates set the goal of being the first to set the newly-created official FAI records. On October 5, at the California Capital Airshow in Sacramento, he successfully set his first Guinness World Record, "Time to Climb to 500 Meters," with a performance of 1:02.58, measured from wheels stopped until the aircraft reaches the required altitude.

The following day, on October 6, 2013, Yates set a second Guinness World Record for "Fastest Electric Airplane" with a run in one direction of the 1 kilometer course of 220.9 mph (355.5 kph), in the other of 212.9 (342.6 kph), with an average 216.9 mph (349 kph):

> So there I was, strapped into the tight cockpit of the electric airplane once again, only this time there were 100,000 people watching me and I had fresh memories of three power failure, dead-stick emergency landings during similar world record flights from the months before. Behind me were the Canadian Thunderbird jet demonstration team waiting to follow me, and in front of me was the empty runway where I had been cleared for takeoff by the California Capital Airshow control tower. I held the brakes tightly as I ran up the 258 horsepower electric motor, checked that I had 450 volts ready to go, tightened my seat belts and let it rip. There was a lot of vibration and a big surge of acceleration

as I was catapulted down the runway. Just as I reached the area where the spectators were gathered along the fence line, I pulled back on the fighter-style joystick controller and flew the plane off the ground at an incredibly steep angle. My instruments showed I was in control and climbing at greater than 2,500 feet per minute. By 1,500 feet the Guinness record was mine. I flew some exhibition banana passes for the crowd, landed and pulled out my American flag, which waved in the wind as I taxied past the fans and back to the pits to prepare for the top speed world record.[3]

Yates then announced plans to set new top speed marks over 3 km and 15 km with another attempt at the 3,000 meter time-to-climb mark. He realized these plans only days later: "Time to Climb to a Height of 3,000 Meters" in a record-setting time of 5 minutes 32 seconds; "Speed Over a 3 km Course" with a 4-pass average speed of 201 mph (324.04 km/hr); "Speed over a 15 km Course" with a 2-pass average speed of 140 mph (225.88 km/hr). Yates's time-to-climb world record performance of 5 minutes 32 seconds demonstrated a sustained rate of climb of 2,000 feet per minute from ground level to 9,843 feet above the ground (3,000 meters). This is a greater performance than most gasoline-powered airplanes and was selected by the National Aeronautic Association (NAA) as the "Most Memorable Aviation Record of 2013."

Chip Yates's exploits in pushing electric vehicle technology earned him recognition as a "Pioneer of Aviation" from the State of California in the form of Assembly Resolution #1740, presented to Yates in Sacramento during Senate and Assembly sessions on August 30, 2012.

Following his electric motorcycle and electric airplane world record campaign, Yates expanded his role in aviation through a number of ventures including winning a U.S. Marine Corps contract to design and build a 1,000-lb. (450-kg) payload disposable drone to resupply troops in harm's way, leading a hybrid-electric aircraft development program, conducting battery pack design for automotive manufacturers, and marketing an aerospace structural titanium 3D printing technology. In November 2017, the Marine Corps Warfighting Laboratory (MCWL) selected the Yates Electrospace Corporation (YEC) Silent

Chip Yates III adapted the technology of his record-setting electric motorcycle to modify this Rutan Long-EZ, in which, on October 6, 2013, watched by 100,000 spectators at the California Capital Airshow in Sacramento, he increased his own world electric air speed record average from 202.6 mph to 216.9 mph (photograph: Tara Larivee/Yates Electrospace).

Chip Yates, 2013 (photograph: Tara Larivee/Yates Electrospace).

Arrow product line of autonomous cargo aircraft for a 12-month flight test program. A fleet of 10 aircraft from 500 to 1,000-pound gross weights will be evaluated.

Meanwhile others went air record-breaking. On June 21, 2014, Gary Davis took off from Greenville Downtown Airport in Greenville, South Carolina, in his ultralight trike, a custom combination of a North Wing Stratus wing matched to an electric-powered trike frame designed by Gary Randall Fishman, and reached a record altitude of 4,660 feet (1,420 meters).

Raphaël Domjan of Yverdon-les-Bains, Switzerland, well known in the nautical world for his solar-powered boat, the *PlanetSolar*, which completed the first circumnavigation of the globe in 2012 without using a drop of fossil fuel, is currently working on a venture he calls *SolarStratos*. This is a twin-seat solar aircraft which Domjan will pilot to a record-breaking altitude of 80,000 feet (25,000 m), requiring 4 hours of constant climb. Designed by Calin Gologan, *SolarStratos* would be a lightweight adaptation of a PC Aero single-seat electrically powered aircraft. Measuring 7.7 meters (26 ft.) long, its 25-meter (82-ft) wingspan and tailplane are covered with 20 m^2 of solar cells (24 percent efficiency) linked to eighty 1 kg Li-Ion (20 kWh) batteries to power a 13.5 kW electric engine. For a fee, a passenger will be able to share the same breathtaking experience. On December 7, the *SolarStratos* (registration HB-SXA) was officially unveiled at Payerne airfield; 300 guests attended including personalities, ambassadors, representatives of local and national authorities, and sponsors.

On May 5, 2017, coordinated by SolarXplorers and financed by numerous sponsors, piloted by Damien Hischier, the 25-m-span Elektra-2 solar electric ultralight (420 kg) aircraft successfully completed its first 7-minute flight at an altitude of 300 meters over Payerne. *SolarStratos* was equipped with a space suit made by Zvezda of Moscow, enabling Domjan to experience a simulated flight to 13,000 meters at a temperature of -55° Celsius; the aircraft will be towed around on the ground by a bio-fueled vehicle. The first pilotless stratospheric flights at an altitude of up to 20 km are planned in 2018, which will culminate in new record altitudes for solar-electric aircraft. It will have a pilot on board and up to 50 kg of payload; no other solar-electric aircraft in the world with a comparable payload has reached this altitude. The *SolarStratos* is equipped with an automatic flight control system developed at the German Aerospace Centre (DLR) Institute of Robotics and Mechatronics in Oberpfaffenhofen.

On September 3, 2017, on its eleventh flight, the 33-ft (10-m) Airbus *Perlan II* carbon-fiber glider reached a new world record altitude of 15,902 meters (9.88 miles). The two pilots, Australian Morgan Sanderock and American Jim Payne, were towed up into the sky above El Calafate, southern Argentina, in the mid-wing, two-seats-in-tandem, pressurized, experimental research glider as designed by Greg Cole and built by Windward Performance at Bend Municipal Airport, Oregon. With its 83.83-ft (25.55-m)-span wing, the *Perlan II* was able to use stratospheric mountain orographic waves to fly up to the stratosphere at over 600 kph. The onboard Li-Po battery enabled the pilots to keep warm using electric socks and vests. Ultimately the *Perlan II* will attempt to reach 90,000 feet, a world record for any wing-supported flight, with or without an engine.

The English have a fine tradition of breaking airspeed records. The e-challenge has been taken up by Rhodesian-born composites guru and flying enthusiast Roger Targett, originally of Severn Valley Sailplanes repair and maintenance center in Gloucestershire, England. Originally Targett called his plane the TEACO Bat, but partly because Internet inquiries often sent inquirers to TESCO, a UK grocery retail company, he changed the name to Electroflight. With Targett's wealth and experience coming from the composites industry, inevitably Electroflight's 200 kW *Electric Lightning P1 E*, once built, will combine advances in carbon composite materials and construction methods with emerging electric motor, control system and energy storage technologies. Its twin propeller contra-rotating propulsion offers unique maneuvering capabilities. Electroflight linked up with Williams Advanced Engineering, with their experience in Formula E racing automobiles in the goal to establish a 300 mph+ record. Two- and four-seater aircraft are envisaged and will be sold through the name Electropulsion based at Nympsfield Aerodrome, near Stonehouse, Gloucestershire, England. By early 2017 the propulsion system design was finalized and the structural test panels had been made and tested. In October 2017, when Cranfield University in Bedfordshire opened its Aerospace Integration Research Centre (AIRC), co-funded by Airbus, Rolls-Royce and the Higher Education Funding Council for England (HEFCE), on the program was the completion of Electroflight's challenger.

Drones have also established records. On February 23, 2016, Dirk Brunner, technical consultant and engineer, set a Guinness World Record when his custom-built quadcopter drone took just 1.3 seconds to accelerate to a speed of 100 kph (60 mph), reaching a maximum climb speed of 189 kph (119 mph), and took a total of 3.871 seconds to reach a height of 100 m (328 feet). Brunner's ascent was accurately measured by an on-board logging barometric altimeter. The record took place above Pasing Model Airfield in Munich, Bayern, Germany.

Where the maximum speed of a racing drone is concerned, on July 27, 2016, George Matus, Jr., of Salt Lake City, manufacturer of the Teal drone, used a production unit to fly at top speed of 85 mph. One year later, on July 13, 2017, a 1.75-lb (0.8-kg) quadcopter, *RacerX*, hand-built by a Drone Racing League team led by Rian Gury, became the fastest drone in the world when it hit a top speed of 179.78 mph (288.6 kph), although it was timed at an official average speed of 165.2 mph (265.87 kph) over a 100-meter distance at Cunningham Park golf course in Queens, upstate New York, observed by the *Guinness Book of World Records*. Its rotors were spinning at up to 46,000 rpm, creating a high-pitched whine. The Drone Racing League reported that prototypes actually burst into flames when hitting its highest point of acceleration due to the amount of power used.

On Thursday, March 23, 2017, taking off from the Dinslaken Schwarze Heide airfield, Walter Extra piloted his 330LE aircraft to a top speed of around 337.50 kph (209.7 mph) over a distance of three kilometers—13.48 kph (8.38 mph) faster than the previous record, set by Chip Yates in 2013. The World Air Sports Federation (FAI) officially recognized the record flight in the category "Electric airplanes with a take-off weight less than 1,000 kilograms." In a slightly modified configuration with an overall weight exceeding one metric ton, test pilot Walter Kampsmann then flew 330LE to a speed of 342.86 kph (213.04 mph).

In October 2017, the author received the following communication from Chip Yates:

> The electric Long-EZ is preparing for test flights in November/December and will be attempting 10 new world records, capturing back a number of records that Siemens took from us recently! I am building also a new electric plane capable of 350 mph and intend to be the first person to break 200 [done], 300, 400, 500, 600 and the speed of sound. I purchased and own 1 megawatt of electric motors, which is 4 × 250kW liquid-cooled custom motors for my record breaking. I also own a 700 volt battery pack for the planes that is over 65 kWh. I have built what we think is the world's most powerful electric test stand, running at ½ megawatt turning 2 contra rotating 3 bladed propellers at my company Yates Electrospace Corporation. We should start test flying in December up in the Mojave Desert.

APPENDIX D

Timeline of Notable Electric Aircraft Flights

1782	The Montgolfier brothers balloons, using "electric smoke."
1881	Tissandier brothers' tethered electric "aerostat" at the International Electrical Exhibition in Paris.
1883	Tissandier brothers in their full-scale electric aerostat above Paris: 20 minutes.
1884	Renard and Krebs in the electric dirigible *La France*: 23 minutes.
1887	Gustave Trouvé's 3-cubic-centimeter tethered electric model helicopter.
1896	The Smith-Maxim electric *Air Ship*: across the United States.
1909	Joseph L. Cannon's tethered electric model airplane in North London: 8 minutes.
1917	PKZ1 hybrid-electric tethered rotorcraft: Fischamend, Austria.
1957	Harold Taplin's electric-powered model aircraft *Radio Queen*: Chalgrove Aerodrome in Oxfordshire: 10 minutes.
1959	Fred Militky's Micromax-engined model electric aircraft *FM 248*: 23 minutes.
1971	Boucher brothers of Astro Flight's first American practical electric-powered model airplane: 29 minutes.
1973	Heino Brditschka's eight test flights in the *MBE-E1*: Wels airport: 15 minutes.
1973	*SkyEye* electric UAV: El Mirage Lake.
1975	Astro Flight's *Sunrise I* solar-powered model airplane: 28 flights over 5,000 ft.
1979	Larry Mauro flight in *Mauro Solar Riser*.
1980	Janice Brown in *Solar Challenger*: 92 minutes.
1981	Steve Ptacek in *Solar Challenger* across the English Channel: 5 hours and 23 minutes.
1983	Günter Rochelt in *Solair I*: 5 hours 41 minutes.
1990	Eric Raymond flies *Sunseeker* across the United States, his two flights covering 496 miles.
1995	AeroVironment's *Pathfinder* solar UAV: 50,000 feet.
1995	Graham Dorrington in *Dirigible-4* over the forests of Malaysia.
1996	Rudolf Voit-Nitschmann in the solar-powered *Icaré II*: 350 km.
1997	Stefan Gehrmann in his NiCad *AE1* motor glider.
1998	AeroVironment's *Pathfinder Plus* UAV: 80,201 feet.
2001	*Helios* UAV: 96,863 feet.
2001	AeroVironment's *Raven* SUAV.
2003	Graham Dorrington his lithium *White Diamond* dirigible.
2003	Axel Lange in the *Antares 20E*, the first single-seater electric production sailplane.

2005	Stefan Gehrmann in his Li-Po *AE1*: 40 minutes.
2005	Alan Cocconi's *SoLong* UAV: 48 hours.
2005	*Zephyr 5* UAV: 6 hours.
2007	Randall B. Fishman in his *ElectraFlyer* trike at Jasper, Tennessee: 60 minutes.
2007	Ivo Boscarol in the Pipistrel *Taurus Electro* in Slovenia: 20 minutes.
2007	Christian Vandamme in the Electravia *Electra* ULM in France: 30 minutes.
2008	Derek Piggot in the Displaced Electric Propulsion *EA9*.
2008	*Zephyr 6* UAV: 87 hours.
2008	Fishman in his *ElectraFlyer Model C* motor glider.
2008	Cecilio Barberan Alonso in the Boeing Fuel Cell Demonstrator, Ocana, Madrid.
2009	Fredrik Müchler, the SAAB *MERA 01* in Sweden.
2009	Eric Raymond in *Sunseeker II* across the Alps: 400 km.
2009	Axel Lange in fuel cell *Antares DLR-H2*.
2009	Gérard Thevenot in hydrogen-powered ultralight "trike" across the English Channel.
2009	Maurizio Cheli in SkySpark: 252 kph (155 mph).
2010	Turin University's fuel cell Jihlaven Rapid 200: Reggio Emilio Airport.
2010	*Zephyr 7*: 336 hours continuous.
2011	*Taurus Electro G4*: 400 miles.
2011	*E-Genius*: 200 miles at speeds over 100 mph.
2012	Jean-Luc Soullier in the *MC30E* :189.87 kph (117.98 mph).
2012	Canada's *Green1* above British Columbia.
2013	Czech's *EvektorSportStar EPOS*: Kunovice airport.
2013	Chip Yates III: 216.9 mph (349 kph): Sacramento.
2013	Siemens-engined DA36 E-Star 2 hybrid motor-glider.
2013	Pierre Chabert and Gerard Feldzer in *Iris Challenger 2* blimp across the English Channel.
2015	Eric and Irena Raymond in their *Sunseeker Duo* across the Alps.
2015	Didier Esteyne in Airbus *E-Fan* (38 minutes); Hugues Duval in Electravia *Cri-Cri* (17 minutes): across the English Channel.
2015	Brazil's *SORA-e* above São Paulo.
2015	*Greased Lightning* hybrid multi-prop.
2015	Bert Rutan's *SkiGull*.
2015–16	Bertrand Piccard and André Borschberg in *Solar Impulse*: around the world: 552 hours total and 43,041 km (26,744 mi).
2016	*Enduro1* quadcopter across the English Channel: 72 minutes.
2016	George Matus Jr.'s Teal drone: 85 mph.
2016	Siemens-engined Extra 330LE: Schwarze Heide Airport.
2016	*Ehang184* PAV: China.
2016	Alexander Zosel in Volocopter VC-200 *White Lady*: Bruchsal.
2017	China's fuel cell RX1E.
2017	*Lilium* PAV: Bavaria.
2017	Hamilton H55 Silent *Twister*: Raron.
2017	Extra *330LE*: 342.86 kph (213.04 mph).
2017	DRL *RacerX* quadcopter: 179.78 mph (288.6 kph).
2017	Grenchen, three e-airplane flypast.
2017/18	Vahana

Chapter Notes

Chapter One

1. Numbers 10:33.
2. Ezekiel 1:4.
3. Deborah M. Coulter-Harris, *The Queen of Sheba: Legend, Literature and Lore* (Jefferson, NC: McFarland, 2013); *Jewish Encyclopaedia*, "Solomon's Carpet."
4. Richard Burton, *The Thousand Nights and a Night*, vol. 13, 1885.
5. E. Feifel, *Pao Phu Tzu* (Nei Phien) Ms 1944.
6. Leonardo da Vinci, *Codex Atlanticus*, fol. 161r-a.
7. Leonardo da Vinci, *Arundel Codex* 279v.
8. The mathematical and philosophical works of the Right Reverend John Wilkins late lord bishop of Chester: to which is prefix'd the author's life, and an account of his works; in two volumes (1708).
9. Kevin Desmond, *Innovators in Battery Technology: Profiles of 93 Influential Electrochemists* (Jefferson, NC: McFarland, 2016), Musschenbroek entry.
10. Desmond, *Innovators*, Franklin entry.
11. Desmond, *Innovators*, Volta entry.
12. Letter from Benjamin Franklin to his friend John Ingenhousz, a Dutch biologist and chemist, dated January 16, 1784 (Franklin Papers).
13. Marion Fulgence, *Les Ballons et Les Voyages Aériens* (Paris: Librairie L. Hachette et Cie. 1869), 284.
14. A Paines engine is a type of internal combustion engine using gunpowder as its fuel.
15. Thomas Edison Papers Project, Rutgers University.
16. This criticism was later endorsed by Henri de Graffigny in his book *La Navigation Aerienne*, published in 1888; the report was reviewed by Edwin T. Teale, "Edison's Early Dream Ships," *Popular Aviation*, June 1933.
17. *The Papers of Thomas A. Edison*, vol. 5 (Baltimore: Johns Hopkins University Press, 2004), pp. 777–778.
18. Kevin Desmond, *Gustave Trouvé: French Electrical Genius* (Jefferson, NC: McFarland, 2015), 58–60.
19. Chris Morgan, *The Shape of Futures Past: The Story of Prediction* (Exeter, UK: Webb & Bower, 1980), 159.
20. Desmond, *Trouvé*, 81.
21. "Les Applications de l'électricité à la navigation aérienne," *La Lumière électrique*, 1884.
22. Desmond, *Innovators*, Renard entry.
23. The Renard series in geometrical progression were later adopted in 1952 in the ISO 3 norm.
24. "Fait Divers," *La Lumière Electrique*, Saturday, March 20, 1886.
25. *La Lumière Electrique*, Saturday, July 24, 1886.
26. Jules Verne, "Robur Le Conquerant," *Journal des débats politiques et littéraires*, June 29 to August 18, 1886, then in book form (Paris: Hetzel, 1886).
27. Verne, *Robur*, Chapter 7, "On Board the Albatross."
28. David Seed, "The Land of the Future: British Accounts of the USA at the Turn of the Nineteenth Century," *European Journal of American Studies* 11, no. 2 (Summer 2016).
29. Illinois historian Howard Scamehorn reported that both Octave Chanute and Albert Francis Zahm "publicly denounced and mathematically proved the fallacy of the vacuum principle"; however, the author does not give his source.
30. Ed Zotti, "19th century Chicago 'aeroplane,'" *The Straight Dope*, July 11, 2010.
31. Clara MacCarald and Kathi McCarthy, "Dr. Martin Braun's Air Ship," *Thousand Islands life.com*, March 13, 2016.
32. Desmond, *Trouvé*, 111.
33. Desmond, *Trouvé*, 132.
34. Hiram Maxim, "Aerial Navigation, the Power Required," *The Century* 42, no. 6 (October 1891): 829–837.
35. Albert Robida, *Le Vingtième siècle; La Vie Electrique* (Paris: La Librairie Illustrée, 1893).
36. George Chetwynd Griffith, *The Angel of the Revolution: A Tale of the Coming Terror* (1893).
37. "The Electric Motor in Aeronautics," *The Electrical Review*, no. 852 (March 23, 1894).
38. "Electricity in Aeronautics," *The Electrical Review*, no. 880 (October 5, 1894).

Chapter Two

1. "To Traverse the Air," *San Francisco Call* 79, no. 91 (February 29, 1896).
2. *Daily Sentinel* 5, no. 40 (January 12, 1898).
3. Garrett P. Serviss, "Edison's Conquest of Mars," *New York Evening Journal*, January 12–February 10, 1898, serialized in 26 parts, with illustrations by P. Gray.
4. U.S. Patent no. 834658 A.
5. International Aeronautical Congress, St. Louis, MO, 1904.
6. *Electrical Review*, April 24, 1903.
7. The Blackpool ride was still in service 100 years later.
8. James Hamilton and Phil Holmes, *Chronic Inventor: Life and Work of Hiram Stevens Maxim, 1840–1916*

(Bexley, UK: Directorate of Education, Libraries & Museums, 1991).
9. "C.E. Taylor Laid to Rest With Aviation Pioneers," *Los Angeles Times*, February 3, 1956.
10. "French Honours for Aviation," *Flight Magazine*, July 31, 1909.
11. R.H. Goddard. "The green notebooks," vol. 1. The Dr. Robert H. Goddard Collection at Clark University Archives, Clark University, Worcester, MA.
12. Robert H. Goddard, "An autobiography," *Astronautics* 4:24–27, 106–109, 1959.
13. Donald W Horner, *The Encyclopedia of Science Fiction*, April 26, 2016.
14. *Those Magnificent Men in their Flying Machines; Or, How I Flew from London to Paris in 25 Hours 11 Minutes* is a 1965 British comedy film from 20th Century Fox.
15. "Unbelievable Flying Objects," Wright-Brothers.org
16. Henri Loyrette, *Henri Eiffel, un Ingénieur et Son Oeuvre* (New York: Rizzoli, 1985).
17. Hermann Scheit and Bobeth, "Test of an air-propeller," *VDI Zeitschrift* 55 (November 4, 1911): 1840–1846.
18. "An Electrically Driven Model Aeroplane," *The Model Engineer and Electrician* 21 (October 21, 1909): 391.
19. "The Wireless War in the Air," Marconi Heritage Group.
20. "Tesla promises Big Things. Planeless, Screwless Air Ship," *New York Sun*, September 12, 1911.
21. "First Night Flight at Nashville in 1910, Recalled," *Tennessean Magazine*, August 8, 1937.
22. Charles Delfus and Henri Boucher, "Histoire de l'Aeronautique," *Illustration*, 1942.
23. "Flying at Hendon," *Flight*, June 21, 1913.
24. C. Cole and E.F. Cheesman, *The Air Defense of Britain 1914–1918* (New York: Putnam, 1984).
25. G. Wallace, *Claude Grahame-White: A Biography* (London: Putnam, 1960), 183.
26. "Flying at Hendon," *Flight*, November 15, 1913.
27. Igor Sikorsky, *Story of the Winged-S: An Autobiography by Igor Sikorsky* (New York: Dodd, Mead, 1938).

Chapter Three

1. "Edison's New Flying Machine," *The Straits Times*, May 20, 1892, 2.
2. Cole and Cheesman, *The Air Defense of Britain*.
3. William Edward Fischer Jr., *The Development of Military Night Aviation to 1919* (Alabama Air University Press, Maxwell Air Force Base, 1998).
4. "Starýplánunikátníholetadlabylnalezenna Novojičínsku," *Českátelevize*, June 3, 2011; and Die k.u.k. Aeronautische Anstalt Fischamend, 2011.
5. Louis Rodier, "Le Plus Puissant Phare d'Aviation," *La Science et La Vie*, no. 61 (March 1922).
6. Subbaraya Shastry, G.R. Josyer, *Vymaanika Shaastra–Aeronautics by Maharshi Bharadwaaja* (Mysore: Coronation Press,1973).
7. H. Oberth, *Wegezur Raumschiffahrt* (Munich and Berlin: Druck und Verlag von R. Oldenbourg, 1929).
8. U.S. Patent 1,865,828 (filed 1929).
9. Communication from David Wilson, Farnborough Air Sciences Trust.
10. Advertisement in *Flight*, June 27, 1929.
11. Arthur M. Young website; Brad McNally, "Rotorcraft Pioneers, Part IV: Arthur M Young—The Genius Behind Bell Helicopters," *Rotorcraft PR*, June 15, 2010.
12. A. Piccirillo, "Electric Aircraft Pioneer: The Focke-Wulf Fw 190," *SAE Technical Paper 965631*, 1996, doi:10.4271/965631.
13. G. Gingell, "Supermarine Spitfire—40 years on," *Royal Aeronautical Society*, 1976.
14. Rob Arndt, "Viktor Schauberger Repulsin A & B (1940–1945)," discaircraft.greyfalcon.us.

Chapter Four

1. "First All-Electric Takes The Air," *Aeromodeller*, September 1957.
2. "Ein Leben für den Modellflug," *Kirchheimer Zeitung*, September 24, 1976.
3. *Aeromodeller Annual 1960-61* (Watford, UK: Model Aeronautical Pr., 1960).
4. *Aeromodeller*, October 1962.
5. Communication from Heidno Brditschka, June 10, 2016.
6. "Sportflugzeuge müssen keinen Lärm Machen," *Kirchheimer Zeitung*, November 10, 1973.
7. Communication from Bob Boucher, December 2016.
8. Idem., Boucher.

Chapter Five

1. Desmond, *Innovators*, Chapin entry.
2. Communication from Bob Boucher, December 2016.
3. Roland Boucher, "Project Sunrise Los Angeles California 1974 Flight of the World's First Solar Powered Aircraft."
4. "A Revolutionary Idea for a Solar-Powered Airplane: Larry Mauro tests his revolutionary idea for a solar-powered airplane," *Mother Earth News*, July/August 1979.
5. Communication from Eric Raymond, November 2016.
6. Communication from Janice Sullivan (née Brown), December 6, 2016.
7. Communication from Steve Ptacek, November 21, 2016.
8. Ptacek.
9. At time of writing, Mrs. Janice Sullivan (née Brown) is a flight instructor at Bakersfield Flying Club, while Steve Ptacek is nearing retirement after flying Boeing 777s for United Airlines. Aged 66, he would like to fly an electric sailplane.
10. "Solair I: Das erste deutsche Solarflugzeug fliegt. Günter Rocheltverwirklichttechnologischen Traum," *aerokurier* (April 1981): 463–465.
11. Communication from Eric Raymond, November 2016.
12. "China launches solar aircraft Wed," *AFP*, August 31, 1994.
13. "Solar-powered UAV flies two days straight," *Machine Design*, August 18, 2005.
14. Communication from Eric Raymond, March 2016.
15. Denise Civiletti, "Luminati will have till at least Nov. 30 to culminate April letter of intent for EPCAL; Catsimatidis still not committed to deal," *Riverhead Local*, November 4, 2017.

Chapter Six

1. Desmond, *Innovators*.
2. Communication from John Emms of Puffin Models, July 2016.
3. Communication from Stefan Gehrmann, November 2016.
4. Communication from Randall B. Fishman, June 2016.
5. Communication from Jonas Larssen, principal engineer, electric power and avionics, Saab Aeronautics, September 25, 2016.
6. Dean Sigler, "Power Up Front," January 11, 2009.
7. Nicole Bertin, "L'avion 100% électrique est né à Royan! Perspective de commercialization en 2017," *Blog*, October 11, 2014.
8. "The first electric-powered amphibia in the world," *Seaplane International*, August 12, 2011.
9. Stephen Trimble, "Rutan enters seaplane market with SkiGull unveiling," *Flightglobal*, August 11, 2015; Pia Bergqvist, "Project Notebook: Retirement Dream Burt Rutan's SkiGull takes shape. Will it be his last airplane?" *Flying*, March 21, 2016.
10. Viknesh Vijayenthiran "Dutch sports car brand Spyker developing electric SUV," *Green Car Reports*, April 4, 2016.
11. "British company launches electric Furio project," aeroexpo.co.uk/news, November 13, 2015.
12. Jon Excell, "UK team eyes solar-powered airship potential," *The Engineer*, March 21, 2015.

Chapter Seven

1. Kevin Desmond, *Electric Boats and Ships: A History* (Jefferson, NC: McFarland, 2017).
2. Communication from Lorne Campbell, September 4, 2016.
3. IEEE History Center, The Institute of Electrical and Electronics Engineers, Inc.
4. Tom Koehler, "Boeing makes history with flights of Fuel Cell Demonstrator Airplane," www.boeing.com, May 1, 2008.
5. "Emission-free, electric flight—DLR presents the HY4 research platform at the 2016 Hannover Trade Fair," April 26, 2016.
6. Hydrogeit, "Fuel Cell Passenger Aircraft for Medium-Distance Flights," *H-2 International*, March 1, 2017.
7. "China becomes 3rd country to test hydrogen-powered plane," *RT Question More*, January 9, 2017.
8. Shani Elitzur et al., "On-board hydrogen production for auxiliary power in passenger aircraft," *International Journal of Hydrogen Energy* (2017).
9. Joshua Barrie, "The UK Tests The First Electric Plane That Can Charge Its Batteries During Flight," businessinsider.com, January 14, 2015.
10. Stephen Trimble, "GE reveals major achievements in hybrid electric propulsion," FlightGlobal.com, August 25, 2017.

16. "University of Maryland Achieves First Flight of a Solar-Powered, Piloted Helicopter," *UMD Right Now*, September 9, 2016.
17. "Researchers have shown ultrathin, air-stable perovskite solar cells for powering unmanned aircraft," *SPIE* 30, August 2016.

Chapter Eight

1. Dave Calderwood, "Pipistrel to make electric aircraft in China," *Flyer*, November 1, 2016.
2. Communication from Taja Boscarol, September 10, 2017.
3. Thomas A. Horne, "Siemens denies e-motors to Pipistrel," Aopa.org, July 7, 2015.
4. Didier Esteyne, "The Cross Channel Story," communication to the author, November 28, 2016.
5. "Russia has created an all-electric aircraft," popmech.ru, April 1, 2014.
6. Dean Sigler, "China's First Certified Electric Airplane Ready for Mass Production," *Sustainable Skies*, March 15, 2016.
7. "Chinese-Hungarian joint venture to manufacture electric-powered aircraft," *Budapest Beacon*, September 13, 2016.
8. Dean Sigler, "Electric Powerplants," *Sustainable Aviation*, September 26, 2016.
9. AE 590: Dr. Mark Moore, NASA Langley Research Center, "The Coming Era of Distributed Electric Propulsion—and What It Means," Department of Aerospace Engineering, University of Illinois at Urbana–Champaign, October 26, 2015.
10. "Flying Whales and Skeleton Technologies partner to develop next generation of large capacity airships," *Graphen-info*, July 4, 2016.
11. Dave Calderwood, "Airbus to show hybrid E-Fan Plus at Oshkosh," *Flyer*, July 22, 2016.
12. "Elektromotorstelltzweifachen Geschwindigkeitsrekord auf," *Siemens*, München, April 4, 2017.
13. Woodrow Bellamy III, "Airbus CTO Sees Electric Aircraft Drastically Changing Commercial Aviation," aviationtoday.com, June 6, 2017.
14. The e-aircraft which flew at Grenchen: Magnus eFusion, Aeros ANT-E, Alisport Silent 2 Electro, IFB eGenius, EGO Trike, RühleATOS Silent Glider, Antares 20 E, ATOS VR-190, Archeopteryx Electro, Icaré II xxl, and the D-14 Phoenix.
15. Logging up 126 hours, Cornelia was the third woman electric aviatrix since Janice Brown flight-tested *Solar Challenger* in 1980 and Irena Raymond logged up 110 hours in *Sunseeker II*.
16. Communication from René Meier, September 18, 2017.
17. "Air Transat to test taxiing system that could cut emissions," *Canadian Press*, January 18, 2017.
18. Alan Boyle, "Zunum Aero emerges from stealth mode with big plans for hybrid electric planes," *Geek-Wire*, April 5, 2017.
19. Mark Molloy, "Plans for London-Paris electric flight in 'next decade' unveiled," *The Telegraph Technology*, March 22, 2017.
20. "Schwarzenegger debuts 'dream' electric Hummer," *Wheels*, September 21, 2017.

Chapter Nine

1. Jean-François Prevéraud, "Le drone a 100 ans," *Vieilletechnologique*, March 3, 2015.
2. David Boddington, *Radio-Controlled Model Aircraft* (Crowood Press, 2004).
3. Kevin Desmond, "I, Nikola Tesla," *Radio Control Boat Modeller*, May/June 1987.
4. B.F. Meissner, *Radiodynamics: The Wireless Control of Torpedoes and Other Mechanisms* (New York: Van

Nostrand, 1916). The FL-boat (Fernlenkboote, literally "remote controlled boat" was a weapon employed by the Imperial German Navy during World War I; it was, however, powered by gasoline engines.

5. Edmond Petit, *New aviation history* (Albin Michel, 1997), and Louis Bonte, *L'histoire des essaisenvol (1914–1940)* (Paris: Editions Larivière, 1975).

6. Steve Zaloga (Teal Group), "A Word With a History," *Defense News*, 2013.

7. Leonardo da Vinci, *Codex Atlanticus B*, fol 83 verso.

8. Kevin Desmond, *Gustave Trouvé, French Electrical Genius* (Jefferson, NC: McFarland, 2015).

9. Mike Hirschberg, "Electric VTOL Wheel of Fortune," *Vertiflite*, March/April 2017.

Chapter Ten

1. Communication from Bob Boucher, December 21, 2016.
2. Communication from Ray Morgan, 2016.
3. Sebastian Anthony, "NASA's electric vertical-takeoff airplane takes first flight, aims to eventually replace the helicopter," *Extreme Tech*, August 20, 2014.
4. Arjan Khapal, "Amazon wants to use lampposts, churches," CNBC, July 2016.
5. "Hackers to start hijacking drones used for deliveries and policing," thejournal.ie, December 3, 2016.
6. Communications from Novadem during July 2016.
7. "Students build electric-powered personal flying machine," *Science Daily*, December 2, 2015.
8. Patent 20160031555.
9. "Lady Gaga pilots first flying dress," *dezeen magazine*, November 11, 2013.
10. Thomas J. Mueller, "On the Birth of Micro Air Vehicles," *International Journal of Micro Air Vehicles* 1, no. 1 (2009).
11. Jennifer Chu, "Miniaturizing the brain of a drone: Method for designing efficient computer chips may get miniature smart drones off the ground," *MIT News*, July 11, 2017.
12. Fatima Bhojani, "For Good or Bad, Intelligent, Swarming Nanobots Are the Next Frontier of Drones," *Motherboard*, May 21, 2014.
13. Andrew Liszewski, "Borrowing a Clever Trick From Birds, This Smart Glider Could One Day Fly Forever Without a Motor," *Gizmodo*, http://gizmodo.com/borrowing-a-clever-trick-from-birds-this-smart-glider-1797896712, August 11, 2017.
14. Dina Spector, "Tiny Flying Robots Are Being Built to Pollinate Crops Instead of Real Bees," *Business Insider*, July 7, 2014; Alexia Erickson, "Robotic Bees are now being built to pollinate crops instead of real bees," *NextStory*, October 5, 2016.
15. Emily Matchar, "Turning Dragonflies Into Drones," The Drone News.com, February 11, 2017.

Chapter Eleven

1. Hannah Kuchler, "Facebook hails first test flight of Aquila internet drone," *Technology*, July 21, 2016.
2. Jordan Mintzer, *Hollywood Reporter*, February 14, 2016.
3. Rachel Swatman, "Intel launches 500 drones into sky and breaks world record in spectacular style," *Guinness World Records*, November 4, 2016.
4. FlightRadar24.
5. Oliver Starr, "Fleye could be your personal autonomous robot drone," *Gizmag*, December 10, 2015.
6. Kerry Reals, "Aircraft engineers turn to biomimicry for greener designs," *Flight Global*, August 9, 2010.
7. Mary-Ann Russon, "Good news for *Star Wars* fans—Flike personal flight vehicle passes first manned test flights," *International Business Times*, June 1, 2015.
8. Dean Sigler, "An Electric Autogyro," *Sustainable Aviation*, July 24, 2015.
9. "Airbus working on flying car," *Motoring*, August 23, 2016.

Chapter Twelve

1. "L'Aéroplane à Moteur de M. Vuia," *L'Aérophile*, September 1906.
2. Loz Blain, "Safety last: Russian hoverbike is equally amazing and horrifying," *New Atlas*, February 21, 2017.

Chapter Thirteen

1. "Boeing Files Patent to Generate Electricity from Airport Noise," *What A Future!*, September 18, 2015.
2. Fabrizio Poli, "The Electric Plane," linkedin.com, November 8, 2014.
3. Jon Hemmerdinger, "Three 'miracles' required for mainstream electric-powered aircraft, says P&W," *Flightglobal*, May 23, 2014.
4. Tom Neuman, "How I Designed a Practical Electric Plane for NASA," *IEEE Spectrum*, May 24, 2016. This article appeared in the June 2016 print issue as "Fly the Electric Skies."
5. Jim Banke, "Students Get Charged Up Designing Future Electric Aircraft," NASA Aeronautics Research Mission Directorate, September 20, 2016.
6. Philippe J. Masson, Department of Mechanical Engineering, University of Houston, "Development of a 3D Sizing Model for All-Superconducting Machines for Turbo-Electric Aircraft Propulsion," *IEEE Transactions on Applied Superconductivity* 23, no. 3 (January 14, 2013).
7. Communication from Daniel Cristóbal, Executive Partnern AXTER AEROSPACE—March 2017.
8. "'Game-changing' electric motors could power aircraft," *Drives & Controls*, March 14, 2017.
9. Dyllan Furness, "Boeing's solar-powered airplane looks ridiculous, but it might just be crazy enough to work," *Digital Trends*, June 15, 2016.
10. "Digital Morphing Wing: Active Wing Shaping Concept Using Composite Lattice-Based Cellular Structures," *Soft Robotics* 4, no. 1 (2017).
11. Ryan O'Hare, "Bat drone takes to the skies! Flying machine uses shape-changing wings inspired by the mammals," *MailOnline*, February 18, 2016.
12. Richard Gray, *MailOnline*, March 6, 2015.
13. David L. Chandler | *MIT News Office* January 6, 2017.
14. Greg Nibler, "Amazon Patents a Flying Warehouse for Drone Delivery," *DT Daily*, December 29, 2016.
15. "Retailer Gets Patent for Mega-Drone," *LiveScience* January 6, 2017.
16. Communication from Lee Cronin, June 15, 2017; Anthony Cutherbertson, "Drones could be 'grown' with chemputer machine," *Newsweek*, May 7, 2016.

17. Alan Tovey, "Rolls-Royce: the future of flight is electric," *The Telegraph*, December 31, 2016.
18. *Deccan Herald*, December 16, 2003; *Times of India*, October 18, 2004.
19. "'Telepathic' microchip could help paraplegics control computers," *The Telegraph*, September 3, 2009.
20. Bin He et al., "Quadcopter control in three-dimensional space using a noninvasive motor imagery-based brain–computer interface," *Journal of Neural Engineering* 10, no. 4 (June 2013).
21. "Using wireless interface, operators control multiple drones by thinking of various tasks," *Science Daily*, July 13, 2016.
22. Jack Stewart, "I Used Only My Mind to Fly a Plane Around Seattle," *Wired*, November 2016.

Appendix B

1. Communication from André Borscberg, December 7, 2016.

2. Borschberg.
3. Idem.
4. Idem.
5. Borschberg.
6. Idem.
7. From information kindly supplied to the author by Julie Conti, Senior Media Relations Officer, EPFL, Lausanne, Switzerland.
8. Miquel Ros "Greener skies: Frenchman prepares for history-making zero-carbon Atlantic flight," CNN, May 13, 2016.

Appendix C

1. Communication from Ray Morgan, December 2016.
2. Communication from Jean-Luc Soullier, August 2, 2017.
3. Communication from Chip Yates, December 2016.

Bibliography

Books

Boucher, Robert J. *A History of Solar Flight.* Astro Flight, Inc. AIAA, 1981.
Chanute, Octave. *Progress in Flying Machines.* New York: The American Engineer and Railroad Journal, 1894.
Cromie, Robert. A Plunge into Space. London & New York: Frederick Warne & Co., 1890.
Desmond, Kevin. *Electric Boats and Ships: A History.* Jefferson, NC: McFarland, 2017.
_____. *Gustave Trouvé: French Electrical Genius.* Jefferson, NC: McFarland, 2015.
_____. *Innovators in Battery Technology: Profiles of 93 Influential Electrochemists.* Jefferson, NC: McFarland, 2016.
De Lafolie, Louis-Guillaume. *La Philosophe sans prétention ou l'Homme rare. Ouvrage physique, chymique, politique et moral, dédié aux savants.* Rouen, France: Clousier, 1775.
Dixon, Charles. *Fifteen Hundred Miles an Hour.* London: Blis78s, Sands & Foster, 1895.
L'Electricité à bord des Aéronefs. Guide Pratique, chapter 2.
Godwin, Francis, Bishop of Hereford. *The Man in the Moone, or a discourse of a voyage thither.* London: John Norton, 1638.
Goodenough, J.B., et al. *Solid-State Chemistry of Energy Conversion and Storage.* Washington, D.C.: American Chemical Society, 1977.
Griffith, George Chetwynd. *The Angel of the Revolution: A Tale of the Coming Terror.* Pearson's Weekly, 1893.
Hamon, Augustin Henri. *La Navigation aérienne.* Paris: Marponet Flammarion, 1885.
Hay, William Delisle. *Three Hundred Years Hence, or, a Voice from Posterity.* London: Newman, 1881.
LeCornu, Joseph. *La Navigation Aérienne: Histoire documentaire et anecdotique.* Paris: Vuibert and Nony, 1903.
Norton, Alice Mary. *Sargasso of Space.* New York City: Gnome Press, 1955.
Parrott, Timothy C. *The Secret Life of Dr. Charles Abbott Smith: San Francisco's 19th Century Airship Inventor.* Privately printed.
Piccard, Bertrand. *Solar Impulse: Objectif tour du monde,* 2015.
Pseudo-Apollodorus. *Epitome of the Biblioteca.* Athens, Greece.
Robida, Albert. *La Vie électrique (Vingtième siècle).* Paris: La Librairie Illustrée, 1893.
Salgari, Emilio. *Le meraviglie del Duemila.* Florence: Bemporad, 1907.
Schoeberl Ernst. *From Sunrise to Solar-Impulse: 34 Years of Solar Powered Flight.* Schweinfurt, Germany: University of Applied Sciences Wuerzburg–Schweinfurt, 2008.
Shastry, Subbaraya. *Vaimānika Shastra.* Trans. by G.R. Josyer, 1952.
Swift, Jonathan. *Travels into Several Remote Nations of the World. In Four Parts. By Lemuel Gulliver, First a Surgeon, and Then a Captain of Several Ships, commonly known as Gulliver's Travels.* London: Charles Bathurst, 1726.
Tissandier, Gaston. *L'Application de l'Électricité à la Navigation aérienne. L'Aérostat électrique à hélice de MM. Albert et Gaston Tissandier.* Note presented to the Society for Encouragement, January 11, 1884. Extract from the *Society Bulletin.* In-40de 16 pages, avec 1 image. Paris: Tremblay, 1884.
Verne, Jules. *Robur le Conquérant.* Paris: Hetzel, 1886.
Wells, Herbert George. *The First Men in the Moon.* London: George Newnes, 1901.

Periodicals

Aerokurier
Aeromodeller
Aérophile
Astronautics
aviationtoday.com
Aviation Week
Canadian Press
Drone News.com
E flight Journal
Electrical Review
The Engineer
Flight Global
Flyer
Flying
Flying Pages
GeekWire
Gizmag
H-2 International
IEEE Spectrum
Illustration
International Journal of Hydrogen Energy
International Journal of Micro Air Vehicles
Kirchheimer Zeitung
Mother Earth News
Motoring
Science Daily

Index

Numbers in **_bold italics_** indicate pages with illustrations

ABC Motors 50
AC Propulsion 98
Advanced Tactics (AT) 241; *Black Knight Transformer* octocopter 241; *Panther* 241
AE-1 108-9
Aeriel 14
Aero Club *see* Royal Aero Club
Aéro-Club de France 121, 174, 283
Aero Electric Aircraft Corporation (AEAC) 100
AeroMobil 237
Aérostable 45
aérostat 16-7, 19, 21, 24, 29, **_31_**, 38, 148, 158, 261; *see also* airship; dirigible
aérostation 15, 22, **_23_**
AeroVironment 82-3, 85, 89, 95, 104, 133, 166, 183-8, **_185_**, 205, 219-20; *Gossamer Albatross I* 82, 85; *Gossamer Albatross II* 82-3; *Gossamer Condor* 82; *Gossamer Penguin* 82-4; *Pathfinder* 184-6, 252, 272; *Pathfinder Plus* **_185_**, 186-7
Aether 253
air-to-ground communication 44, 49
Airbus 134-5, 143-5, **_145-6_**, 148, 151, 157, 161, 163-4, 167-8, 189-90, 212-3, 219, 224-7, **_225_**, 229, 246, 248, 254, 257, 259, 262, 289-90; *A3 Vahana* 262; *E-Fan* 143-5, **_145_**, **_146_**, 212; *E-Fan 1.1* 161; *E-Fan 2.0* 144; *E-Fan 4.0* 144; *E-Fan Plus* 161; *E-Fan X* 161, 163-4; *Perlan II* 289
AirEnergy 107-9, **_109_**
AirQuadOne 226, 240
airship 9, **_10_**, 19, **_20_**, **_21_**-2, 22-5,
27, 31-2, 35-6, 39, 41-2, 44, 49, 53, 55, 99, 105, 111, 118, 127-8, 158-160, 170-1, 177, 250, 253, 256, 258, 260-1, 281-2; *see also* aerostat; dirigible
Alauda Mark 1 Airspeeder quadcopter 242
Alpha Electro 142-144, 147, 153, 165, 169, 262
Alpha Electro G2 143
Amazon 191-3, 211, 218, 221, 258
Ampère Distributed Electric Propulsion airplane 22
Ampère electric boat 22
amphibian aircraft 102-3, 124-5, 252
Ancient Egypt 5-7
Antares motor-gliders 122, 127, 141; *Antares 20E* 109, **_110_**, 122, 127, 135, 138; *Antares 23E* 109, 122, 138, 162; *Antares DLR-H2* 134; *Antares E2* 138, 165; *Antares H2* **_135_**, **_137_**, 138; *Antares H3* 135
Apis 2 motor-glider 126
Apple, Vincent G. 40
Aquila solar drone 212
Archaeopteryx microlift glider **_165_**
Archaeopteryx solar plane 82-3
Arco Solar 82-3
Arcus-E glider 122
Ark of the Covenant 5-6
Astro Flight Incorporated 72-3, 75-6, **_77_**, 78, 83, **_85_**, 86, 182; *Sunrise I* 75-8 **_77_**; *Sunrise II* 76, 78, 83, 95, 182; *see also* Boucher, Robert; Boucher, Roland
AT&T 49, 217
AtlantikSolar 211

Atos hang glider 118
Auguste Piccard submarine 270
Aurora Flight Sciences 156, 187, 228, 248
Autogiro 179
Autonomous Flying Microbots 208-9
Autoplane 52
Autovolantor 236
Avery, William 37
Avro 43, 60, 140; *Avro 504* 50, 140; *504K* 58; *Lancaster* 60; *RJ* 164
Avroplane model airplane 43

balloon 11, 13, 16-17, 19, 21-4, 28-31, 34-5, 51, 92, 111, 127, 159, 177, 188-9, 256-7, 259, 270-1, 273
Barton, Frank 38
batteries 10-1, 13-4, 16, 19, 22-5, 28, 25, 40-4, 48, 53, 59-61, 65-6, 68-9, 75, 76-81, 86, 89, 92, 94-95, 98-9, 101, 105-12, 116, 119, 122-5, 128-9, 133-6, 138-9, 142-5, 147-8, 150-2, 155-6, 161, 164, 168-70, 173, 182, 186-8, 190, 193, 196-7, 199, 202-7, 210, 214, 218, 220-1, 227-8, 238-241, 245-6, 248-50, 255-6, 258-9, 261-2, 267, 272-4, 276, 281, 283-4, 286, 289-90
battery brands: Altair 236; EnerDel ; EPS 30; Exxon 106; Graupner/Ripmax 118; Honeywell 182; Kokam 109, 118, 122, 124, 126, 128, 147, 252, 283; Kreisel 123; Opitpower 212; OXIS 128; Panacis 101; Panasonic 168; Renata 152; Saft 109, 140; Sakti3 128; Sanyo **_96_**, 98, 107; Sion Power

189; Sony 107; Thunder Sky 111; Varta 68, *70*; Yardney 182; Yuasa 105; Yuneec 199
battery types: aluminium-air 169, 252; chlorine-zinc flow *23*; dichromate of potash 19; dry cell 52; electro-voltaic pocket 14; flow 24; lead-acid 16, *18*, 59, 63, 105, 111, 182; lithium 76, 97, 100, 106, 111, *113*, 143, 158, 169, 189, 201, 209, 223, 236; lithium-air 169, 245; lithium-ion (Li-Ion) 97, 101, 107, 109, 128, 134, 140, 158, 169, 206, 213, 239, 286: lithium-ion-phosphate 119; lithium-metal 158, 261; lithium-nickel-cobalt-manganese-oxide 203; lithium-polymer (Li-Po) 99, 107, *109*, 111, 115, 117–8, 121–4, 126, 128, 136, 143, 154, 199, 204, 208–9, 220, 223, 252, 273, 281, 289; lithium-sulphur (Li-S) 128, 190; lithium-thionyl-chloride 107, 182; nickel-cadmium (Ni-Cad) 68, *70*, 72–3, 79–80, 84, 95, 97, 108, 110, 182, 187; nickel-iron 182; nickel-metal hydride (NiMh) *108*–109; rechargeable 106–7, 111, 201, 204; silicon-solar 75; silver-zinc 65, 72, 107, 182–3; zinc-air 169
Bebop drone 196
BEHA (Bio-Electric Hybrid Aircraft) 253
Bell Aircraft Corporation 57, 228
Bell Helicoptor 228
Bell Telephone Laboratories 75
Berblinger Prize 94, 97, 122
Bionic Bat 95
Biplane 37, 42, 45, *46*, 49, 74, 79, 101, 124, 127, 129, 140, 174, 231, 253, 286
BiPod 367 239
Bizjak, Stanley 61, *62*, 63, 130
Blade CX RTF micro helicopter 111
Blériot, Louis Charles Joseph 42–4, 46, 144, 171
blimp 104, 127, 160, 187, 258, 260
BLItz 247
Boeing 133–4, 151, 155–7, 167–8, 182, 187, 213, 226, 229, 244, 246, 248, 251–2, 255, 261, 285; *B24H Liberator* 61; *BWB* 155; *F-15* 139; *17E Flying Fortress* 61; *737* 140, 183, 245; *747* 129, 140; *757* 155; *787 Dreamliner* 121, 158, 239, 272

Borschberg, André 152, 270–75, *275*, 276–80
Boscarol, Ivo 115–*16*, 142, *163*
Boscarol, Taja 143
Bosch 48, 68, *70*, 79–80, 223
Boucher, Captain Max 171, 174
Boucher, Robert "Bob" 65, 69, 71–3, *77*, 77–8, 84, 86, 88–9, 93, 95, 182; *see also* Astro Flight
Boucher, Roland 65, 69, 71, 75, 76–6, *77*, 78, 84, 86, 182; *see also* Astro Flight
Brditschka, Heinrich 65, 67–9
Brditschka, Heinrich "Heino," Jr. 67, *70*
British Air Ministry 54, 64, 124
Brown, Janice 84, *87*, 87–9
Brown, William C. 66
Bruce, Eric Henry Stuart 28–9
Bruner, Donald L. 54
Bruss, Helmut 65, 78
BURD 95
BURD-II 95
Bye, George 100–2; Bye Aerospace 102, 262; Bye Energy 100

Cabot, Godfrey L. 53
Caihong solar UAV 229
Campbell, John Wood, Jr. 74
Campbell, Prof. Peter Carmont 28
Cannon, Joseph L. 44
Central Intelligence Agency (CIA) 213
Centurion solar aircraft 184, 186
Centurion UAV 128
Cessna 69, 72, 90, 100–1, 155; *150* 68, 89; *180* 72
Chanute, Octave 37
Chillson, Charles W. 58–9
Chopra, Naman 240
Chretien, Pascal 204
Chrysalis 95
CityHawk 227
Civil Aviation Administration of China 147
Civil Aviation Authority (CAA) 143, 201
Civil Aviation Authority (ENAC) 197
Civil Aviation Authority of Singapore 226
Civil Aviation Safety Authority 228
Cloud Dancer motor-glider 111
Cobra model glider 65
Cocconi, Alan 98
Comparative Aircraft Flight Efficiency (CAFE) Foundation 99, 106, *117*, 125–6
Concept EP model helicopter 110

Concept 30 model helicopter 110
Consolidated Aircraft Corporation 59
Cormorant 226
Cri-Cri 123–4, 283–4; *E-Cristaline Cri-Cri* 145–6, *146*; *Green Cri-Cri* 124, 143
Curtiss, Glenn 52, 125, 140
Curtiss Airplane and Motor Company 58, 140, 180–1; biplane 45, 174; *Model L* 52; *N-9* seaplane 52, 174; OXX engine 52
Curtiss Flying School 173
Curtiss-Wright Limited *58*, 58–9
Custer, Willard R. 252

D-14 Phoenix motor-glider *166*
Daedalus 7–8, 95, 97
Daimler-Chrysler 134, 273; Austro-Daimler *51*–2; Daimler Benz 133; Daimler 223
da Vinci, Leonardo 8, 175, 220
De Havilland 80, 174; *DH98 Mosquito* 80
De Havilland, Geoffrey 174
Defense Advanced Research Projects Agency (DARPA) 72–3, 75–6, 78, 156, 160, 187, 193, 205, 209, 241, 249, 260; ICARUS program 193
de Lafolie, Louis-Guillaume 11, *12*
DeLorean, Paul 240–1
die Krähe motor-glider 67
Dietrich, Carl 236
Diodon inflatable drone 220
dirigible *23*, 23–25, 49, 74, 159; *see also* aerostat; airship
Dirigible-4 105, 127
Dirigicycle 28
DJI 197–8, 213, 262; *Phantom* 197; *Phantom 2* 197; *Spark* 262–3; *see also* Wang, Frank
Domjan, Raphaël 289
Dorrington, Graham E. 105, 111, 118, *119*, 127
Douglas Aircraft Company 59
Drone Racing League 268, 290
Du Pont 82–6, *88*, 95
Duval, Hugues 145, *146*, 283, 286

E-Fan see Airbus
E-Fenix paratrike 118
e-Flight Waiex 123
e-Genius motor-glider *see* Magnus
The e-Spirit of St Louis 126
e-volo volocopter 2221, 223
EA-9 118
Eagle *224*, 225

Index

Edison, Thomas Alva 15, *16*, 37–39, 48, 190, 263
eFusion see Magnus
EHang 184 quadcopter 199–200, **200**, 216, 262
Eiffel, Gustav 42–43, 45
Electra model aircraft 66
ElectraFlyer 112, *113*; *ElectraFlyer C* 112
Electravia 118, 121, 124, 145, 283; *Electro Trike* 118
electric heating 52, 56
electric landing gear 158
Electric Lightning P1 127
electric lights and lighting 35, 45, **46**, 47, 49–54, 63, 115, 142, 157, 171, 173, 216–7, 220, 278; communication 28
electric propeller 30, 42–3, **58**, 59, 76, **85**, 86, 96, 121, 125, 139
electric radio 49, 55, 63, 76, 144, 173, 201, 205
Electric Swift 119
The Electrical Experimenter 52, 129
electrical heating 51–2, 56,
The Electrical Review 32, 34
Electro-fli model aircraft 73
ElectroFlight (formerly *TEACO Bat*) 290
ElectroLight 2 motor-glider 145
Elektra One 101, 123, 153
Elektra One Solar motor-glider 103, 123
Elektra-2 289
Elektro-UHU motor-glider 65
Elektroflug FM model airplane
Emerson Electric 61
Empress Motor Car and Aviation Company 64
Enduro 1 quadcopter 212
English Electric Aviation Limited 64
Eraole 281
Esteyne, Didier 124, 143–4, **145–6**, 161, 212
Etlantic microlight 281
Euroglider 257
European Aeronautic Defence and Space Company (EADS) 124, 245
European Space Agency ESA 223, 271, 282
EViation 169–70, 250, 261; *Alice ER* 169; *Orca* 169
Experimental Aircraft Association (EAA) 112, 240
Extra, Walter 151, 153, 161–2, 290
Extra Aircraft Construction 151, 162, *162*; *EA* 151; *300 Elektro* 162; *330LE* 153, 161
Exxon 106

Facebook 103, 212
fairground rides **38**, 39, 56
Faradair 252
Faulhaber, Dr. Ing. Fritz 65
Federal Aviation Authority (FAA) 79, 101–2, 114, 126, 197, 200–1, 214–5, 220–1, 223, 237, 249
Federal Office of Civil Aviation (FOCA) 153, 165
Firefly 145
"The First Men in the Moon" 37
Fishman, Randall B. 111–2, *113*–4, 289
Flair 94
Flair 30 ultralight glider 94
Flash Falcon 256
Fleye 219
Flightcraft 140
Flike tricopter 220
Fly-B 241
The Flyer 40
FM241 65
FM248 65
Focke-Wulf FW 190 59
Fournier RF-4 model glider 72–3
La France airship **22–3**, 24, 105
Franklin, Benjamin 10–11, 13
Furnival, Capt. J.M. 49

G2 Gyrocycle 242
General Electric GE 32, 60, 139, 155, 183
Gerrand Industries Limited 64
GL-10 Greased Lightning 190
Glassock, Richard 227,
Global Hawk 189, 213
Global Observer 133
GM EV-1 98
GM Impact 98
Goddard, Robert H. 41
Gonsales, Domingo 8
Goodenough, Prof. John B. 106
Google 99, 117, 142, 191, 199, 211, 219, 221, 239–40, 260, 262, 274, 277
Gossamer see AeroVironment
Grahame-White, Claude 45
Grahame-White Flying School 45, *46*
Grandseigne, Robert 45
Graphene 135, 158, 257–8
Graupner 65, **66–7**, 73, 78, 116, 257; *Silentius FM254* 65, *67*
Graupner, Johannes 65
Green 1 121
Green Pioneer 98
Green Wing E430 198
GreenWing International 198–9

Gyrodrive 237
Gyroplane No. I 176
Gyroplane No. II 176

Halo airship 253
Hamilton, Charles K. 45
Hamilton aEro **152**
Hamilton H55 153
HB-3 motor-glider 67–8
HB-21 motor-glider 67–8
Hélicostat 177
Helios 184, 186, 229
Heliotech **77**
Heliotek (later Spectrolab) 75–6, 78 83–4; 78 83–4
Heliplats 133
Hemera airship 253
Hercules octocopter 190
Hexacopter 197–8
Hi-Fly model airplane 67
High Altitude Solar (HALSOL) 183
Hirobo Electric Corporation 202; *Eagle* model helicopter 203; *HX-1 BIT* model helicopter 203
Hogan, Edward 28
Honda 130, 139
Horizon X2 125
Horner, Donald W. 42
HorseFly octocopter 192
Hovercraft 90, 139, 235, 237
HoverMast-100 214
Huazhi Hu 199, **200**
Hussain, Begum Rokeya Sakhawat 40
HY4 135–6
hybrid-electric 129, **131**, 138–9, **141**, 142–4, 148–9, 151, 162, 164, 168, 190, 228, 231, 237, 240–1, 250, 255, 262, 269, 287
hydrogen powered flight 16, 19, 22, 24–5, 40, 49–51, 103, 111, 129, 131–6, 138, 142, 147, 156–7, 171, 184, 186, 213, 227, 243, 246–8, 255–6
HyFish 134
Hyperblimp 253
HyperMach 244

Icaré 97
Icaré II 97
Icarus 7, 97, 266
ICARUS rescue project 211
ignition system 40, 48, 174,
IHI (formerly Ishikawajima-Harima Heavy Industries Company Limited) 157
Inozemtsev, Alexander 147
Intel Shooting Star drone 216
Ion 247
ion engine 42
Ion Tiger 247–8

Iris Challenger 2 blimp 127
Isis airship 160

Joby 262
Joby Aviation 154, 203; *S2* 203
Joby Energy Incorporated 203

Kalt 30 Baron Whisper 110
Kelleher, Christopher Charles 188–9; *Zephyr* 188–90, 229
Kettering, Charles F. 174
Kipling, Rudyard 41
Kite 10, 37, 171, 260
Klein, Štefan 237
Kramer, Dale 125
Krebs, Arthur Constantin 22–4
Kremer, Hanoch "Henry" 80, 82; Kremer Prize 81–2, 94–5; Kremer Speed Challenge 95

Lady Gaga (aka Stefani Joanne Angelina Germanotta) 204–5, 216
LAK 17 122
Lange, Axel **108**, 109, **110**, 122, 127, 134, **135**, **137**, 138, 140, **264**; Lange Aviation GmbH 109, 165
laser-powered airplane 132, 244
Lavrand, Anne 118
Lazair 125
LCA60T airship 158–9
Lektro 140
Levavasseur, Léon 24, 40
Leyden jar 9–11
LF 20 motor-glider 108–9
Li, Danny H.Y. 97–8
LightCraft 132, 244
Lilium 223–4, **224**, 225, **265**
Lindbergh, Charles 106, 114, 270, 279, 286
Lindbergh, Erik 106, 122, 126, 140, 240
Lindbergh prizes and awards 106, 114, 117, 122–3, 140, 221
Lippisch, Prof. Alexander 63, 65, 250
LISA Aeroplane Company 103
Lissaman, Dr. Peter B.S. 82, 183, 187
Lockheed Aircraft Corporation *see* Lockheed Martin
Lockheed Martin 59, 75, 83, 98, 139, 160, 182, 187, 241, 254–5; *Electra* 91; *HALE-D* 98–9
Lotus 154
Low, Archibald Mongomery 173
Luftwaffe 59, 63, 65, 69
Luminati Aerospace 103
Lung Biotechnology Tier 1 helicopter 249
Lynch, Cedric 111, 118, 284

MacCready, Marshall 84
MacCready, Parker 90
MacCready, Dr. Paul Beattie 72, 81–4, **87**, 88–91, 95–6, 133, 183, 205, 211–2, 272, 280
Magnus Aircraft 150; *eFusion* 150, 162, 166; *eFusion 212* 150–1; *e-Genius* motor-glider 99, 117, 162, 165, **166**
Malibu 72
Marconi Company 24, 44, 49, 55
Massachusetts Institute of Technology (MIT) 55, 74, 95, 103, 125, 193, 197–8, 203, 206–7, 225, 236, 239, 245, 254, 257, 261, 271
Mauro, Larry 79–80
Mauro Solar Riser 79–80
Maxim, Sir Hiram Stevens 29–30, 34–5, **38**, 39–40, 56
Maxwell X-57 154–6, 228, 261
MBE-1 68–9, 107
Meier, Hans Justus 69
Meier, Lorenz 198
Meier, René **164**
Michaelis, Dominic 92
micro-drones 190, 194, 198, 205–10, 215, 255
Militky, Alfred 65, **66-7**, 67–9, 73, 78, **79**, 107, 122, 182, 257, 280; Militky Cup 67, 78
MinAir 94
Minerva 13
Ministry of Defence (MOD) 188–90
model aircraft 5, 14, 22, 43–5, 57, 75–6, 78–9, **79**, 81, 92, 95, 104, 107, 110–1, 122, 124, 132, 133, 134, 170–1, 179, 182, 203, 217–8, 227, 257, 259
Modular Approach to Hybrid-Electric Propulsion Architecture (MAHEPA) 143
Moller, Paul Sandler 236
Monarch B 95
Monnett, John 122–3; *Moni* motor-glider 112, 122
Monocopter 255
Montgolfier, Jacques-Etienne 11, 13
Montgolfier, Joseph-Michel 11, 13
Moog Incorporated 101–2
Moonspinner 216
Moreau, Albert 45
Morgan, Ray 83–4, 86, 89–91, 183, 186–7
motor glider 65, 67–8, 72, 108–9, 111–2, 115, 119, 122–4, 126, 133–4, 140, 45, 148, 154, 170
Mott, Samuel D. 38
Mouette E-Trike 136

Multibody Advanced Airship for Transport (MAAT) 128
Musculair 94
Musculair II 94, 96
Musgrave, Major Herbert 48
Musk, Elon 128, 238–9, 246; *see also* Tesla

N3-X 248
NASA (National Aeronautics and Space Administration) 84–85, **86**, 95, 116–7, 125, 130, 132–4, **150**, 154–6, 181–2, 184, **185**, 186–7, 190, 201–3, 214, 220–1, 223, 225, 228, 236, 239, 244, 246–50, 254, 261, 264, 272, 277, 282
La Navette Bretonne 145, **146**
NeoXcraft 240
Neuman, Tom 246
Nixie camera drone 219
North, André 75
Northcliffe, Baron Alfred Lord 43–4
Northrop Grumman Corporation (previously Northrop Corporation) 72, 100, 107, 160, 182, 215, 249
Norton, Alice Mary (aka Andre Norton) 75
Nulli Secundus I 44

Observer 100
Octocopter 147, 165, 190, 192, 199, 215, 227, 240–1
Oemichen, Etienne Edmond 177, **178**, 220

P-Plane 203
Panthera see Pipistrel
paragliders and paragliding 152, 253, 256
Parrot 194–8, 217–8, 265
Partior Q-1 247
Passenger Drone 228–9
passenger drones 199, 226, 262
Pathfinder see AeroVironment
Perlan II see Airbus
Phantom drone *see* DJI
Piccard, Antoine Auguste 270
Piccard, Bertrand 270–5, **275**, 276–80
Piccard, Jacques 270
Pipistrel 99, 115, **116–7**, 116–8, 127, 136, **141**, 141–4, 147, **150**, 152–4, 160, 162, **163**, 169, 228, 262; Electro Taurus **150**, 154; *Panthera* **141**, 141–2, 150; *Panthera Electro* 142; *Panthera Hybrid* 142; *Sinus* 115; *Taurus Electro G2* **141**, 142; *Taurus Electro G4* 115–6, **117**
PKZ 52, 214; *Virus* 115
PKZ1 **51**, 52

Plan Bee see Autonomous Flying Microbots
PLIMP drone 259–60
Pointer 187–8
Pratt and Whitney (P&W) 60, 74, 139, 245
Predator drone 213
Progress Eagle 255
Project Zero 202
Ptacek, Steven R. 88–90, 144, 187
Puffin 203, 223, 228

quadcopter *51*, 52, 176, 178, **180**, 180–1, 192, 194–7, 199, 206, 209, 212, 214–7, 219, 227, 239, 241–2, 262–3, 267, 290
quadrotor 207, 209, 215
Quantix VTOL *219*

Radio Queen 64–5
RAND Corporation 205
Raven 187–8, 219
Raymond, Eric 95–100, **96**, 272, 275
Raymond, Irena **100**
Raytheon 67
Renard, Louis-Marie-Joseph-Charles, Clement "Charles" *22*, 22–4, 30
The Richard C. du Pont Memorial Trophy 81
Rigid No 9 49
Robert, Etienne-Gaspard (aka Robertson) 13
Robobee see Autonomous Flying Microbots
Rochelt, Günther 92–6, **92–4**
Rochelt, Holger 94
Roe, Edwin Alliott Verdon 43–4
Rolls Royce 155–6, 224, 240, 246, 259, 290
Rosenbaum, Bruno 53
Royal Aero Club 43, 140, 188
Royal Aeronautical Society 20, 89, 92, 95
Royal Air Force (RAF) 49, 56, 59, 60, 89–90, 93, 118, *119*, 188
Royal Aircraft Establishment 56, 174
Royal Flying Corps 46, 48–50, 64, 173,
Royal Naval Air Service 50
Royal Society 9, 11, 28
RQ-12 Wasp drone 188
RQ-20A Puma drone 188
Rutan, Burt 125; *Long-EZ* 286 **287**; *Quickie* 283
RX1E 136, 147, **148**
RX1E-A 262

SAAB 119, **120**, 121
Sadowski, Herbert 69

Salgari, Emilio 41
Santos-Dumont, Alberto 121
Sastry, Ann Marie 128
ScanEagle drone 213–4
Schauberger, Viktor 63
Scheit, Prof. Hermann 43
science fiction 9, 17, 25, 29–30, 31, **33**, 36, 40–2, 74–5, 231–3
Scorpion 3 241
Seagull glider 136
Seeley, Brien 106, 126,
Shenzhen Dajiang Innovation Technology see DJI
Short brothers 140, 235
Shrike drone 188
Siemens 19, **20**, 30, 82, 144, **148**, 148–53, 161–2, **162–3**, 165, 170, 221, 226, 262, 290; *330LE* 151–3, 161, 290
Sikorsky Aircraft Corporation 46, 203–4, 246; *S-22 Ilya-Muromets* 46–7
Sikorsky Prize 103
Silencer electric glider 67
Silent Club AE1 107
Silent Falcon 190
Silent Targa 107
Silver Fox model airplane 92
SkiGull 125
Sky Sapience 214
Sky Voyage 256
Skybender UAV 211
Skydrive 238
SkyEye UAV 192
Skylark EH-1 110
SkySpark 283
Smartbird 208
SmartFish 134
SmartFlyer 165
Smartflyer Challenge 164, 166
Smith, Charles Abbott 35–6
Snowstorm 201
Soaring 97
SolAero Technologies Corporation 102
Solair I 92–4, **93**; see also Rochelt, Gunther
Solair II **94**; see also Rochelt, Gunther
The Solar Aircraft Company Limited 74
solar balloon 92
Solar Challenger **85–8**, 82–92, 93, 95, 144, 183, 187, 212, 272, 282
Solar Flight 77, 96–7, 100
solar glider *77*, 98
solar helicopter 103
Solar Impulse 152, 270–281, **275**, **277–8**, **280**
Solar MS-1 biplane 74
Solar One 81
Solar Ship Incorporated 151, 160; *Zenship* 151

Solara 50 drone 211
Solara 60 drone 211
Solaris 78, 92
SolarStratos 289
SoLong 98
Solution F helicopter 204
Sonex Aircraft 122–3; *Waiex* 122
SonicStar 244
SORA-e 121
Soullier, Jean-Luc 281, 283–5, **284**
Souricette 118
SpaceshipOne 125
Spark drone see DJI
Spectrolab see Heliotek
Sperry, Elmer 52–3, 56, 173
Sperry, Lawrence 173–4
Sperry Instruments Corporation 61, 174
SportStar EPOS 121–2
SpykerAero 127
Stampa, Ulrich 69
Starduster 95
StratoAirNet 102
Stratobus airship 159–60, 262
Stringfellow, John **14**, 14–5
SUGAR (Subsonic Ultra Green Aircraft Research) *Volt* 248
Sun Power Corporation 98, 185
SunFlyer 100–1
SunFlyer 2 102, 162
SunFlyer 4 102
Sunraycer 184
Sunrise see Astro Flight
Sunseeker Duo 99
Sunseeker I **96**,
Sunseeker II 97, 99, **100**
Surefly quadcopter 227
Swift 118–9
Swift, Jonathan 9, **10**, 11
Synergy 252

Tailwind hybrid UAV 190
TailWind-E 249
TailWind-H 249
Tampier, René 231–2, **232**
Taplin, Col. Harold John "Taps" 64
Taurus Electro see Pipistrel
Taylor, Charles E. 40
Taylor, George Crosland 43
Taylor, Moulton B. "Molt" 235
Terrafugia **236**, 236–7
Tesla 128, 158, 164, 166–8, 170, 224, 228, 238, 240, 246, 259; see also Musk, Elon
Tesla, Nikola 44–5, 55, 132, 171, **172**, 179, 259
Thompson, Archie F. 56
Tissandier brothers (Albert and Gaston) 16–7, 19, **20–1**, 21, 28–30, 148, 158, 280, 282
To, Freddie 80–1

Tomažič, Tine 115–6, *163*, 165
Toyota 130, 238, 246, 262, 265
Transformable HOvering Rotorcraft (THOR) 255
tricoptor 220, 233
TriFan 102
triplane 15, 42–3, 52, 252
Trouvé, Gustav 16–7, *18*, 19, 24, 29, 31, 34, 48
Tupolev 134, 147; *Tu-154* 132; *Tu-155* 132; *Tu-204* 147; *Tu-214* 147; *Tu-214E* 147; *Tu-2016* 133
Turnbull, Wallace Rupert *57*, 57–8

Uber 228, 300
Ultralight Flying Machines 79
United Aircraft Corporation (UAC) 147
United States Air Force (USAF) 58, *61*, 84, 95, 100, 132, 160–1, 174, 181, 188, 213, 241, 247, 254, 282
United States Army 178
United States Army Air Corps 54, 56
United States Navy 107

Vaculik, Juraj 237
Vahana 266
Vanderlip, Edward G. 181
van Musschenbroek, Pieter 9
Vapor 246
VerdeGo 240
Verne, Jules 25, *26*, 29
von Petróczy, Stephan *51*, 51–2, 215
Votec Evolaris 153
Votec 221 153
Voyager 125
Vuia I 231

Walker, Frederick 38
Walt Disney Corporation 216
Wang, Frank 197; *see also* DJI
"The War of the Worlds" 36
WATTsUP 127
Wellner, Georg 30
Wells, Herbert George 36–7
Westinghouse Electrical Corporation 32, 53, *62*, 129–30, *131*
Wheatley, William B. 59–60
WheelTug 167–8
Whisper octocopter 165
The White Diamond airship 111
The White Lady 222
Williams, John 110
wireless communication 29, 44, 48–50, 53, 173, 201, 228
wireless positioning 55
World War I 45, 47–53, 64, 80, 126, 129, 140, 171, 173–4, 177, 194
World War II *58*, 59, *60*, 60–65, 67, 69, 80–1, 128, 175, 179, 201, 233
Wright, Orville 40, 74, 253
Wright, Wilbur 40
Wright brothers 38, 40, 43, 121, 162, 169, 227, 240, 278; *see also* Wright, Orville; Wright, Wilbur
Wright Electric 169, 262
Wright Flyer 140, 169, 278

Yates, William Morrison "Chip" 153, 161, 169, 285–7, *287-8*, 289
Ypselon GT 153
Yuneec International Company 112, 126, 162, 198–9; *eSpyder* 126, 198; GW430 147

Zee.Aero 225, 239–40
Zephyr see Kelleher, Christopher Charles
Zeppelin 50, 53
Znidarsic, Luke 122
Znidarsic, Matija 122
Zunino, Eric 194, *195*
Zunino, Pascal 194, *195*
Zunum 168, 261; *Aero* 168

www.ingramcontent.com/pod-product-compliance
Ingram Content Group UK Ltd.
Pitfield, Milton Keynes, MK11 3LW, UK
UKHW050541150426
5217IPUK00026B/2031